# THE KERR SPACETIME

## Rotating Black Holes in General Relativity

Rotating black holes, as described by the Kerr spacetime, are the key to understanding the most violent and energetic phenomena in the Universe, from the core collapse of massive supernova explosions producing powerful bursts of gamma rays, to supermassive black hole engines that power quasars and other active galactic nuclei.

This book is a unique, comprehensive overview of the Kerr spacetime, with original contributions and historical accounts from researchers who have pioneered the theory and observation of black holes, and Roy Kerr's own description of his 1963 discovery. It covers all aspects of rotating black holes, from mathematical relativity to astrophysical applications and observations, and current theoretical frontiers. This book provides an excellent introduction and survey of the Kerr spacetime for researchers and graduate students across the spectrum of observational and theoretical astrophysics, general relativity and high-energy physics.

DAVID WILTSHIRE is a Senior Lecturer in the Department of Physics and Astronomy at the University of Canterbury, New Zealand. His research has spanned many areas in general relativity and cosmology, including black holes in higher dimensional gravity, brane worlds, quantum cosmology, dark energy and the averaging problem in inhomogeneous cosmology.

MATT VISSER is Professor of Mathematics at the Victoria University of Wellington, New Zealand. He has published widely in the areas of general relativity, quantum field theory, and theoretical cosmology. He is best known for his contributions to the theory of traversable wormholes, chronology protection, and analogue spacetimes.

SUSAN SCOTT is Associate Professor in the Centre for Gravitational Physics at the Australian National University. She is well known for her contributions to mathematical relativity and cosmology, and is currently President of the Australasian Society for General Relativity and Gravitation.

# THE KERR SPACETIME

## Rotating Black Holes in General Relativity

*Edited by*

DAVID L. WILTSHIRE
*University of Canterbury, Christchurch*

MATT VISSER
*Victoria University of Wellington*

SUSAN M. SCOTT
*Australian National University, Canberra*

CAMBRIDGE UNIVERSITY PRESS
Cambridge, New York, Melbourne, Madrid, Cape Town, Singapore, São Paulo, Delhi

Cambridge University Press
The Edinburgh Building, Cambridge CB2 8RU, UK

Published in the United States of America by Cambridge University Press, New York

www.cambridge.org
Information on this title: www.cambridge.org/9780521885126



First published 2009

Printed in the United Kingdom at the University Press, Cambridge

*A catalogue record for this publication is available from the British Library*

*Library of Congress Cataloguing in Publication data*
The Kerr spacetime : rotating black holes in general relativity / edited by David L. Wiltshire, Matt Visser, Susan Scott.
p. cm.
Includes index.
ISBN 978-0-521-88512-6
1. Kerr black holes. 2. Black holes (Astronomy) – Mathematical models. 3. Space and time – Mathematical models. 4. Kerr, R. P. (Roy P.) I. Wiltshire, David L., 1956– II. Visser, Matt. III. Scott, Susan, 1956– IV. Title.
QB843.B55K47 2009
523.8′875 – dc22 2008045980

ISBN 978-0-521-88512-6 hardback

# Contents

# Contributors

**Steve Carlip**
*Physics Department, University of California at Davis, CA 95616, USA*

**Brandon Carter**
*Observatoire de Paris–Meudon, Place Jules Janssen, F-92195 Meudon, France*

**Andrew C. Fabian**
*Institute of Astronomy, University of Cambridge, Madingley Road, Cambridge CB3 0HA, UK*

**Gary T. Horowitz**
*Physics Department, University of California at Santa Barbara, CA 93107, USA*

**Roy P. Kerr**
*Department of Physics and Astronomy* and *Department of Mathematics and Statistics, University of Canterbury, Private Bag 4180, Christchurch 8140, New Zealand*

**Benjamin R. Lewis**
*Centre for Gravitational Physics, Department of Physics, Australian National University, Canberra ACT 0200, Australia*

**Fulvio Melia**
*Department of Physics and Steward Observatory, University of Arizona, Tucson, Arizona 85721, USA*

**Giovanni Miniutti**
*Institute of Astronomy, University of Cambridge, Madingley Road, Cambridge CB3 0HA, UK*

**Roger Penrose**
*Mathematical Institute, Oxford University, 24-29 St Giles St, Oxford OX1 3LB, UK*

**Maurice H. P. M. van Putten**
*Kavli Institute for Astrophysics and Space Research, Massachusetts Institute of Technology 77 Massachusetts Avenue, 37-287, Cambridge, MA 02139, USA*

**David C. Robinson**
*Mathematics Department, King's College London, Strand, London WC2R 2LS, UK*

**Remo Ruffini**
*Dipartimento di Fisica, Università di Roma "La Sapienza", Piazzale Aldo Moro 5, I-00185 Roma, Italy;* and *ICRANet, Piazzale della Repubblica 10, I-65122 Pescara, Italy*

**Susan M. Scott**
*Centre for Gravitational Physics, Department of Physics, Australian National University, Canberra ACT 0200, Australia*

**Matt Visser**
*School of Mathematics, Statistics, and Computer Science, Victoria University of Wellington, PO Box 600, Wellington, New Zealand*

# Foreword

It is an amazing fact that there are roughly $10^{20}$ rotating black holes in the observable universe, and that the spacetime near each of them is, to a very good approximation, given by an exact solution of Einstein's vacuum field equations discovered in 1963 by Roy Kerr. The Kerr spacetime is a defining feature of modern astrophysics. It is becoming increasingly important as the basis for understanding astrophysical processes from core collapse supernovae which produce gamma-ray bursts while forming black holes to the supermassive black hole engines that power quasars. These processes are the most violent and energetic phenomena in the universe since the Big Bang, and the key to understanding them is the Kerr geometry.

Yet the man behind this remarkable solution has remained somewhat enigmatic. When one of the contributors to this book heard that a conference was to be held to celebrate Roy's 70th birthday in 2004, he was suprised to learn that Roy was still alive! The idea of such a conference was suggested to me by another of my former teachers, Brian Wybourne, shortly before Brian passed away late in 2003. This book, which contains the invited lectures of the 2004 *Kerr Fest*, is the lasting result.

As someone who took courses from Roy at the University of Canterbury, one personal anecdote is in order. Roy amused his students with his laid back style and laconic humour, while always impressing upon us how to see quickly to the point in complex calculations, to back-track and fix our mistakes. He was also disdainful of needless bureaucracy, and took it as a badge of honour that he was the last person to deliver his exam scripts to the university Registry for printing. "I can't see why the Registry always want the exams so early," he complained. "In Texas we just used to write the exams on the board. Look, don't worry about the exam. I just rehash the old questions from past years."

In one lecture that year, after filling a blackboard with calculations that weren't getting closer to the expected solution, Roy wrote

$$\cdots = \text{mess}$$

"Look, at this point I think I'd go off to the pub", he sighed. Standing back to look at the board for a minute, he enlightened us further. "Ah, I screwed up a factor of $ct$ in the second line... I can't be bothered doing all that again. Just take down a note: *The stupid bastard screwed up the ct*."

When we came to sit the exam for that course, some last minute rushing in its preparation became evident. The questions may just have been "rehashed", but when one put the values in, the algebra rapidly became very formidable, for question after question. One student left halfway through in an obvious state of distress. While wrestling with one particular question, which two pages of algebra made me suspect was actually not analytically soluble, the value of Roy's teaching became clear in a flash of inspiration. "$\cdots =$ mess", I wrote, "I can't be bothered solving this equation, but if I was to proceed here is the method I would use:... "

This anecdote not only illustrates the fun we had as Roy's students, but serves as a useful analogy for the state of understanding in 1963 about the problem of solving Einstein's equations for the exterior field of a rotating mass. After some decades of work the consensus of the experts was that the problem equalled a "mess", which most general relativists had given up on. The status of "black holes" as possible physical objects in the universe was also disputed. The term "black hole" itself had yet to be coined by Wheeler in 1967. Despite the work of Oppenheimer and Snyder on gravitational collapse in 1939, and the eventual understanding of the properties of the horizon in Schwarzschild's 1916 solution, everyone knew that realistic bodies rotated. Thus back in 1963 it was still a possibility that the Schwarzschild horizon was a mathematical curiosity, which might not survive the perturbation of adding angular momentum. Roy's solution changed all that. He established black holes as possible entities in the physical universe, and recent decades of many astrophysical observations have confirmed that nature has made good use of them.

Roy's brilliance in achieving what nobody had for decades, by pen and paper before the days of algebraic computing, was really a consequence of his insights into dealing with "messes". He understood deeply the importance of simplifying symmetry principles, of differential and integral constraints, and how to quickly sort out the relevant degrees of freedom from the irrelevant ones. The Kerr solution is a monumental legacy to Roy's incisive insight.

David Wiltshire

# Preface

The book is organized into three parts.

- *General relativity.* Wherein the contributors discuss the classical physics of rotating black holes. There is an orientational overview of the Kerr spacetime, followed by Roy Kerr's own account of its discovery, twistorial applications, constants of the motion, the uniqueness theorems and visualization of the Kerr geometry.
- *Astrophysics.* Observational evidence for rotating black holes, supermassive black holes, merger and collapse.
- *Quantum gravity.* Theoretical frontiers, black hole entropy and quantum physics, higher dimensional black strings and black rings.

The contributions encompass by and large the invited lectures of the *Kerr Fest: a Symposium on Black Holes in Astrophysics, General Relativity and Quantum Gravity*, held at Roy Kerr's home institution, the University of Canterbury, in Christchurch, New Zealand, in August 2004 to celebrate Roy's 70th birthday a few months beforehand on 16 May. As one of the original speakers, Zoltan Perjés, sadly passed away a few months after the Symposium, one of Roy's former colleagues who was unable to attend the Symposium, Roger Penrose, kindly agreed to submit an article in his place.

As appendices, we also reprint Roy Kerr's renowned *Physical Review Letter* of 1963, at less than two pages, stunning in its impact and brevity, along with his conference proceeding article presented at the *First Texas Symposium on Relativistic Astrophysics* in December 1963, in which the black hole property of the solution was first described.

There should be something in this book for almost everyone – there are relatively few books one can turn to to get significant coverage of the Kerr spacetime, and most of the relevant material is scattered around the scientific literature in small bits and pieces. We hope that this book will serve as a coherent starting point for those interested in more technical details, and as a broad overview for those who are interested in the current state of play.

David Wiltshire — *Christchurch*
Matt Visser — *Wellington*
Susan Scott — *Canberra*
*January 2008*

Roy Patrick Kerr, 2007.

# Illustrations

# Part I

## General relativity:<br>Classical studies of the Kerr geometry

# 1

# The Kerr spacetime – a brief introduction

Matt Visser

## 1.1 Background

The Kerr spacetime has now been with us for some 45 years (Kerr 1963, 1965). It was discovered in 1963 through an intellectual *tour de force*, and continues to provide highly non-trivial and challenging mathematical and physical problems to this day. This chapter provides a brief introduction to the Kerr spacetime and rotating black holes, touching on the most common coordinate representations of the spacetime metric and the key features of the geometry – the presence of horizons and ergospheres. The coverage is by no means complete, and serves chiefly to orient oneself when reading subsequent chapters.

The final form of Albert Einstein's general theory of relativity was developed in November 1915 (Einstein 1915, Hilbert 1915), and within two months Karl Schwarzschild (working with one of the slightly earlier versions of the theory) had already solved the field equations that determine the exact spacetime geometry of a non-rotating "point particle" (Schwarzschild 1916a). It was relatively quickly realized, via Birkhoff's uniqueness theorem (Birkhoff 1923, Jebsen 1921, Deser and Franklin 2005, Johansen and Ravndal 2006), that the spacetime geometry in the vacuum region outside any localized spherically symmetric source is equivalent, up to a possible coordinate transformation, to a portion of the Schwarzschild geometry – and so of direct physical interest to modelling the spacetime geometry surrounding and exterior to idealized non-rotating spherical stars and planets. (In counterpoint, for modelling the *interior* of a finite-size spherically symmetric source, Schwarzschild's "constant density star" is a useful first approximation (Schwarzschild 1916b). This is often referred to as Schwarzschild's "interior" solution, which is potentially confusing as it is an utterly distinct physical

The Kerr Spacetime: Rotating Black Holes in General Relativity. Ed. David L. Wiltshire, Matt Visser and Susan M. Scott. Published by Cambridge University Press. 

spacetime solving the Einstein equations in the presence of a specified distribution of matter.)

Considerably more slowly, only after intense debate was it realized that the "inward" analytic extension of Schwarzschild's "exterior" solution represents a non-rotating black hole, the endpoint of stellar collapse (Oppenheimer and Snyder 1939). In the most common form (Schwarzschild coordinates, also known as curvature coordinates), which is not always the most useful form for understanding the physics, the Schwarzschild geometry is described by the line element

$$\mathrm{d}s^2 = -\left[1 - \frac{2m}{r}\right] \mathrm{d}t^2 + \frac{\mathrm{d}r^2}{1 - 2m/r} + r^2(\mathrm{d}\theta^2 + \sin^2\theta\, \mathrm{d}\phi^2), \tag{1.1}$$

where the parameter $m$ is the physical mass of the central object.

But astrophysically, we know that stars (and for that matter planets) rotate, and from the weak-field approximation to the Einstein equations we even know the approximate form of the metric at large distances from a stationary isolated body of mass $m$ and angular momentum $J$ (Thirring and Lense 1918; Misner, Thorne, and Wheeler 1973; Adler, Bazin, and Schiffer 1975; D'Inverno 1992; Hartle 2003; Carroll 2004; Pfister 2005). In suitable coordinates:

$$\begin{aligned} \mathrm{d}s^2 = &-\left[1 - \frac{2m}{r} + O\left(\frac{1}{r^2}\right)\right] \mathrm{d}t^2 - \left[\frac{4J\sin^2\theta}{r} + O\left(\frac{1}{r^2}\right)\right] \mathrm{d}\phi\, \mathrm{d}t \\ &+ \left[1 + \frac{2m}{r} + O\left(\frac{1}{r^2}\right)\right] [\mathrm{d}r^2 + r^2(\mathrm{d}\theta^2 + \sin^2\theta\, \mathrm{d}\phi^2)]. \end{aligned} \tag{1.2}$$

This approximate metric is perfectly adequate for almost all solar system tests of general relativity, but there certainly are well-known astrophysical situations (e.g. neutron stars) for which this approximation is inadequate – and so a "strong field" solution is physically called for. Furthermore, if a rotating star were to undergo gravitational collapse, then the resulting black hole would be expected to retain at least some fraction of its initial angular momentum – thus suggesting on physical grounds that somehow there should be an extension of the Schwarzschild geometry to the situation where the central body carries angular momentum.

Physicists and mathematicians looked for such a solution for many years, and had almost given up hope, until the Kerr solution was discovered in 1963 (Kerr 1963) – some 48 years after the Einstein field equations were first developed. From the weak-field asymptotic result we can already see that angular momentum destroys spherical symmetry, and this lack of spherical symmetry makes the calculations *much* more difficult. It is not that the basic principles are all that different, but simply that the algebraic complexity of the computations is so high that relatively few physicists or mathematicians have the fortitude to carry them through to completion.

Indeed it is easy to both derive and check the Schwarzschild solution by hand, but for the Kerr spacetime the situation is rather different. For instance in Chandrasekhar's magnum opus on black holes (Chandrasekhar 1998), only part of which is devoted to the Kerr spacetime, he is moved to comment:

> The treatment of the perturbations of the Kerr space-time in this chapter has been prolixius in its complexity. Perhaps, at a later time, the complexity will be unravelled by deeper insights. But meantime, the analysis has led us into a realm of the rococo: splendorous, joyful, and immensely ornate.

More generally, Chandrasekhar also comments:

> The nature of developments simply does not allow a presentation that can be followed in detail with modest effort: the reductions that are required to go from one step to another are often very elaborate and, on occasion, may require as many as ten, twenty, or even fifty pages.

Of course the Kerr spacetime was not constructed *ex nihilo*. Some of Roy Kerr's early thoughts on this and related matters can be found in (Kerr 1959; Kerr and Goldberg 1961), and over the years he has periodically revisited this theme (Goldberg and Kerr 1964; Kerr and Schild 1965; Debney, Kerr, and Schild 1969; Kerr and Debney 1970; Kerr and Wilson 1977; Weir and Kerr 1977; Fackerell and Kerr 1991; Burinskii and Kerr 1995).

For practical and efficient computation in the Kerr spacetime many researchers will prefer to use general symbolic manipulation packages such as Maple, Mathematica, or more specialized packages such as GR-tensor. When used with *care* and *discretion*, symbolic manipulation tools can greatly aid physical understanding and insight.[1]

Because of the complexity of calculations involving the Kerr spacetime there is relatively little textbook coverage dedicated to this topic. An early discussion can be found in the textbook by Adler, Bazin, and Schiffer (1975 second edition). The only dedicated single-topic textbook I know of is that by O'Neill (1995). There are also comparatively brief discussions in the research monograph by Hawking and Ellis (1975), and the standard textbooks by Misner, Thorne, and Wheeler (1973), D'Inverno (1992), Hartle (2003), and Carroll (2004). One should particularly note the 60-page chapter appearing in the very recent textbook by Plebański and Krasiński (2006). An extensive and highly technical discussion of Kerr black holes is given in Chandrasekhar (1998), while an exhaustive discussion of the class of spacetimes described by Kerr–Schild metrics is presented in the book "Exact Solutions to Einstein's Field Equations" (Stephani *et al.*, 2002).

[1] For instance, the standard distribution of Maple makes some unusual choices for its sign conventions. The signs of the Einstein tensor, Ricci tensor, and Ricci scalar (though *not* the Riemann tensor and Weyl tensor) are opposite to what most physicists and mathematicians would expect.

To orient the reader I will now provide some general discussion, and explicitly present the line element for the Kerr spacetime in its most commonly used coordinate systems. (Of course the physics cannot depend on the coordinate system, but specific computations can sometimes be simplified by choosing an appropriate coordinate chart.)

## 1.2 No Birkhoff theorem

Physically, it must be emphasized that there is no Birkhoff theorem for rotating spacetimes – it is *not* true that the spacetime geometry in the vacuum region outside a generic rotating star (or planet) is a part of the Kerr geometry. The best result one can obtain is the much milder statement that outside a rotating star (or planet) the geometry asymptotically approaches Kerr geometry.

The basic problem is that in the Kerr geometry all the multipole moments are very closely related to each other – whereas in real physical stars (or planets) the mass quadrupole, octopole, and higher moments of the mass distribution can in principle be independently specified. Of course from electromagnetism you will remember that higher $n$-pole fields fall off as $1/r^{2+n}$, so that far away from the object the lowest multipoles dominate – it is in this *asymptotic* sense that the Kerr geometry is relevant for rotating stars or planets.

On the other hand, if the star (or planet) gravitationally collapses – then classically a black hole can be formed. In this case there *are* a number of powerful uniqueness theorems which guarantee the direct physical relevance of the Kerr spacetime, but as the unique exact solution corresponding to stationary rotating black holes, (as opposed to merely being an asymptotic solution to the far field of rotating stars or planets).

## 1.3 Kerr's original coordinates

The very first version of the Kerr spacetime geometry to be explicitly written down in the literature was the line element (Kerr 1963)

$$\begin{aligned} \mathrm{d}s^2 = &-\left[1 - \frac{2mr}{r^2 + a^2\cos^2\theta}\right] (\mathrm{d}u + a\sin^2\theta\ \mathrm{d}\phi)^2 \\ &+ 2(\mathrm{d}u + a\sin^2\theta\ \mathrm{d}\phi)\,(\mathrm{d}r + a\sin^2\theta\ \mathrm{d}\phi) \\ &+ (r^2 + a^2\cos^2\theta)\,(\mathrm{d}\theta^2 + \sin^2\theta\ \mathrm{d}\phi^2). \end{aligned} \tag{1.3}$$

The key features of this spacetime geometry are:

- Using symbolic manipulation software it is easy to verify that this manifold is Ricci flat, $R_{ab} = 0$, and so satisfies the vacuum Einstein field equations. Verifying this by hand is at best tedious.
- There are three off-diagonal terms in the metric – which is one of the features that makes computations tedious.
- By considering (for instance) the $g_{uu}$ component of the metric, it is clear that for $m \neq 0$ there is (at the very least) a coordinate singularity located at $r^2 + a^2 \cos^2\theta = 0$, that is:

$$r = 0; \qquad \theta = \pi/2. \tag{1.4}$$

  We shall soon see that this is actually a curvature singularity. In these particular coordinates there are no other obvious coordinate singularities.
- Since the line element is independent of both $u$ and $\phi$ we immediately deduce the existence of two Killing vectors. Ordering the coordinates as $(u, r, \theta, \phi)$ the two Killing vectors are

$$U^a = (1, 0, 0, 0); \qquad R^a = (0, 0, 0, 1). \tag{1.5}$$

  Any constant-coefficient linear combination of these Killing vectors will again be a Killing vector.
- Setting $a \to 0$ the line element reduces to

$$\mathrm{d}s^2 \to -\left[1 - \frac{2m}{r}\right] \mathrm{d}u^2 + 2\,\mathrm{d}u\,\mathrm{d}r + r^2\,(\mathrm{d}\theta^2 + \sin^2\theta\,\mathrm{d}\phi^2), \tag{1.6}$$

  which is the Schwarzschild geometry in the so-called "advanced Eddington–Finkelstein coordinates". Based on this, by analogy the line element (1.3) is often called the advanced Eddington–Finkelstein form of the Kerr spacetime. Furthermore since we know that $r = 0$ is a curvature singularity in the Schwarzschild geometry, this strongly suggests (but does not yet prove) that the singularity in the Kerr spacetime at $(r = 0, \theta = \pi/2)$ is a curvature singularity.
- Setting $m \to 0$ the line element reduces to

$$\begin{aligned} \mathrm{d}s^2 \to \mathrm{d}s_0^2 = &-(\mathrm{d}u + a\sin^2\theta\,\mathrm{d}\phi)^2 \\ &+ 2(\mathrm{d}u + a\sin^2\theta\,\mathrm{d}\phi)\,(\mathrm{d}r + a\sin^2\theta\,\mathrm{d}\phi) \\ &+ (r^2 + a^2\cos^2\theta)\,(\mathrm{d}\theta^2 + \sin^2\theta\,\mathrm{d}\phi^2), \end{aligned} \tag{1.7}$$

  which is actually (*but certainly not obviously*) flat Minkowski spacetime in disguise. This is most easily seen by using symbolic manipulation software to verify that for this simplified line element the Riemann tensor is identically zero: $R_{abcd} \to 0$.
- For the general situation, $m \neq 0 \neq a$, all the non-zero components of the Riemann tensor contain at least one factor of $m$.

- Indeed, in a suitably chosen orthonormal basis the result can be shown to be even stronger: All the non-zero components of the Riemann tensor are then proportional to $m$:

$$R_{\hat{a}\hat{b}\hat{c}\hat{d}} \propto m. \tag{1.8}$$

(This point will be discussed more fully below, in the section on the rational polynomial form of the Kerr metric. See also the discussion in Plebański and Krasiński (2006).)

- Furthermore, the only non-trivial quadratic curvature invariant is

$$\begin{aligned} R_{abcd}\, R^{abcd} &= C_{abcd}\, C^{abcd} \\ &= \frac{48m^2(r^2 - a^2\cos^2\theta)\,[(r^2 + a^2\cos^2\theta)^2 - 16r^2a^2\cos^2\theta]}{(r^2 + a^2\cos^2\theta)^6}, \end{aligned} \tag{1.9}$$

guaranteeing that the singularity located at

$$r = 0; \qquad \theta = \pi/2, \tag{1.10}$$

is actually a curvature singularity. (We would have strongly suspected this by considering the $a \to 0$ case above.)

- In terms of the $m = 0$ line element we can put the line element into manifestly Kerr–Schild form by writing

$$\mathrm{d}s^2 = \mathrm{d}s_0^2 + \frac{2mr}{r^2 + a^2\cos^2\theta}\,(\mathrm{d}u + a\sin^2\theta\,\mathrm{d}\phi)^2, \tag{1.11}$$

or the equivalent

$$g_{ab} = (g_0)_{ab} + \frac{2mr}{r^2 + a^2\cos^2\theta}\,\ell_a\,\ell_b, \tag{1.12}$$

where we define

$$(g_0)_{ab} = \begin{bmatrix} -1 & 1 & 0 & 0 \\ 1 & 0 & 0 & a\sin^2\theta \\ 0 & 0 & r^2 + a^2\cos^2\theta & 0 \\ 0 & a\sin^2\theta & 0 & (r^2 + a^2)\sin^2\theta \end{bmatrix}, \tag{1.13}$$

and

$$\ell_a = (1, 0, 0, a\sin^2\theta). \tag{1.14}$$

- It is then easy to check that $\ell_a$ is a null vector, with respect to both $g_{ab}$ and $(g_0)_{ab}$, and that

$$g^{ab} = (g_0)^{ab} - \frac{2mr}{r^2 + a^2\cos^2\theta}\,\ell^a\,\ell^b, \tag{1.15}$$

where

$$(g_0)^{ab} = \frac{1}{r^2 + a^2 \cos^2\theta} \begin{bmatrix} a^2 \sin^2\theta & r^2 + a^2 & 0 & -a \\ r^2 + a^2 & r^2 + a^2 & 0 & -a \\ 0 & 0 & 1 & 0 \\ -a & -a & 0 & (\sin^2\theta)^{-1} \end{bmatrix}, \tag{1.16}$$

and

$$\ell^a = (0, 1, 0, 0). \tag{1.17}$$

- The determinant of the metric takes on a remarkably simple form

$$\det(g_{ab}) = -(r^2 + a^2 \cos^2\theta)^2 \sin^2\theta = \det([g_0]_{ab}), \tag{1.18}$$

where the $m$ dependence has cancelled. (This is a side effect of the fact that the metric is of the Kerr–Schild form.)

- At the curvature singularity we have

$$\mathrm{d}s_0^2\big|_{\text{singularity}} = -\mathrm{d}u^2 + a^2\, \mathrm{d}\phi^2, \tag{1.19}$$

showing that, in terms of the "background" geometry specified by the disguised Minkowski spacetime with metric $(g_0)_{ab}$, the curvature singularity is a "ring". Of course in terms of the "full" geometry, specified by the physical metric $g_{ab}$, the intrinsic geometry of the curvature singularity is, unavoidably and by definition, singular.

- The null vector field $\ell^a$ is an affinely parameterized null geodesic:

$$\ell^a\, \nabla_a \ell^b = 0. \tag{1.20}$$

- More generally

$$\ell_{(a;b)} = \begin{bmatrix} 0 & 0 & 0 & 0 \\ 0 & 0 & 0 & 0 \\ 0 & 0 & r & 0 \\ 0 & 0 & 0 & r\sin^2\theta \end{bmatrix}_{ab} - \frac{m(r^2 - a^2 \cos^2\theta)}{(r^2 + a^2 \cos^2\theta)^2}\, \ell_a\, \ell_b. \tag{1.21}$$

(And it is easy to see that this automatically implies the null vector field $\ell^a$ is an affinely parameterized null geodesic.)

- The divergence of the null vector field $\ell^a$ is also particularly simple

$$\nabla_a \ell^a = \frac{2r}{r^2 + a^2\, \cos^2\theta}. \tag{1.22}$$

- Furthermore, with the results we already have it is easy to calculate

$$\ell_{(a;b)}\, \ell^{(a;b)} = \ell_{(a;b)} g^{bc} \ell_{(c;d)} g^{da} = \ell_{(a;b)} [g_0]^{bc} \ell_{(c;d)} [g_0]^{da} = \frac{2r^2}{(r^2 + a^2 \cos^2\theta)^2}, \tag{1.23}$$

whence

$$\frac{\ell_{(a;b)}\,\ell^{(a;b)}}{2} - \frac{(\nabla_a \ell^a)^2}{4} = 0. \tag{1.24}$$

This invariant condition implies that the null vector field $\ell^a$ is "shear free".

- Define a one-form $\ell$ by

$$\ell = \ell_a \, \mathrm{d}x^a = \mathrm{d}u + a\sin^2\theta \, \mathrm{d}\phi, \tag{1.25}$$

then

$$\mathrm{d}\ell = a\sin 2\theta \; \mathrm{d}\theta \wedge \mathrm{d}\phi \neq 0, \tag{1.26}$$

but implying

$$\mathrm{d}\ell \wedge \mathrm{d}\ell = 0. \tag{1.27}$$

- Similarly

$$\ell \wedge \mathrm{d}\ell = a\sin 2\theta \; \mathrm{d}u \wedge \mathrm{d}\theta \wedge \mathrm{d}\phi \neq 0, \tag{1.28}$$

but in terms of the Hodge-star we have the invariant relation

$$*(\ell \wedge \mathrm{d}\ell) = -\frac{2a\cos\theta}{r^2 + a^2\cos^2\theta} \; \ell, \tag{1.29}$$

or in component notation

$$\epsilon^{abcd}(\ell_b \, \ell_{c,d}) = -\frac{2a\cos\theta}{r^2 + a^2\cos^2\theta} \; \ell^a. \tag{1.30}$$

This allows one to pick off the so-called "twist" as

$$\omega = -\frac{a\cos\theta}{r^2 + a^2\cos^2\theta}. \tag{1.31}$$

This list of properties is a quick, but certainly not exhaustive, survey of the key features of the spacetime that can be established by direct computation in this particular coordinate system.

### 1.4 Kerr–Schild "Cartesian" coordinates

The second version of the Kerr line element presented in the original article (Kerr 1963), also discussed in the early follow-up conference contribution (Kerr 1965), was defined in terms of "Cartesian" coordinates $(t, x, y, z)$:

$$ds^2 = -dt^2 + dx^2 + dy^2 + dz^2 + \frac{2mr^3}{r^4 + a^2z^2}\left[dt + \frac{r(x\,dx + y\,dy)}{a^2 + r^2} + \frac{a(y\,dx - x\,dy)}{a^2 + r^2} + \frac{z}{r}\,dz\right]^2, \tag{1.32}$$

subject to $r(x, y, z)$, which is now a dependent function, not a coordinate, being implicitly determined by:

$$x^2 + y^2 + z^2 = r^2 + a^2\left[1 - \frac{z^2}{r^2}\right]. \tag{1.33}$$

- The coordinate transformation used in going from (1.3) to (1.32) is

$$t = u - r; \qquad x + iy = (r - ia)\,e^{i\phi}\sin\theta; \qquad z = r\cos\theta. \tag{1.34}$$

Sometimes it is more convenient to explicitly write

$$x = (r\cos\phi + a\sin\phi)\sin\theta = \sqrt{r^2 + a^2}\,\sin\theta\cos[\phi - \arctan(a/r)]; \tag{1.35}$$

$$y = (r\sin\phi - a\cos\phi)\sin\theta = \sqrt{r^2 + a^2}\,\sin\theta\sin[\phi - \arctan(a/r)]; \tag{1.36}$$

and so deduce

$$\frac{x^2 + y^2}{\sin^2\theta} - \frac{z^2}{\cos^2\theta} = a^2, \tag{1.37}$$

or the equivalent

$$\frac{x^2 + y^2}{r^2 + a^2} + \frac{z^2}{r^2} = 1. \tag{1.38}$$

- The $m \to 0$ limit is now manifestly Minkowski space

$$ds^2 \to ds_0^2 = -dt^2 + dx^2 + dy^2 + dz^2. \tag{1.39}$$

Of course the coordinate transformation (1.34) used in going from (1.3) to (1.32) is also responsible for taking (1.7) to (1.39).

- The $a \to 0$ limit is

$$ds^2 \to -dt^2 + dx^2 + dy^2 + dz^2 + \frac{2m}{r}\left[dt + \frac{(x\,dx + y\,dy + z\,dz)}{r}\right]^2, \tag{1.40}$$

now with

$$r = \sqrt{x^2 + y^2 + z^2}. \tag{1.41}$$

After a change of coordinates this can also be written as

$$\mathrm{d}s^2 \to -\mathrm{d}t^2 + \mathrm{d}r^2 + r^2(\mathrm{d}\theta^2 + \sin^2\theta\, \mathrm{d}\phi^2) + \frac{2m}{r}\,[\mathrm{d}t + \mathrm{d}r]^2, \tag{1.42}$$

which is perhaps more readily recognized as the Schwarzschild spacetime. In fact if we set $u = t + r$ then we regain the Schwarzschild line element in advanced Eddington–Finkelstein coordinates; compare with Equation (1.6).

- The full $m \neq 0 \neq a$ metric is now manifestly of the Kerr–Schild form

$$g_{ab} = \eta_{ab} + \frac{2mr^3}{r^4 + a^2 z^2}\,\ell_a\,\ell_b, \tag{1.43}$$

where now

$$\ell_a = \left(1, \frac{rx + ay}{r^2 + a^2}, \frac{ry - ax}{r^2 + a^2}, \frac{z}{r}\right). \tag{1.44}$$

Here $\ell_a$ is a null vector with respect to both $g_{ab}$ and $\eta_{ab}$ and

$$g^{ab} = \eta^{ab} - \frac{2mr^3}{r^4 + a^2 z^2}\,\ell^a\,\ell^b, \tag{1.45}$$

with

$$\ell^a = \left(-1, \frac{rx + ay}{r^2 + a^2}, \frac{ry - ax}{r^2 + a^2}, \frac{z}{r}\right). \tag{1.46}$$

- Define $R^2 = x^2 + y^2 + z^2$ then explicitly

$$r(x, y, z) = \sqrt{\frac{R^2 - a^2 + \sqrt{(R^2 - a^2)^2 + 4a^2 z^2}}{2}}, \tag{1.47}$$

where the positive root has been chosen so that $r \to R$ at large distances. Because of this relatively complicated expression for $r(x, y, z)$, direct evaluation of the Ricci tensor via symbolic manipulation packages is prohibitively expensive in terms of computer resources – it is better to check the $R_{ab} = 0$ in some other coordinate system, and then appeal to the known trivial transformation law for the zero tensor.

- Consider the quadratic curvature invariant $R_{abcd}\,R^{abcd} = C_{abcd}\,C^{abcd}$, which we have already seen exhibits a curvature singularity at $(r = 0, \theta = \pi/2)$. In this new coordinate system the curvature singularity is located at

$$x^2 + y^2 = a^2; \qquad z = 0. \tag{1.48}$$

Again we recognize this as occurring on a “ring”, now in the $(x, y, z)$ “Cartesian” background space.

- In these coordinates the existence of a time-translation Killing vector, with components

$$K^a = (1, 0, 0, 0) \tag{1.49}$$

is obvious. Less obvious is what happens to the azimuthal (rotational) Killing vector, which now takes the form

$$R^a = (0, 0, y, -x). \tag{1.50}$$

- Note that in these coordinates

$$g^{tt} = -1 - \frac{2mr^3}{r^4 + a^2 z^2}, \tag{1.51}$$

and consequently

$$g^{ab}\, \nabla_a t\, \nabla_b t = -1 - \frac{2mr^3}{r^4 + a^2 z^2}. \tag{1.52}$$

Thus $\nabla t$ is certainly a timelike vector in the region $r > 0$, implying that this portion of the manifold is "stably causal", and that if one restricts attention to the region $r > 0$ there is no possibility of forming closed timelike curves. However, if one chooses to work with the maximal analytic extension of the Kerr spacetime, then the region $r < 0$ does make sense (at least mathematically), and certainly does contain closed timelike curves. (See for instance the discussion in Hawking and Ellis (1975).) Many (most?) relativists would argue that this $r < 0$ portion of the maximally extended Kerr spacetime is purely of mathematical interest and not physically relevant to astrophysical black holes.

So there is a quite definite trade-off in going to Cartesian coordinates – some parts of the geometry are easier to understand, others are more obscure.

## 1.5 Boyer–Lindquist coordinates

Boyer–Lindquist coordinates are best motivated in two stages: First, consider a slightly different but completely equivalent form of the metric which follows from Kerr's original "advanced Eddington–Finkelstein" form (1.3) via the coordinate substitution

$$u = t + r, \tag{1.53}$$

in which case

$$\begin{aligned} ds^2 &= -dt^2 \\ &\quad + dr^2 + 2a \sin^2\theta\, dr\, d\phi + (r^2 + a^2\cos^2\theta)\, d\theta^2 + (r^2 + a^2)\sin^2\theta\, d\phi^2 \\ &\quad + \frac{2mr}{r^2 + a^2\cos^2\theta}\,(dt + dr + a\sin^2\theta\, d\phi)^2. \end{aligned} \tag{1.54}$$

Here the second line is again simply flat 3-space in disguise. An advantage of this coordinate system is that $t$ can naturally be thought of as a time coordinate – at least at large distances near spatial infinity. There are, however, still three off-diagonal terms in the metric so this is not yet any great advance on the original form (1.3). One can easily consider the limits $m \to 0$, $a \to 0$, and the decomposition of this metric into Kerr–Schild form, but there are no real surprises.

Second, it is now extremely useful to perform a further *m-dependent* coordinate transformation, which will put the line element into Boyer–Lindquist form:

$$t = t_{BL} + 2m \int \frac{r\,\mathrm{d}r}{r^2 - 2mr + a^2}; \qquad \phi = -\phi_{BL} - a \int \frac{\mathrm{d}r}{r^2 - 2mr + a^2}; \tag{1.55}$$

$$r = r_{BL}; \qquad \theta = \theta_{BL}. \tag{1.56}$$

Making the transformation, and dropping the $BL$ subscript, the Kerr line-element now takes the form:

$$\begin{aligned} \mathrm{d}s^2 = &- \left[1 - \frac{2mr}{r^2 + a^2\cos^2\theta}\right] \mathrm{d}t^2 - \frac{4mra\sin^2\theta}{r^2 + a^2\cos^2\theta}\,\mathrm{d}t\,\mathrm{d}\phi \\ &+ \left[\frac{r^2 + a^2\cos^2\theta}{r^2 - 2mr + a^2}\right] \mathrm{d}r^2 + (r^2 + a^2\cos^2\theta)\,\mathrm{d}\theta^2 \\ &+ \left[r^2 + a^2 + \frac{2mra^2\sin^2\theta}{r^2 + a^2\cos^2\theta}\right] \sin^2\theta\,\mathrm{d}\phi^2. \end{aligned} \tag{1.57}$$

- These Boyer–Lindquist coordinates are particularly useful in that they minimize the number of off-diagonal components of the metric – there is now only *one* off-diagonal component. We shall subsequently see that this helps particularly in analysing the asymptotic behaviour, and in trying to understand the key difference between an “event horizon” and an “ergosphere”.
- Another particularly useful feature is that the asymptotic ($r \to \infty$) behaviour in Boyer–Lindquist coordinates is

$$\begin{aligned} \mathrm{d}s^2 = &- \left[1 - \frac{2m}{r} + O\left(\frac{1}{r^3}\right)\right] \mathrm{d}t^2 - \left[\frac{4ma\sin^2\theta}{r} + O\left(\frac{1}{r^3}\right)\right] \mathrm{d}\phi\,\mathrm{d}t \\ &+ \left[1 + \frac{2m}{r} + O\left(\frac{1}{r^2}\right)\right] [\mathrm{d}r^2 + r^2(\mathrm{d}\theta^2 + \sin^2\theta\,\mathrm{d}\phi^2)]. \end{aligned} \tag{1.58}$$

  From this we conclude that $m$ is indeed the mass and $J = ma$ is indeed the angular momentum.
- If $a \to 0$ the Boyer–Lindquist line element reproduces the Schwarzschild line element in standard Schwarzschild curvature coordinates.

- If $m \to 0$ the Boyer–Lindquist line element reduces to

$$\begin{aligned} \mathrm{d}s^2 \to &-\mathrm{d}t^2 + \frac{r^2 + a^2 \cos\theta^2}{r^2 + a^2}\, \mathrm{d}r^2 \\ &+ (r^2 + a^2 \cos\theta^2)\, \mathrm{d}\theta^2 + (r^2 + a^2)\, \sin^2\theta\, \mathrm{d}\phi^2. \end{aligned} \tag{1.59}$$

This is flat Minkowski space in so-called "oblate spheroidal" coordinates, and you can relate them to the usual Cartesian coordinates of Euclidean 3-space by defining

$$x = \sqrt{r^2 + a^2}\, \sin\theta\, \cos\phi; \tag{1.60}$$

$$y = \sqrt{r^2 + a^2}\, \sin\theta\, \sin\phi; \tag{1.61}$$

$$z = r\, \cos\theta. \tag{1.62}$$

- One can re-write the Boyer–Lindquist line element as

$$\begin{aligned} \mathrm{d}s^2 = &-\mathrm{d}t^2 + \frac{r^2 + a^2 \cos\theta^2}{r^2 + a^2}\, \mathrm{d}r^2 \\ &+ (r^2 + a^2 \cos\theta^2)\, \mathrm{d}\theta^2 + (r^2 + a^2)\, \sin^2\theta\, \mathrm{d}\phi^2 \\ &+ \frac{2m}{r} \left\{ \frac{[\mathrm{d}t - a \sin^2\theta\, \mathrm{d}\phi]^2}{(1 + a^2 \cos^2\theta/r^2)} + \frac{(1 + a^2 \cos^2\theta/r^2)\, \mathrm{d}r^2}{1 - 2m/r + a^2/r^2} \right\}. \end{aligned} \tag{1.63}$$

In view of the previous comment, this makes it clear that the Kerr geometry is of the form (flat Minkowski space) + (distortion). Note, however, that this is *not* a Kerr–Schild decomposition for the Boyer–Lindquist form of the Kerr line element.

- Of course there will still be a Kerr–Schild decomposition of the metric

$$g_{ab} = (g_0)_{ab} + \frac{2m\,r}{r^2 + a^2 \cos^2\theta}\, \ell_a\, \ell_b, \tag{1.64}$$

but its form in Boyer–Lindquist coordinates is not so obvious. In these coordinates we have

$$\ell = \mathrm{d}t + \frac{r^2 + a^2 \cos^2\theta}{r^2 - 2mr + a^2}\, \mathrm{d}r - a \sin^2\theta\, \mathrm{d}\phi, \tag{1.65}$$

or the equivalent

$$\ell_a = \left(1, +\frac{r^2 + a^2 \cos^2\theta}{r^2 - 2mr + a^2}, 0, -a \sin^2\theta\right). \tag{1.66}$$

Perhaps more surprisingly, at least at first glance, because of the $m$-dependent coordinate transformation used in going to Boyer–Lindquist coordinates the "background" metric

$(g_0)_{ab}$ takes on the form:

$$\begin{bmatrix} -1 & -\frac{2mr}{r^2-2mr+a^2} & 0 & 0 \\ -\frac{2mr}{r^2-2mr+a^2} & \frac{(r^2+a^2\cos^2\theta)(r^2-4mr+a^2)}{(r^2-2mr+a^2)^2} & 0 & \frac{2mar\sin^2\theta}{r^2-2mr+a^2} \\ 0 & 0 & r^2+a^2\cos^2\theta & 0 \\ 0 & \frac{2mar\sin^2\theta}{r^2-2mr+a^2} & 0 & (r^2+a^2)\sin^2\theta \end{bmatrix}. \tag{1.67}$$

This is (arguably) the least transparent form of representing flat Minkowski space that one is likely to encounter in any reasonably natural setting. (That this is again flat Minkowski space in disguise can be verified by using symbolic manipulation packages to compute the Riemann tensor and checking that it is identically zero.) In short, while Boyer–Lindquist coordinates are well adapted to some purposes, they are very ill-adapted to probing the Kerr–Schild decomposition.

- In these coordinates the time translation Killing vector is

$$K^a = (1, 0, 0, 0), \tag{1.68}$$

while the azimuthal Killing vector is

$$R^a = (0, 0, 0, 1). \tag{1.69}$$

- In Boyer–Lindquist coordinates, (because the $r$ and $\theta$ coordinates have not been modified, and because the Jacobian determinants arising from the coordinate transformations (1.53) and (1.55)–(1.56) are both unity), the determinant of the metric again takes the relatively simple $m$-independent form

$$\det(g_{ab}) = -\sin^2\theta\,(r^2+a^2\cos^2\theta)^2 = \det([g_0]_{ab}). \tag{1.70}$$

- In Boyer–Lindquist coordinates, (because the $r$ and $\theta$ coordinates have not been modified), the invariant quantity $R_{abcd}\,R^{abcd}$ looks identical to that calculated for the line element (1.3). Namely:

$$R_{abcd}\,R^{abcd} = \frac{48m^2(r^2-a^2\cos^2\theta)\,[(r^2+a^2\cos^2\theta)^2-16r^2a^2\cos^2\theta)]}{(r^2+a^2\cos^2\theta)^6}. \tag{1.71}$$

- On the axis of rotation ($\theta = 0, \theta = \pi$), the Boyer–Lindquist line element reduces to:

$$\mathrm{d}s^2\Big|_{\text{on-axis}} = -\left[\frac{1-2m/r+a^2/r^2}{1+a^2/r^2}\right]\mathrm{d}t^2 + \left[\frac{1+a^2/r^2}{1-2m/r+a^2/r^2}\right]\mathrm{d}r^2. \tag{1.72}$$

This observation is useful in that it suggests that the on-axis geometry (and in particular the on-axis causal structure) is *qualitatively* similar to that of the Reissner–Nordström geometry.

- On the equator ($\theta = \pi/2$) one has

$$\begin{aligned} \mathrm{d}s^2\big|_{\text{equator}} = &-\left[1 - \frac{2m}{r}\right]\mathrm{d}t^2 - \frac{4ma}{r}\,\mathrm{d}t\,\mathrm{d}\phi \\ &+ \frac{\mathrm{d}r^2}{1 - 2m/r + a^2/r^2} + \left[r^2 + a^2 + \frac{2ma^2}{r}\right]\mathrm{d}\phi^2. \end{aligned} \tag{1.73}$$

Alternatively

$$\begin{aligned} \mathrm{d}s^2\big|_{\text{equator}} = &-\mathrm{d}t^2 + \frac{\mathrm{d}r^2}{1 - 2m/r + a^2/r^2} + (r^2 + a^2)\,\mathrm{d}\phi^2 \\ &+ \frac{2m}{r}\,(\mathrm{d}t + a\,\mathrm{d}\phi)^2. \end{aligned} \tag{1.74}$$

This geometry is still rather complicated and, in contrast to the on-axis geometry, has no "simple" analogue. (See, for example, Visser and Weinfurtner (2005).)

## 1.6 "Rational polynomial" coordinates

Starting from Boyer–Lindquist coordinates, we can define a new coordinate by $\chi = \cos\theta$ so that $\chi \in [-1, 1]$. Then $\sin\theta = \sqrt{1 - \chi^2}$ and

$$\mathrm{d}\chi = \sin\theta\,\mathrm{d}\theta \qquad \Rightarrow \qquad \mathrm{d}\theta = \frac{\mathrm{d}\chi}{\sqrt{1 - \chi^2}}. \tag{1.75}$$

In terms of these $(t, r, \chi, \phi)$ coordinates the Boyer–Lindquist version of the Kerr spacetime becomes

$$\begin{aligned} \mathrm{d}s^2 = &-\left\{1 - \frac{2mr}{r^2 + a^2\chi^2}\right\}\mathrm{d}t^2 - \frac{4amr(1 - \chi^2)}{r^2 + a^2\chi^2}\,\mathrm{d}\phi\,\mathrm{d}t \\ &+ \frac{r^2 + a^2\chi^2}{r^2 - 2mr + a^2}\,\mathrm{d}r^2 + (r^2 + a^2\chi^2)\,\frac{\mathrm{d}\chi^2}{1 - \chi^2} \\ &+ (1 - \chi^2)\left\{r^2 + a^2 + \frac{2ma^2r(1 - \chi^2)}{r^2 + a^2\chi^2}\right\}\mathrm{d}\phi^2. \end{aligned} \tag{1.76}$$

This version of the Kerr metric has the virtue that all the metric components are now simple rational polynomials of the coordinates – the elimination of trigonometric functions makes computer-based symbolic computations much less resource intensive – some calculations speed up by factors of 100 or more.

- For instance, the quadratic curvature invariant

$$R^{abcd}\, R_{abcd} = \frac{48(r^2 - a^2\chi^2)[(r^2 + a^2\chi^2)^2 - 16r^2a^2\chi^2]}{(r^2 + a^2\chi^2)^6}, \tag{1.77}$$

can (on a "modern" laptop, *c.* 2008) now be extracted in a fraction of a second – as opposed to a minute or more when trigonometric functions are involved.[2]

- More generally, consider the (inverse) tetrad

$$e^A_{\ a} = \begin{bmatrix} e^0_{\ a} \\ e^1_{\ a} \\ e^2_{\ a} \\ e^3_{\ a} \end{bmatrix} \tag{1.78}$$

specified by

$$\begin{bmatrix} \sqrt{\frac{r^2-2mr+a^2}{r^2+a^2\chi^2}} & 0 & 0 & -a(1-\chi^2)\sqrt{\frac{r^2-2mr+a^2}{r^2+a^2\chi^2}} \\ 0 & \sqrt{\frac{r^2+a^2\chi^2}{r^2-2mr+a^2}} & 0 & 0 \\ 0 & 0 & \sqrt{\frac{r^2+a^2\chi^2}{1-\chi^2}} & 0 \\ -a\sqrt{\frac{1-\chi^2}{r^2+a^2}} & 0 & 0 & \sqrt{\frac{1-\chi^2}{r^2+a^2}}(r^2+a^2) \end{bmatrix}, \tag{1.79}$$

which at worst contains square roots of rational polynomials. Then it is easy to check

$$g_{ab}\, e^A_{\ a}\, e^B_{\ b} = \eta^{AB}; \qquad \eta_{AB}\, e^A_{\ a}\, e^B_{\ b} = g_{ab}. \tag{1.80}$$

- Furthermore, the associated tetrad is

$$e_A^{\ a} = \begin{bmatrix} e_0^{\ a} \\ e_1^{\ a} \\ e_2^{\ a} \\ e_3^{\ a} \end{bmatrix} \tag{1.81}$$

specified by

$$\begin{bmatrix} \frac{r^2+a^2}{\sqrt{(r^2-2mr+a^2)(r^2+a^2\chi^2)}} & 0 & 0 & \frac{a}{\sqrt{(r^2-2mr+a^2)(r^2+a^2\chi^2)}} \\ 0 & \sqrt{\frac{r^2+a^2\chi^2}{r^2-2mr+a^2}} & 0 & 0 \\ 0 & 0 & \sqrt{\frac{r^2+a^2\chi^2}{1-\chi^2}} & 0 \\ a\sqrt{\frac{1-\chi^2}{r^2+a^2}} & 0 & 0 & \frac{1}{\sqrt{(1-\chi^2)(r^2+a^2)}} \end{bmatrix}. \tag{1.82}$$

[2] Other computer-based symbolic manipulation approaches to calculating the curvature invariants have also been explicitly discussed in Lake (2003, 2004).

- In this orthonormal basis the non-zero components of the Riemann tensor take on only two distinct (and rather simple) values:

$$\begin{aligned} R_{0101} &= -2R_{0202} = -2R_{0303} = 2R_{1212} = 2R_{1313} = -R_{2323} \\ &= -\frac{2mr(r^2 - 3a^2\chi^2)}{(r^2 + a^2\chi^2)^3}, \end{aligned} \tag{1.83}$$

and

$$R_{0123} = R_{0213} = -R_{0312} = \frac{2ma\chi(3r^2 - a^2\chi^2)}{(r^2 + a^2\chi^2)^3}. \tag{1.84}$$

(See the related discussion in Plebański and Krasiński 2006.)

- As promised these orthonormal components of the Riemann tensor are now linear in $m$. Furthermore, note that the only place where *any* of the orthonormal components becomes infinite is where $r = 0$ and $\chi = 0$ – thus verifying that the ring singularity we identified by looking at the scalar $R_{abcs}\ R^{abcd}$ is indeed the whole story – there are no other curvature singularities hiding elsewhere in the Kerr spacetime.
- On the equator, $\chi = 0$, the only non-zero parts of the Riemann tensor are particularly simple

$$\begin{aligned} R_{0101} &= -2R_{0202} = -2R_{0303} = 2R_{1212} = 2R_{1313} = -R_{2323} \\ &= -\frac{2m}{r^3}, \end{aligned} \tag{1.85}$$

which are the same as would arise in the Schwarzschild geometry.

- On the axis of rotation, $\chi = \pm 1$, we have

$$\begin{aligned} R_{0101} &= -2R_{0202} = -2R_{0303} = 2R_{1212} = 2R_{1313} = -R_{2323} \\ &= -\frac{2mr(r^2 - 3a^2)}{(r^2 + a^2)^3}, \end{aligned} \tag{1.86}$$

and

$$R_{0123} = R_{0213} = -R_{0312} = \pm\frac{2ma(3r^2 - a^2)}{(r^2 + a^2)^3}. \tag{1.87}$$

## 1.7 Doran coordinates

The coordinates relatively recently introduced by Chris Doran, see (Doran 2000), give yet another view on the Kerr spacetime. This coordinate system was specifically developed to be as close as possible to the Painlevé–Gullstrand form of the Schwarzschild line element (Painlevé 1921, Gullstrand 1922, Visser 1998), a form that has become more popular over the last decade with continued developments in the "analogue spacetime" programme.

Specifically, one takes the original Eddington–Finkelstein form of the Kerr line element (1.3) and performs the $m$-dependent substitutions

$$du = dt + \frac{dr}{1 + \sqrt{2mr/(r^2 + a^2)}}; \tag{1.88}$$

$$d\phi = d\phi_{\text{Doran}} + \frac{a\, dr}{r^2 + a^2 + \sqrt{2mr(r^2 + a^2)}}. \tag{1.89}$$

After dropping the subscript "Doran", in the new $(t, r, \theta, \phi)$ coordinates Doran's version of the Kerr line element takes the form (Doran 2000):

$$\begin{aligned} ds^2 = & -dt^2 + (r^2 + a^2 \cos^2\theta)\, d\theta^2 + (r^2 + a^2) \sin^2\theta\, d\phi^2 \\ & + \left[\frac{r^2 + a^2\cos^2\theta}{r^2 + a^2}\right] \left\{ dr + \frac{\sqrt{2mr(r^2 + a^2)}}{r^2 + a^2\cos^2\theta} (dt - a \sin^2\theta\, d\phi) \right\}^2 . \end{aligned} \tag{1.90}$$

Key features of this line element are:

- As $a \to 0$ one obtains

$$ds^2 \to -dt^2 + \left\{ dr + \sqrt{\frac{2m}{r}}\, dt \right\}^2 + r^2(d\theta^2 + \sin^2\theta\, d\phi^2), \tag{1.91}$$

  which is simply Schwarzschild's geometry in Painlevé–Gullstrand form (Painlevé 1921, Gullstrand 1922, Visser 1998).
- As $m \to 0$ one obtains

$$\begin{aligned} ds^2 \to ds_0^2 = & -dt^2 + \left[\frac{r^2 + a^2\cos^2\theta}{r^2 + a^2}\right] dr^2 \\ & + (r^2 + a^2\cos^2\theta)\, d\theta^2 + (r^2 + a^2)\sin^2\theta\, d\phi^2, \end{aligned} \tag{1.92}$$

  which is flat Minkowski space in oblate spheroidal coordinates.
- Verifying that the line element is Ricci flat is best done with a symbolic manipulation package.
- Again, since the $r$ and $\theta$ coordinates have not changed, the explicit formulae for $\det(g_{ab})$ and $R^{abcd}\, R_{abcd}$ are unaltered from those obtained in the Eddington–Finkelstein or Boyer–Lindquist coordinates.
- A nice feature of the Doran form of the metric is that for the contravariant inverse metric

$$g^{tt} = -1. \tag{1.93}$$

In the language of the ADM formalism, the Doran coordinates slice the Kerr spacetime in such a way that the "lapse" is everywhere unity. This can be phrased more invariantly as the statement

$$g^{ab} \nabla_a t \nabla_b t = -1, \tag{1.94}$$

implying

$$g^{ab} \nabla_a t \nabla_c \nabla_b t = 0, \tag{1.95}$$

whence

$$(\nabla_a t) g^{ab} \nabla_b (\nabla_c t) = 0. \tag{1.96}$$

That is, the vector field specified by

$$W^a = \nabla^a t = g^{ab} \nabla_b t = g^{ta} \tag{1.97}$$

is an affinely parameterized timelike geodesic. Integral curves of the vector $W^a = \nabla^a t$ (in the Doran coordinate system) are thus the closest analogue the Kerr spacetime has to the notion of a set of "inertial frames" that are initially at rest at infinity, and are then permitted to free-fall towards the singularity.

- Because of this unit lapse feature, it might at first seem that in Doran coordinates the Kerr spacetime is stably causal and that causal pathologies are thereby forbidden. However, the coordinate transformation leading to the Doran form of the metric breaks down in the region $r < 0$ of the maximally extended Kerr spacetime. (The coordinate transformation becomes complex.) So there is in fact complete agreement between Doran and Kerr–Schild forms of the metric: The region $r > 0$ is stably causal, and timelike curves confined to lie in the region $r > 0$ cannot close back on themselves.

There is also a "Cartesian" form of Doran's line element in $(t, x, y, z)$ coordinates. Slightly modifying the presentation of (Doran 2000), the line element is given by

$$ds^2 = -dt^2 + dx^2 + dy^2 + dz^2 + (F^2 V_a V_b + F[V_a S_b + S_a V_b]) dx^a dx^b, \tag{1.98}$$

where

$$F = \sqrt{\frac{2mr}{r^2 + a^2}} \tag{1.99}$$

and where $r(x, y, z)$ is the same quantity as appeared in the Kerr–Schild coordinates, *cf.* Equation (1.33). Furthermore

$$V_a = \sqrt{\frac{r^2(r^2 + a^2)}{r^2 + a^2 z^2}} \left(1, \frac{ay}{r^2 + a^2}, \frac{-ax}{r^2 + a^2}, 0\right), \tag{1.100}$$

and

$$S_a = \sqrt{\frac{r^2(r^2+a^2)}{r^4+a^2z^2}} \left(0, \frac{rx}{r^2+a^2}, \frac{ry}{r^2+a^2}, \frac{z}{r}\right). \tag{1.101}$$

It is then easy to check that $V$ and $S$ are orthonormal with respect to the "background" Minkowski metric $\eta_{ab}$:

$$\eta^{ab}\, V_a V_b = -1; \qquad \eta^{ab}\, V_a S_b = 0; \qquad \eta^{ab}\, S_a S_b = +1. \tag{1.102}$$

These vectors are *not* orthonormal with respect to the full metric $g_{ab}$ but are nevertheless useful for investigating principal null congruences (Doran 2000).

## 1.8 Other coordinates?

We have so far seen various common coordinate systems that have been used over the past 44 years to probe the Kerr geometry. As an open-ended exercise, one could find (either via Google, the arXiv, the library, or original research) as many different coordinate systems for Kerr as one could stomach. Which of these coordinate systems is the "nicest"? This is by no means obvious. It is conceivable (if unlikely) that other coordinate systems may be found in the future that might make computations in the Kerr spacetime significantly simpler.

## 1.9 Horizons

To briefly survey the key properties of the horizons and ergospheres occurring in the Kerr spacetime (see Figure 1.1), let us concentrate on using the Boyer–Lindquist coordinates. First, consider the components of the metric in these coordinates. The metric components have singularities when either

$$r^2 + a^2 \sin^2\theta = 0, \qquad \text{that is,} \qquad r = 0 \quad \text{and} \quad \theta = \pi/2, \tag{1.103}$$

or

$$1 - 2m/r + a^2/r^2 = 0, \qquad \text{that is,} \qquad r = r_\pm \equiv m \pm \sqrt{m^2 - a^2}. \tag{1.104}$$

The first of these possibilities corresponds to what we have already seen is a real physical curvature singularity, since $R_{abcd}\, R^{abcd} \to \infty$ there. In contrast $R_{abcd}\, R^{abcd}$ remains finite at $r_\pm$.

In fact the second option above ($r = r_\pm$) corresponds to *all* orthonormal curvature components and *all* curvature invariants being finite – it is a coordinate singularity but not a curvature singularity. Furthermore as $a \to 0$ we have the

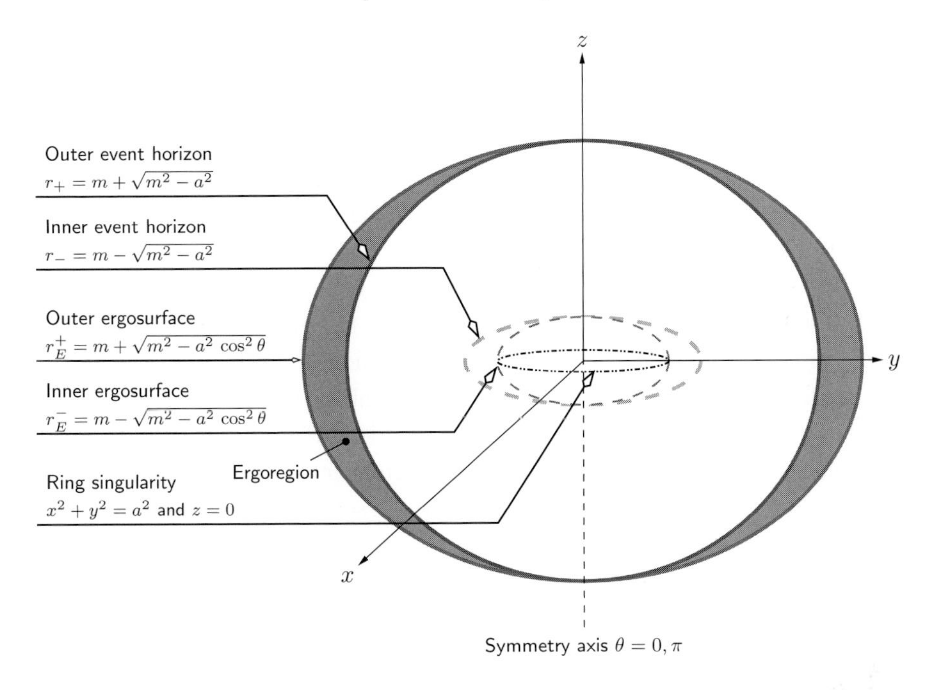


Fig. 1.1. Schematic location of the horizons, ergosurfaces, and curvature singularity in the Kerr spacetime.

smooth limit that $r_\pm \to 2m$, the location of the horizon in the Schwarzschild geometry. We will therefore tentatively identify $r_\pm$ as the locations of the inner and outer horizons. (I shall restrict our attention to the case $a < m$ to avoid having to deal with either the extremal case $a = m$ or the naked singularities that occur for $a > m$.)

Consider now the set of three-dimensional surfaces formed by fixing the $r$ coordinate to take on some specified fixed value $r = r_*$, and letting the other three coordinates $(t, \theta, \phi)$ run over their respective ranges. On these three-dimensional surfaces the induced metric is found by formally setting $\mathrm{d}r \to 0$ and $r \to r_*$ so that

$$\begin{aligned} \mathrm{d}s^2_{\text{3-surface}} &= -\mathrm{d}t^2 + (r_*^2 + a^2\cos\theta^2)\,\mathrm{d}\theta^2 + (r_*^2 + a^2)\,\sin^2\theta\,\mathrm{d}\phi^2 \\ &\quad + \frac{2m}{r_*}\frac{\left[\mathrm{d}t - a\sin^2\theta\,\mathrm{d}\phi\right]^2}{(1 + a^2\cos^2\theta/r_*^2)}. \end{aligned} \tag{1.105}$$

Now we could calculate the determinant of this 3-metric directly. Alternatively we could use the the block-diagonal properties of the metric in Boyer–Lindquist coordinates plus the fact that for the full $(3+1)$-dimensional metric we have already observed

$$\det(g_{ab}) = -\sin^2\theta\,(r^2 + a^2\cos^2\theta)^2, \tag{1.106}$$

to allow us to deduce that on the 3-surface $r = r_*$:

$$\det(g_{ij})_{\text{3-surface}} = -\sin^2\theta\,(r_*^2 + a^2\cos^2\theta)\,(r_*^2 - 2mr_* + a^2). \tag{1.107}$$

Now for $r_* > r_+$ or $r_* < r_-$ the determinant $\det(g_{ij})_{\text{3-surface}}$ is negative, which is a necessary condition for these 3-surfaces to be $(2+1)$ dimensional (two space plus one time dimension). For $r_* \in (r_-, r_+)$ the determinant $\det(g_{ij})_{\text{3-surface}}$ is positive, indicating the lack of any time dimension – the label "$t$" is now misleading and "$t$" actually denotes a spacelike direction. Specifically at what we shall soon see are the inner and outer horizons, $r_* = r_\pm$, the determinant is zero – indicating that the induced 3-metric $[g_{ij}]_{\text{3-surface}}$ is singular (in the sense of being represented by a singular matrix, *not* in the sense that there is a curvature singularity) everywhere on both of these 3-surfaces.

In particular, since $[g_{ij}]$ is a singular matrix, then at each point in either of the 3-surfaces at $r = r_\pm$ there will be some 3-vector $L^i$ that lies in the 3-surface $r = r_\pm$ such that

$$[g_{ij}]\,L^i = 0, \tag{1.108}$$

which implies in particular

$$[g_{ij}]\,L^i\,L^j = 0. \tag{1.109}$$

Now promote the 3-vector $L^i$ [which lives in three-dimensional $(t, \theta, \phi)$ space] to a 4-vector by tacking on an extra coefficient that has value zero:

$$L^i \to L^a = (L^t, 0, L^\theta, L^\phi). \tag{1.110}$$

Then in the $(3+1)$-dimensional sense we have

$$g_{ab}\,L^a\,L^b = 0 \qquad (L^a \text{ only defined at } r = r_\pm), \tag{1.111}$$

i.e. there is a set of curves, described by the vector $L^a$, that lie precisely on the 3-surfaces $r = r_\pm$ and which do not leave those 3-surfaces. Furthermore on the surfaces $r = r_\pm$ these curves are null curves and $L^a$ is a null vector. Note that these null vector fields $L^a$ are defined only on the inner and outer horizons $r = r_\pm$ and that they are quite distinct from the null vector field $\ell^a$ occurring in the Kerr–Schild decomposition of the metric, that vector field being defined throughout the entire spacetime.

Physically, the vector fields $L^a$ correspond to photon "orbits" that skim along the surface of the inner and outer horizons without either falling in or escaping to infinity. You should then be able to easily convince yourself that the outer horizon is an "event horizon" ("absolute horizon") in the sense of being the boundary of the region from which null curves do not escape to infinity, and we shall often concentrate discussion on the outer horizon $r_+$.

Indeed if we define quantities $\Omega_\pm$ by

$$\Omega_\pm = \frac{a}{2mr_\pm} = \frac{a}{r_\pm^2 + a^2}, \tag{1.112}$$

and now define

$$L^a_\pm = (L^t_\pm, 0, 0, L^\phi_\pm) = (1, 0, 0, \Omega_\pm), \tag{1.113}$$

then it is easy to check that

$$g_{ab}(r_\pm)\, L^a_\pm\, L^b_\pm = 0 \tag{1.114}$$

at the 3-surfaces $r = r_\pm$ respectively. That is, the spacetime curves

$$X(t) = (t, r(t), \theta(t), \phi(r)) = (t,\ r_\pm,\ \theta_0,\ \phi_0 + \Omega_\pm\, t) \tag{1.115}$$

are an explicit set of null curves (so they could represent photon trajectories) that skim along the 3-surfaces $r = r_\pm$, the outer and inner horizons.

In fact $r_+$ is the "event horizon" of the Kerr black hole, and $\Omega_+$ is a constant over the 3-surface $r = r_+$. This is a consequence of the "rigidity theorems", and the quantity $\Omega_+$ is referred to as the "angular velocity" of the event horizon. The event horizon "rotates" *as though* it were a solid body. (Somewhat counter-intuitively, the inner horizon also rotates as though it were a solid body, but with a different angular velocity $\Omega_-$.)

Because the inner and outer horizons are specified by the simple condition $r = r_\pm$, you might be tempted to deduce that the horizons are "spherical". Disabuse yourself of this notion. We have already looked at the induced geometry of the 3-surface $r = r_\pm$ and found the induced metric to be described by a singular matrix. Now let's additionally throw away the $t$ direction, and look at the 2-surface $r = r_\pm, t = 0$, which is a two-dimensional "constant time" slice through the horizon. Following our previous discussion for 3-surfaces the intrinsic 2-geometry of this slice is described by Smarr (1973)

$$ds^2_{\text{2-surface}} = (r_\pm^2 + a^2\cos^2\theta)\, d\theta^2 + \left(\frac{4m^2 r_\pm^2}{r_\pm^2 + a^2\cos^2\theta}\right) \sin^2\theta\, d\phi^2, \tag{1.116}$$

or equivalently

$$\mathrm{d}s^2_{\text{2-surface}} = (r_\pm^2 + a^2 \cos^2\theta)\,\mathrm{d}\theta^2 + \left(\frac{[r_\pm^2 + a^2]^2}{r_\pm^2 + a^2\cos^2\theta}\right) \sin^2\theta\,\mathrm{d}\phi^2. \tag{1.117}$$

So while the horizons are *topologically* spherical, they are emphatically not *geometrically* spherical. In fact the area of the horizons is (Smarr 1973)

$$A_H^\pm = 4\pi(r_\pm^2 + a^2) = 8\pi(m^2 \pm \sqrt{m^4 - m^2 a^2}) = 8\pi(m^2 \pm \sqrt{m^4 - L^2}). \tag{1.118}$$

Furthermore, the Ricci scalar is (Smarr 1973)

$$R_{\text{2-surface}} = \frac{2(r_\pm^2 + a^2)\,(r_\pm^2 - 3a^2\cos^2\theta)}{(r_\pm^2 + a^2\cos^2\theta)^3}. \tag{1.119}$$

At the equator

$$R_{\text{2-surface}} \to \frac{2(r_\pm^2 + a^2)}{r_\pm^4} > 0, \tag{1.120}$$

but at the poles

$$R_{\text{2-surface}} \to \frac{2(r_\pm^2 + a^2)\,(r_\pm^2 - 3a^2)}{(r_\pm^2 + a^2)^3}. \tag{1.121}$$

So the intrinsic curvature of the outer horizon (the event horizon) can actually be *negative* near the axis of rotation if $3a^2 > r_+^2$, corresponding to $a > (\sqrt{3}/2)\,m$ – and this happens *before* one reaches the extremal case $a = m$ where the event horizon vanishes and a naked singularity forms.

The Ricci 2-scalar of the outer horizon is negative for

$$\theta < \cos^{-1}\left(\frac{r_+}{\sqrt{3}a}\right), \quad \text{or} \quad \theta > \pi - \cos^{-1}\left(\frac{r_+}{\sqrt{3}a}\right), \tag{1.122}$$

which only has a non-vacuous range if $a > (\sqrt{3}/2)\,m$. In contrast, on the inner horizon the Ricci 2-scalar goes negative for

$$\theta < \cos^{-1}\left(\frac{r_-}{\sqrt{3}a}\right), \quad \text{or} \quad \theta > \pi - \cos^{-1}\left(\frac{r_-}{\sqrt{3}a}\right), \tag{1.123}$$

which always includes the region immediately surrounding the axis of rotation. In fact for slowly rotating Kerr spacetimes, $a \ll m$, the Ricci scalar goes negative on the inner horizon for the rather large range

$$\theta \lesssim \pi/2 - \frac{1}{2\sqrt{3}}\frac{a}{m}; \qquad \theta \gtrsim \pi/2 - \frac{1}{2\sqrt{3}}\frac{a}{m}. \tag{1.124}$$

In contrast to the intrinsic geometry of the horizon, in terms of the Cartesian coordinates of the Kerr–Schild line element the location of the horizon is given by

$$x^2 + y^2 + \left(\frac{r_\pm^2 + a^2}{r_\pm^2}\right) z^2 = r_\pm^2 + a^2, \tag{1.125}$$

which implies

$$x^2 + y^2 + \left(\frac{2m}{r_\pm}\right) z^2 = 2mr_\pm. \tag{1.126}$$

So in terms of the background geometry $(g_0)_{ab}$ the event horizons are a much simpler pair of oblate ellipsoids. But like the statement that the singularity is a "ring", this is not a statement about the intrinsic geometry – it is instead a statement about the mathematically convenient but fictitious flat Minkowski space that is so useful in analysing the Kerr–Schild form of the Kerr spacetime. The semi-major axes of the oblate ellipsoids are

$$S_x = S_y = \sqrt{2mr_\pm} \le 2m \tag{1.127}$$

and

$$S_z = r_\pm \le \sqrt{2mr_\pm} \le 2m. \tag{1.128}$$

The eccentricity is

$$e_\pm = \sqrt{1 - \frac{S_z^2}{S_x^2}} = \sqrt{1 - \frac{r_\pm}{2m}}. \tag{1.129}$$

The volume interior to the horizons, with respect to the background Minkowski metric, is easily calculated to be

$$V_0^\pm = \frac{4\pi}{3}(2m)r_\pm^2. \tag{1.130}$$

The surface area of the horizons, with respect to the background Minkowski metric, can then be calculated using standard formulae due to Lagrange:

$$A_0^\pm = 2\pi(2m)r_\pm \left[1 + \frac{1 - e_\pm^2}{2e_\pm} \ln\left(\frac{1 + e_\pm}{1 - e_\pm}\right)\right]. \tag{1.131}$$

This, however, is not the intrinsic surface area, and is not the quantity relevant for the second law of black hole mechanics – it is instead an illustration and warning of the fact that while the background Minkowski metric is often an aid to visualization, this background should not be taken seriously as an intrinsic part of the physics.

## 1.10 Ergospheres

There is a new concept for rotating black holes, the "ergosphere", that does not arise for non-rotating black holes. Suppose we have a rocket ship and turn on its engines, and move so as to try to "stand still" at a fixed point in the coordinate system – that is, suppose we try to follow the world line:

$$X(t) = (t, r(t), \theta(t), \phi(r)) = (t, r_0, \theta_0, \phi_0)\,. \tag{1.132}$$

Are there locations in the spacetime for which it is impossible to "stand still" (in this coordinate dependent sense)? Now the tangent vector to the world line of an observer who is "standing still" is

$$T^a = \frac{\mathrm{d}X^a(t)}{\mathrm{d}t} = (1, 0, 0, 0), \tag{1.133}$$

and a necessary condition for a physical observer to be standing still is that his 4-trajectory should be timelike. That is, we need

$$g(T, T) < 0. \tag{1.134}$$

But

$$g(T, T) = g_{ab}\, T^a\, T^b = g_{tt}, \tag{1.135}$$

so in the specific case of the Kerr geometry (in Boyer–Lindquist coordinates)

$$g(T, T) = -\left(1 - \frac{2mr}{r^2 + a^2 \cos^2\theta}\right). \tag{1.136}$$

But the RHS becomes positive once

$$r^2 - 2mr + a^2 \cos^2\theta < 0. \tag{1.137}$$

That is, defining

$$r_E^{\pm}(m, a, \theta) \equiv m \pm \sqrt{m^2 - a^2 \cos^2\theta}, \tag{1.138}$$

the RHS becomes positive once

$$r_E^-(m, a, \theta) < r < r_E^+(m, a, \theta). \tag{1.139}$$

The surfaces $r = r_E^\pm(m, a, \theta)$, between which it is impossible to stand still, are known as the "stationary limit" surfaces.

Compare this with the location of the event horizons

$$r_\pm \equiv m \pm \sqrt{m^2 - a^2}. \tag{1.140}$$

We see that

$$2m \geq r_E^+(m, a, \theta) \geq r_+ \geq m \geq r_- \geq r_E^-(m, a, \theta) \geq 0. \tag{1.141}$$

In fact $r_E^+(m, a, \theta) \geq r_+$ with equality only at $\theta = 0$ and $\theta = \pi$ (corresponding to the axis of rotation). Similarly $r_E^-(m, a, \theta) \leq r_-$ with equality only at the axis of rotation. (This inner stationary limit surface touches the inner horizon on the axis of rotation, but then plunges down to the curvature singularity $r = 0$ at the equator, $\theta = \pi/2$.)

Restricting attention to the "outer" region: There is a region between the outer stationary limit surface and the outer event horizon in which it is impossible to "stand still", but it is still possible to escape to infinity. This region is known as the "ergosphere". Note that as rotation is switched off, $a \to 0$, the stationary limit surface moves to lie on top of the event horizon and the ergosphere disappears. (Sometimes one sees authors refer to the stationary limit surface as the "ergosurface", and to refer to the ergosphere as the "ergoregion". Some authors furthermore use the word "ergoregion" to refer to the entire region between the stationary limit surfaces – including the black hole region located between the horizons.)

- By setting $r = r_E^\pm(m, a, \theta)$ and then replacing

$$\mathrm{d}r \to \mathrm{d}r_E^\pm = \left(\frac{\mathrm{d}r_E^\pm}{\mathrm{d}\theta}\right) \mathrm{d}\theta = \pm \frac{a^2 \cos\theta \sin\theta}{\sqrt{m^2 - a^2\cos^2\theta}}\, \mathrm{d}\theta \tag{1.142}$$

one can find the intrinsic induced 3-geometry on the ergosurface. One form of the result is

$$\begin{aligned} \mathrm{d}s^2_{\text{3-geometry}} &= -2a\sin^2\theta\, \mathrm{d}t\, \mathrm{d}\phi \\ &\quad + \frac{2m^3 r_E^\pm}{m^2 - a^2\cos^2\theta}\, \mathrm{d}\theta^2 + 2\left[m r_E^\pm + a^2\sin^2\theta\right]\sin^2\theta\, \mathrm{d}\phi^2. \end{aligned} \tag{1.143}$$

This result is actually quite horrid, since if we wish to be explicit

$$\begin{aligned} ds^2_{\text{3-geometry}} &= -2a \sin^2\theta \; dt \; d\phi \\ &+ \frac{2m^3(m \pm \sqrt{m^2 - a^2\cos^2\theta})}{m^2 - a^2\cos^2\theta} \, d\theta^2 \\ &+ 2\left[m(m \pm \sqrt{m^2 - a^2\cos^2\theta}) + a^2\sin^2\theta\right] \sin^2\theta \; d\phi^2. \end{aligned} \tag{1.144}$$

- By now additionally setting $dt = 0$ one can find the intrinsic induced 2-geometry on an instantaneous constant time-slice of the ergosurface (Pelavas, Neary, and Lake 2001). One obtains

$$\begin{aligned} ds^2_{\text{2-geometry}} &= +\frac{2m^3(m \pm \sqrt{m^2 - a^2\cos^2\theta})}{m^2 - a^2\cos^2\theta} \, d\theta^2 \\ &+ 2\left[m(m \pm \sqrt{m^2 - a^2\cos^2\theta}) + a^2\sin^2\theta\right] \sin^2\theta \; d\phi^2. \end{aligned}$$

- The intrinsic area of the ergosurface can then be evaluated as an explicit expression involving two incomplete elliptic integrals (Pelavas, Neary, and Lake 2001). A more tractable approximate expression for the area of the outer ergosurface is (Pelavas, Neary, and Lake 2001):

$$A_E^+ = 4\pi \left[(2m)^2 + a^2 + \frac{3a^4}{20m^2} + \frac{33a^6}{280m^4} + \frac{191a^8}{2880m^6} + \mathcal{O}\left(\frac{a^{10}}{m^8}\right)\right]. \tag{1.145}$$

- The two-dimensional Ricci scalar is easily calculated but is again quite horrid (Pelavas, Neary, and Lake 2001). A tractable approximation is

$$R_E^+ = \frac{1}{2m^2}\left[1 + \frac{3(1 - 6\cos^2\theta)a^2}{4m^2} - \frac{3(1 - 5\cos^2\theta + 3\cos^4\theta)a^4}{8m^4} + \mathcal{O}\left(\frac{a^6}{m^6}\right)\right]. \tag{1.146}$$

On the equator of the outer ergosurface there is a simple exact result that

$$R_E^+(\theta = \pi/2) = \frac{4m^2 + 5a^2}{4m^2(2m^2 + a^2)}. \tag{1.147}$$

At the poles the Ricci scalar has a delta function contribution coming from conical singularities at the north and south poles of the ergosurface. Near the north pole the metric takes the form

$$ds^2 = K\left[d\theta^2 + \left(1 - \frac{a^2}{m^2}\right)\theta^2 \, d\phi^2 + \mathcal{O}(\theta^4)\right], \tag{1.148}$$

where $K$ is an irrelevant constant.

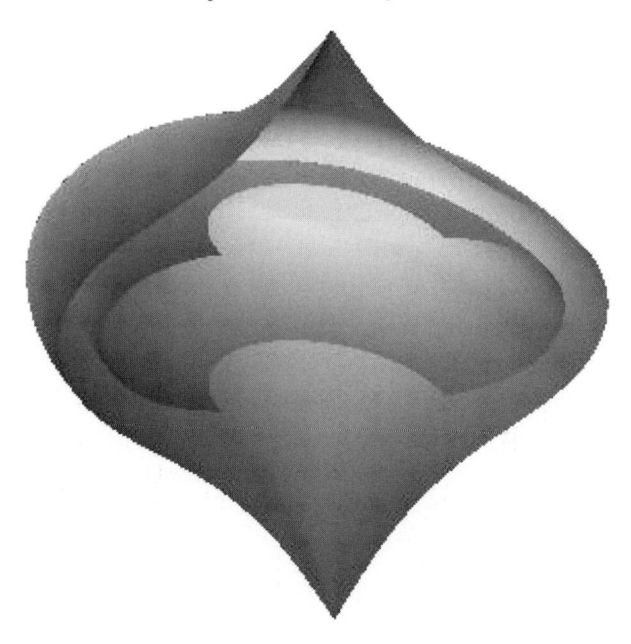

Fig. 1.2. Isometric embeddings in Euclidean space of the outer ergosurface and a portion of the outer horizon. For these images $a = 0.90\ m$. A cut to show the (partial) outer horizon, which is not isometrically embeddable around the poles, is shown. The polar radius of the ergosurface diverges for $a \to m$ as the conical singularities at the poles become more pronounced.

- An isometric embedding of the outer ergosurface and a portion of the outer horizon in Euclidean 3-space is presented in Fig. 1.2. Note the conical singularities at the north and south poles.
- The same techniques applied to the inner ergosurface yield

$$A_E^- = 4\pi \left[ \frac{2\sqrt{2}-1}{3} a^2 + \frac{12\sqrt{2} - 13a^4}{20m^2} + \frac{292\sqrt{2} - 283a^6}{280m^4} + \mathcal{O}\left(\frac{a^8}{m^6}\right) \right], \tag{1.149}$$

and

$$R_E^- = -\frac{2(5 - 2\cos^2\theta)}{(2-\cos^2\theta)^2 a^2} - \frac{76 - 198\cos^2\theta + 128\cos^4\theta - 25\cos^6\theta}{2(8 - 12\cos^2\theta + 6\cos^4\theta - \cos^6\theta)m^2} + \mathcal{O}\left(\frac{a^2}{m^4}\right). \tag{1.150}$$

- At the equator, the inner ergosurface touches the physical singularity. In terms of the intrinsic geometry of the inner ergosurface this shows up as the two-dimensional Ricci scalar becoming infinite. The metric takes the form

$$\mathrm{d}s^2 = a^2(\theta - \pi/2)^2\,\mathrm{d}\theta^2 + a^2\{2 - 3(\theta - \pi/2)^2\}\,\mathrm{d}\phi^2 + \mathcal{O}[(\theta - \pi/2)^4], \tag{1.151}$$

and after a change of variables can be written

$$ds^2 = dh^2 + \{2a^2 - 6a|h|\} \, d\phi^2 + \mathcal{O}[h^2]. \tag{1.152}$$

At the poles the inner ergosurface has conical singularities of the same type as the outer ergosurface. See Equation (1.148).

- Working now in Kerr–Schild Cartesian coordinates, since the ergosurface is defined by the coordinate condition $g_{tt} = 0$, we see that this occurs at

$$r^4 - 2mr^3 + a^2 z^2 = 0. \tag{1.153}$$

But recall that $r(x, y, z)$ is a rather complicated function of the Cartesian coordinates, so even in the "background" geometry the ergosurface is quite tricky to analyse. In fact it is better to describe the ergosurface parametrically by observing

$$\sqrt{x_E^2 + y_E^2} = \sqrt{r_E(\theta)^2 + a^2} \, \sin\theta; \qquad z_E = r_E(\theta) \, \cos\theta. \tag{1.154}$$

That is

$$\sqrt{x_E^2 + y_E^2} = \left( \left[ m \pm \sqrt{m^2 - a^2 \cos^2\theta} \right]^2 + a^2 \right)^{1/2} \sin\theta; \tag{1.155}$$

$$z_E = (m \pm \sqrt{m^2 - a^2 \cos^2\theta}) \, \cos\theta. \tag{1.156}$$

So again we see that the ergosurfaces are quite tricky to work with. See Fig. 1.3 for a polar slice through the Kerr spacetime in these Cartesian coordinates.

Putting these particular mathematical issues aside: physically and astrophysically, it is extremely important to realize that (assuming validity of the Einstein equations, which is certainly extremely reasonable given their current level of experimental and observational support) you should trust in the existence of the ergosurface and event horizon (that is, the outer ergosurface and outer horizon), and the region immediately below the event horizon.

However, you should *not* physically trust in the inner horizon or the inner ergosurface. Although they are certainly there as mathematical solutions of the exact vacuum Einstein equations, there are good physics reasons to suspect that the region at and inside the inner horizon, which can be shown to be a Cauchy horizon, is grossly unstable – even classically – and unlikely to form in any real astrophysical collapse.

Aside from issues of stability, note that although the causal pathologies (closed timelike curves) in the Kerr spacetime have their genesis in the maximally extended $r < 0$ region, the effects of these causal pathologies can reach out into part of the $r > 0$ region, in fact out to the inner horizon at $r = r_-$ (Hawking and Ellis 1975) – so the inner horizon is also a chronology horizon for the maximally extended Kerr

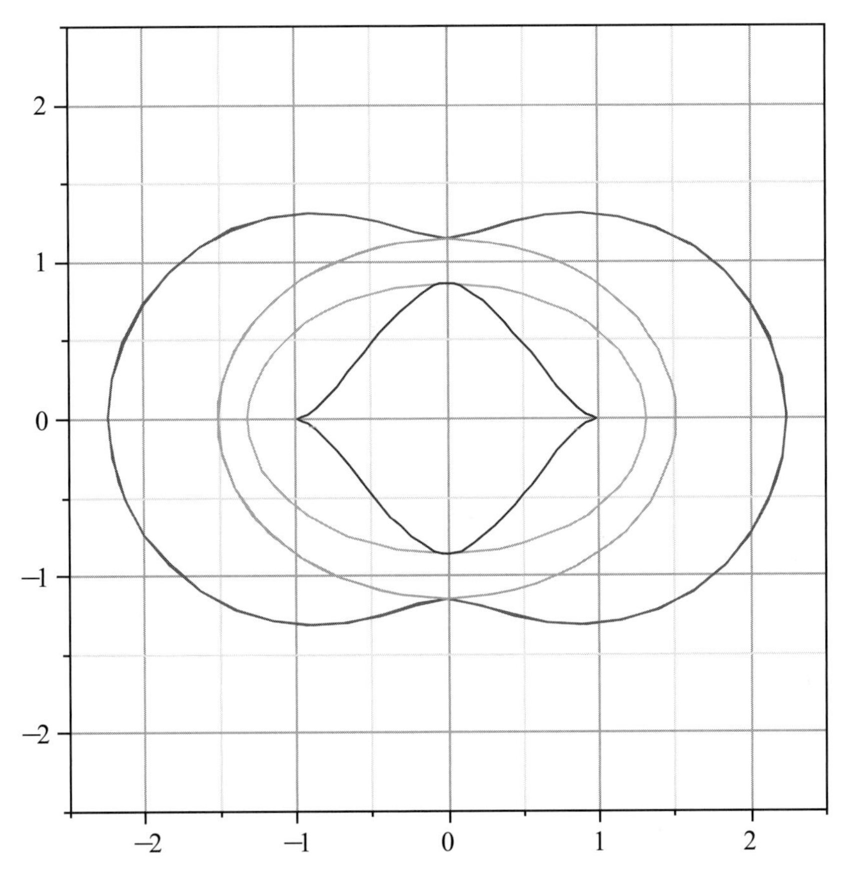


Fig. 1.3. Polar slice through the Kerr spacetime in Cartesian Kerr–Schild coordinates. Location of the horizons, ergosurfaces, and curvature singularity is shown for $a = 0.99\,m$ and $m = 1$. Note that the inner and outer horizons are ellipses in these coordinates, while the inner and outer ergosurfaces are more complicated. The curvature singularity lies at the kink in the inner ergosurface.

spacetime. Just what does go on deep inside a classical or semiclassical black hole formed in real astrophysical collapse is still being debated – see for instance the literature regarding "mass inflation" for some ideas (Poisson and Israel 1989; Balbinot and Poisson 1993). For astrophysical purposes it is certainly safe to discard the $r < 0$ region, and almost all relativists would agree that it is safe to discard the entire region inside the inner horizon $r < r_-$.

## 1.11 Killing vectors

By considering the Killing vectors of the Kerr spacetime it is possible to develop more invariant characterizations of the ergosurfaces and horizons. The two obvious

Killing vectors are the time translation Killing vector $K^a$ and the azimuthal Killing vector $R^a$; in addition any constant coefficient linear combination $a\, K^a + b\, R^a$ is also a Killing vector – and this exhausts the set of all Killing vectors of the Kerr spacetime.

The time translation Killing vector $K^a$ is singled out as being the unique Killing vector that approaches a unit timelike vector at spatial infinity. The vanishing of the norm

$$g_{ab}\, K^a\, K^b = 0 \tag{1.157}$$

is an invariant characterization of the ergosurfaces – the ergosurfaces are where the time translation Killing vector becomes null.

Similarly, the horizon is a null 3-surface (since it is defined by a set of photon trajectories) that is further characterized by the fact that it is invariant (since the photons neither fall into nor escape from the black hole). This implies that there is *some* Killing vector $a\, K^a + b\, R^a$ that becomes null on the horizon (a result that holds in any stationary spacetime, and leads to the "rigidity theorem"). Without loss of generality we can set $a \to 1$ and define $b \to \Omega_H$ so that the horizon is invariantly characterized by the condition

$$g_{ab}\, (K^a + \Omega_H\, R^a)\, (K^b + \Omega_H\, R^b) = 0. \tag{1.158}$$

It is then a matter of computation to verify that this current definition of $\Omega_H$ coincides with that in the previous discussion.

## 1.12 Comments

There are a (large) number of things I have not discussed in this brief introduction, but I have to cut things off somewhere. In particular, you can go to several textbooks, the original literature, and the rest of this book to see discussions of:

- The Kerr–Newman geometry (a charged rotating black hole).
- Carter–Penrose diagrams of causal structure.
- Maximal analytic extensions of the Kerr spacetime.
- Achronal regions and chronology horizons (time travel) in the idealized Kerr spacetime.
- Extremal and naked Kerr spacetimes; cosmic censorship.
- Particle and photon orbits in the Kerr black hole.
- Geodesic completeness.
- Physically reasonable sources and possible "interior solutions" for the Kerr spacetime.
- Penrose process – energy extraction from a rotating black hole.
- Black hole uniqueness theorems.
- Singularity theorems guaranteeing the formation of black holes in certain circumstances.

- The classic laws of black hole mechanics; area increase theorem.
- Black hole thermodynamics (Hawking temperature and Bekenstein entropy).

These and other issues continue to provoke considerable ongoing research – and I hope this brief introduction will serve to orient the reader, whet one's appetite, and provoke interest in the rest of this book.

## Acknowledgments

Supported by the Marsden Fund administered by the Royal Society of New Zealand. Figure 1.1 courtesy of Silke Weinfurtner. Figure 1.2 courtesy of Kayll Lake. Additionally, I wish to acknowledge useful comments from Silke Weinfurtner, Kayll Lake, Bartolome Alles, and Roy Kerr.

## References

Adler, R. J., Bazin, M. & Schiffer, M. (1975), *Introduction to General Relativity*, Second edition (McGraw-Hill, New York).
[It is important to acquire the 1975 second edition, the 1965 first edition does not contain any discussion of the Kerr spacetime.]

Balbinot, R. & Poisson, E. (1993), "Mass inflation: the semiclassical regime", *Phys. Rev. Lett.* **70**, 13.

Birkhoff, G. D. (1923), *Relativity and Modern Physics* (Harvard University Press, Cambridge).

Burinskii, A. & Kerr, R. P. (1995), "Nonstationary Kerr congruences", arXiv:gr-qc/9501012.

Carroll, S. (2004), *An Introduction to General Relativity: Spacetime and Geometry* (Addison Wesley, San Francisco).

Chandrasekhar, S. (1998), *The Mathematical Theory of Black Holes* (Oxford University Press, Oxford).

Debney, G. C., Kerr, R. P. & Schild, A. (1969), "Solutions of the Einstein and Einstein–Maxwell equations", *J. Math. Phys.* **10**, 1842.

Deser, S. & Franklin, J. (2005), "Schwarzschild and Birkhoff *à la* Weyl", *Am. J. Phys.* **73**, 261 [arXiv:gr-qc/0408067].

D'Inverno, R. (1992), *Introducing Einstein's Relativity* (Oxford University Press, Oxford).

Doran, C. (2000), "A new form of the Kerr solution", *Phys. Rev.* **D 61**, 067503 [arXiv:gr-qc/9910099].

Einstein, A. (1915), "Zur allgemeinen Relativitatstheorie", *Sitzungsber. Preuss. Akad. Wiss.*, 778; Addendum ibid., 799.

Fackerell, E. D. & Kerr, R. P. (1991), "Einstein vacuum field equations with a single non-null Killing vector", *Gen. Rel. Grav.* **23**, 861.

Goldberg, J. N. & Kerr, R. P. (1964), "Asymptotic properties of the electromagnetic field", *J. Math. Phys.* **5**, 172.

Gullstrand, A. (1922), "Allgemeine Lösung des statischen Einkörperproblems in der Einsteinschen Gravitationstheorie", *Ark. Mat. Astron. Fys.* **16**(8), 1.

Hartle, J. (2003), *Gravity: an Introduction to Einstein's General Relativity* (Addison Wesley, San Francisco).

Hawking, S. W. & Ellis, G. F. R. (1975), *The Large Scale Structure of Space-time* (Cambridge University Press, Cambridge).

Hilbert, D. (1915), "Die Grundlagen der Physik (Erste Mitteilung)", *Nachr. Gesell. Wiss. Göttingen Math. Phys. Kl.*, 395.

Jebsen, J. T. (1921), "Über die allgemeinen kugelsymmetrischen Lösungen der Einsteinschen Gravitationsgleichungen im Vakuum", *Ark. Mat. Ast. Fys.* **15**(18), 1.

Johansen, N. V. & Ravndal, F. (2006), "On the discovery of Birkhoff's theorem", *Gen. Rel. Grav.* **38**, 537 [arXiv: physics/0508163].

Kerr, R. P. (1959), "The Lorentz-covariant approximation method in general relativity", *Nuovo Cimento* **13**, 469.

Kerr, R. P. & Goldberg, J. N. (1961), "Einstein spaces with four-parameter holonomy groups", *J. Math. Phys.* **2**, 332.

Kerr, R. P. (1963), "Gravitational field of a spinning mass as an example of algebraically special metrics", *Phys. Rev. Lett.* **11**, 237.

Kerr, R. P. (1965), "Gravitational collapse and rotation", published in: *Quasi-stellar sources and gravitational collapse: including the proceedings of the First Texas Symposium on Relativistic Astrophysics*, ed. Robinson, I., Schild, A. & Schücking, E. L. (University of Chicago Press, Chicago), pp. 99–102.
The conference was held in Austin, Texas, on 16–18 December 1963.

Kerr, R. P. & Schild, A. (1965), "Some algebraically degenerate solutions of Einstein's gravitational field equations", *Proc. Symp. Appl. Math.* **17**, 199.

Kerr, R. P. & Debney, G. C. (1970), "Einstein spaces with symmetry groups", *J. Math. Phys.* **11**, 2807.

Kerr, R. P. & Wilson, W. B. (1977), "Singularities in the Kerr–Schild metrics", *Proceedings of the 8th International Conference on General Relativity and Gravitation*, 7–12 August 1977, Waterloo, Ontario, Canada, p. 378.

Lake, K. (2003), "Comment on negative squares of the Weyl tensor", *Gen. Rel. Grav.* **35**, 2271 [arXiv:gr-qc/0302087].

Lake, K. (2004), "Differential invariants of the Kerr vacuum", *Gen. Rel. Grav.* **36**, 1159 [arXiv:gr-qc/0308038].

Misner, C. A., Thorne, K. & Wheeler, J. A. (1973), *Gravitation* (W. H. Freeman, San Francisco).

O'Neill, B. (1995), *Geometry of Kerr Black Holes* (A. K. Peters, Wellesly, MA).

Oppenheimer, J. R. & Snyder, H. (1939), "On continued gravitational contraction", *Phys. Rev.* **56**, 455.

Painlevé, P. (1921), "La mécanique classique et la theorie de la relativité", *C. R. Acad. Sci.* **173**, 677.

Pelavas, N., Neary, N. & Lake, K. (2001), "Properties of the instantaneous ergo surface of a Kerr black hole", *Class. Quantum Grav.* **18**, 1319 [arXiv:gr-qc/0012052].

Pfister, H. (2005), "On the history of the so-called Lense–Thirring effect", http://philsci-archive.pitt.edu/archive/00002681/01/lense.pdf

Plebański, J. & Krasiński, A. (2006), *An Introduction to General Relativity and Cosmology* (Cambridge University Press, Cambridge).

Poisson, E. & Israel, W. (1989), "Inner-horizon instability and mass inflation in black holes", *Phys. Rev. Lett.* **63**, 1663–6.

Schwarzschild, K. (1916a), "Über das Gravitationsfeld eines Massenpunktes nach der Einsteinschen Theorie", *Sitzungsber. Preuss. Akad. Wiss.*, 189.

Schwarzschild, K. (1916b), "Über das Gravitationsfeld einer Kugel aus inkompressibler Flussigkeit nach der Einsteinschen Theorie", *Sitzungsber. Preuss. Akad. Wiss.*, 424.

Smarr, L. (1973), "Surface geometry of charged rotating black holes", *Phys. Rev.* **D 7**, 269.

Stephani, H., Kramer, D., MacCallum, M. A. H., Hoenselaers, C. & Herlt, E. (2002), *Exact Solutions of Einstein's Field Equations* (Cambridge University Press, Cambridge).

Thirring, H. & Lense, J. (1918), "Über den Einfluss der Eigenrotation der Zentralkörper auf die Bewegung der Planeten und Monde nach der Einsteinschen Gravitationstheorie", *Phys. Z.* **19**, 156;
English translation by B. Mashoon, F. W. Hehl & D. S. Theiss, "On the influence of the proper rotations of central bodies on the motions of planets and moons in Einstein's theory of gravity", *Gen. Rel. Grav.* **16**, 727 (1984).

Visser, M. (1998), "Acoustic black holes: horizons, ergospheres, and Hawking radiation", *Class. Quantum Grav.* **15**, 1767 [arXiv:gr-qc/9712010].

Visser, M. & Weinfurtner, S. Ch. (2005), "Vortex geometry for the equatorial slice of the Kerr black hole", *Class. Quantum Grav.* **22**, 2493 [arXiv:gr-qc/0409014].

Weir, G. & Kerr, R. P. (1977), "Diverging type-D metrics", *Proc. Roy. Soc. Lond.*, **A 355**, 31.

# 2

# Discovering the Kerr and Kerr–Schild metrics

Roy P. Kerr

## 2.1 Introduction

The story of this metric begins with a paper by Alexei Zinovievich Petrov (1954) where the simultaneous invariants and canonical forms for the metric and conformal tensor are calculated at a general point in an Einstein space. This paper took a while to be appreciated in the West, probably because the Kazan State University journal was not readily available, but Felix Pirani (1957) used it as the foundation of an article on gravitational radiation theory. He analysed gravitational shock waves, calculated the possible jumps in the Riemann tensor across the wave fronts, and related these to the Petrov types.

I was a graduate student at Cambridge, from 1955 to 1958. In my last year I was invited to attend the relativity seminars at Kings College in London, including one by Felix Pirani on his 1957 paper. At the time I thought that he was stretching when he proposed that radiation was type N, and I said so, a rather stupid thing for a graduate student with no real supervisor to do.[1] It seemed obvious that a superposition of type N solutions would not itself be type N, and that gravitational waves near a macroscopic body would be of general type, not type N.

Perhaps I did Felix an injustice. His conclusions may have been oversimplified but his paper had some very positive consequences. Andrzej Trautman computed the asymptotic properties of the Weyl tensor for outgoing radiation by generalizing Sommerfeld's work on electromagnetic radiation, confirming that the far field is type N. Bondi, van der Burg and Metzner (1962) then introduced appropriate null coordinates to study gravitational radiation in the far zone and related this to the results of Petrov and Pirani.

[1] My nominal supervisor was a particle physicist who had no interest in general relativity.

The Kerr Spacetime: Rotating Black Holes in General Relativity. Ed. David L. Wiltshire, Matt Visser and Susan M. Scott. Published by Cambridge University Press. 

In 1958 I went to Syracuse University as a research associate of Peter Bergmann. While there I was invited to join Joshua Goldberg at the Aeronautical Research Laboratory in Dayton Ohio.[2] There was another relativist at the lab, Dr Joseph Schell, who had studied Einstein's unified field theory under Vaclav Hlavaty. Josh was about to go on study leave to Europe for a few months, and did not want to leave Joe by himself.

Before he left, Josh and I became interested in the new methods that were entering general relativity from differential geometry at that time. We did not have a copy of Petrov's paper so our first project was to re-derive his classification using projective geometry, something which was being done by many other people throughout the world at that time. In each empty Einstein space, $\mathfrak{E}$, the conformal tensor determines four null "eigenvectors" at each point. The metric is called algebraically special (AS)[3] if two of these eigenvectors coincide. This vector is then called a principal null vector (PNV) and the field of these is called a principal null "congruence".

After this, we used a tetrad formulation to study vacuum Einstein spaces with degenerate holonomy groups (Goldberg and Kerr (1961), Kerr and Goldberg (1961)). The tetrad used consisted of two null vectors and two *real* orthogonal space-like vectors,

$$\mathrm{d}s^2 = (\omega^1)^2 + (\omega^2)^2 + 2\omega^3\omega^4.$$

We proved that the holonomy group must be an even-dimensional subgroup of the Lorentz group at each point, and that if its dimension is less than six then coordinates can be chosen so that the metric has the following form:

$$\mathrm{d}s^2 = \mathrm{d}x^2 + \mathrm{d}y^2 + 2\,\mathrm{d}u\left(\mathrm{d}v + \rho\,\mathrm{d}x + \tfrac{1}{2}(\omega - \rho_{,x}v)\,\mathrm{d}u\right),$$

where both $\rho$ and $\omega$ are independent of $v$,[4] and

$$\begin{aligned} \rho_{,xx} + \rho_{,yy} &= 0 \\ \omega_{,xx} + \omega_{,yy} &= 2\rho_{,ux} - 2\rho\rho_{,xx} - (\rho_{,x})^2 + (\rho_{,y})^2 \end{aligned}$$

This coordinate system was not quite uniquely defined. If $\rho$ is bilinear in $x$ and $y$ then it can be transformed to zero, giving the well-known plane-fronted wave

[2] There is a claim spread on internet that we were employed to develop an antigravity engine to power spaceships. This is rubbish! The main reason why the US Air Force had created a General Relativity section was probably to show the navy that they could also do pure research. The only real use that the USAF made of us was when some crackpot sent them a proposal for antigravity or for converting rotary motion inside a spaceship to a translational driving system. These proposals typically used Newton's equations to prove non-conservation of momentum for some classical system.

[3] The term algebraically degenerate is sometimes used instead.

[4] The simple way that the coordinate $v$ appears was to prove typical of these algebraically special metrics.

solutions. These are type N, and have two-dimensional holonomy groups. The more general metrics are type III with four-dimensional holonomy groups.

In September 1961 Joshua joined Hermann Bondi, Andrzej Trautman, Ray Sachs and others at King's College in London. By this time it was well known that all such AS spaces possess a null congruence whose vectors are both geodesic and shearfree. These are the degenerate "eigenvectors" of the conformal tensor at each point. Andrzej suggested to Josh and Ray how they might prove the converse. This led to the celebrated Goldberg–Sachs theorem (see Goldberg and Sachs (1962)).

**Theorem 2.1** *A vacuum metric is algebraically special if and only if it contains a geodesic and shearfree null congruence.*

Either properties of the congruence, being geodesic and shear-free, or a property of the conformal tensor, algebraic degeneracy, could be considered fundamental with the others following from the Goldberg–Sachs theorem. It is likely that most thought that the algebra was fundamental, but I believe that Ivor Robinson and Andrzej Trautman (1962) were correct when they emphasized the properties of the congruence instead. They showed that for any Einstein space with a shear-free null congruence which is also hypersurface orthogonal there are coordinates for which

$$ds^2 = 2r^2 P^{-2}\, d\zeta\, d\bar{\zeta} - 2\, du\, dr - (\Delta \ln P - 2r(\ln P)_{,u} - 2m(u)/r)\, du^2,$$

where $\zeta$ is a complex coordinate, $\zeta = (x + iy)/\sqrt{2}$, say, so that

$$2\, d\zeta\, d\bar{\zeta} = dx^2 + dy^2.$$

The one remaining field equation is,

$$\Delta\Delta(\ln P) + 12m(\ln P)_{,u} - 4m_{,u} = 0, \quad \Delta = 2P^2 \partial_\zeta \partial_{\bar{\zeta}}.$$

The PNV[5] is $\boldsymbol{k} = k^\mu \partial_\mu = \partial_r$, where $r$ is an affine parameter along the rays. The corresponding differential form is $k = k_\mu dx^\mu = du$, so that $\boldsymbol{k}$ is the normal to the surfaces of constant $u$. The coordinate $u$ is a retarded time, the surfaces of constant $r$, $u$ are distorted spheres with metric $ds^2 = 2r^2 P^{-2}\, d\zeta\, d\bar{\zeta}$ and the parameter $m(u)$ is loosely connected with the system's mass. This gives the complete solution to the Robinson–Trautman problem.[6] In 1962 Goldberg and I attended a month-long

[5] The letter $\boldsymbol{k}$ will be used throughout this article to denote the PNV.

[6] In the study of exact solutions, "solving" a problem usually means introducing a useful coordinate system, solving the easier Einstein equations and replacing the ten components of the metric tensor with a smaller number of functions, preferably of less than four variables. These will then have to satisfy a residual set of differential equations, the harder ones, which usually have no known complete solution. For example, the remaining field equation for the Robinson–Trautman metrics is highly non-linear and has no general solution.

Fig. 2.1. Ivor Robinson and Andrzej Trautman constructed all Einstein spaces possessing a hypersurface orthogonal shearfree congruence. Whereas Bondi and his colleagues were looking at spaces with these properties asymptotically, far from any sources, Robinson and Trautman went a step further, constructing exact solutions. (Images courtesy of Andrzej Trautman and the photographer, Marek Holzman.)

meeting in Santa Barbara. It was designed to get mathematicians and relativists talking to each other. Perhaps the physicists learned a lot about more modern mathematical techniques, but I doubt that the geometers learned much from the relativists. All that aside, I met Alfred Schild at this conference. He had just persuaded the Texas state legislators to finance a Center for Relativity at the University of Texas, and had arranged for an outstanding group of relativists to join. These included Roger Penrose and Ray Sachs, but neither could come immediately and so I was invited to visit for the 1962–1963 academic year.

After Santa Barbara, Goldberg and I flew to a conference held at Jablonna near Warsaw. This was the third precursor to the triennial meetings of the GRG society and therefore it could be called GR3. Robinson and Trautman (1964) presented a paper on *"Exact Degenerate Solutions"* at this conference. They spoke about their well-known solution and also showed that when the rays are not hypersurface orthogonal coordinates can be chosen so that

$$ds^2 = -P^2[(d\xi - ak)^2 + (d\eta - bk)^2] + 2\,d\rho\,k + ck^2,$$

Fig. 2.2. Ezra T. Newman, who with T. Unti and L. A. Tambourino, studied the field equations for diverging and rotating algebraically special Einstein spaces.

where, as usual, $k$ is the PNV. Its components, $k_\alpha$, are independent of $\rho$, but $a$, $b$, $c$ and $P$ may be functions of all four coordinates.

I was playing around with the structure of the Einstein equations during 1962, using the new (to physicists) methods of tetrads and differential forms. I had written out the equations for the curvature using a complex null tetrad and self-dual bivectors, and then studied their integrability conditions. In particular, I was interested in the same problem that Robinson and Trautman were investigating where $k$ was not a gradient, i.e. twisting, but there was a major road block in my way. Alan Thompson had also come to Austin that year and was also interested in these methods. Although there seemed to be no reason why there should not be many algebraically special spaces, Alan kept quoting a result from a preprint of a paper by Newman, Tambourino and Unti (1963) in which they had "proved" that the only possible space with a diverging and rotating PNV is NUT space, a one parameter generalization of the Schwarzschild metric. They derived this result using the new Newman–Penrose (N–P) spinor formalism (Newman and Penrose, 1962). Their equations were essentially the same as those obtained by people such as myself using self-dual bivectors: only the names are different. I could not understand how the equations that I was studying could possibly lead to their claimed result, but could only presume it must be so since I did not have a copy of their paper.

In the spring of 1963 Alan obtained a preprint of this paper and loaned it to me. I thumbed through it quickly, trying to see where their hunt for solutions had died. The N–P formalism assigns a different Greek letter to each component of the connection, so I did not try to read it carefully, just rushed ahead until I found what appeared to be the key equation,

$$\tfrac{1}{3}(n_1 + n_2 + n_3)a^2 = 0, \tag{2.1}$$

where the $n_i$ were all small integers, between $-4$ and $+4$. Their sum was not zero so this gave $a = 0$. I had no idea what $a$ represented, but its vanishing seemed to be disastrous and so I looked more carefully to see where this equation was coming from. Three of the previous equations, each involving first derivatives of some of the connection components, had been differentiated and then added together. All the second derivatives cancelled identically and most of the other terms were eliminated using other N–P equations, leaving (2.1).

The mistake that Newman *et al.* made was that they did not notice that they were simply recalculating one component of the Bianchi identities by adding together the appropriate derivatives of three of their curvature equations, and then simplifying the result by using some of their other equations, undifferentiated. The result should have agreed with one of their derived Bianchi identities involving derivatives of the components of the conformal tensor, the $\Psi_i$ functions, giving

$$n_1 + n_2 + n_3 \equiv 0. \tag{2.2}$$

In effect, they rediscovered one component of the identities, but with numerical errors. The real fault was the way the N–P formalism confuses the Bianchi identities with the derived equations involving derivatives of the $\Psi_i$ variables.

Alan Thompson and myself were living in adjoining apartments, so I dashed next door and told him that their result was incorrect. Although it was unnecessary, we recalculated the first of the three terms, $n_1$, obtained a different result to the one in the preprint, and verified that (2.2) was now satisfied. Once this blockage was out of the way, I was then able to continue with what I had been doing and derive the metric and field equations for twisting algebraically special spaces. The coordinates I constructed turned out to be essentially the same as the ones given in Robinson and Trautman (1962). This shows that they are the "natural" coordinates for this problem since the methods used by them were very different to those used by me. Ivor loathed the use of such things as N–P or rotation coefficients, and Andrzej and he had a nice way of proving the existence of their canonical complex coordinates $\zeta$ and $\bar{\zeta}$. I found this same result from one of the Cartan equations, as will be shown in the next section, but I have no doubt that their method is more elegant. Ivor explained it to me on more than one occasion, but unfortunately I never understood what he was saying![7]

[7] While writing this article I read their 1962 paper and finally understood how they derived their coordinates. It only took me 45 years.

At this point I presented the results at a monthly Relativity conference held at the Steven's Institute in Hoboken, N. J. When I told Ted Newman that (2.1) should have been identically zero, he said that they knew that $n_1$ was incorrect, but that the value for $n_2$ given in the preprint was a misprint and so (2.2) was still not satisfied. I replied that since the sum had to be zero the final term, $n_3$ must also be incorrect. Alan and I recalculated it that evening, confirming that (2.2) was satisfied.[8]

## 2.2 Discovery of the Kerr metric

When I realized that the attempt by Newman *et al.* to find all rotating AS spaces had foundered and that Robinson and Trautman appeared to have stopped with the static ones, I rushed headlong into the search for these metrics.

Why was the problem so interesting to me? Schwarzschild, by far the most significant physical solution known at that time, has an event horizon. A spherically symmetric star that collapses inside this is forever lost to us, but it was not known whether angular momentum could stop this collapse to a black hole. Unfortunately, there was no known metric for a rotating star. Schwarzschild was an example of the Robinson–Trautman metrics, none of which could contain a rotating source as they were all hypersurface orthogonal. Many had tried to solve the Einstein equations assuming a stationary and axially symmetric metric, but none had succeeded in finding any physically significant rotating solutions. The equations for such metrics are complicated non-linear PDEs in two variables. What was needed was some extra condition that would reduce these to ODEs, and this might be the assumption that the metric is AS.

The notation used in the rest of this chapter is fairly standard. There were two competing formalisms being used around 1960, complex tetrads and spinors. I used the former,[9] Newman *et al.* the latter. The derived equations are essentially identical, but each approach has some advantages. Spinors make the the Petrov classification trivial once it has been shown that a tensor with the symmetries of the conformal tensor corresponds to a completely symmetric spinor, $\Psi_{ABCD}$. The standard notation for the components of this tensor is

$$\Psi_0 = \Psi_{0000}, \quad \Psi_1 = \Psi_{0001}, \ldots \; \Psi_4 = \Psi_{1111}.$$

If $\zeta^A$ is an arbitrary spinor then the equation

$$\Psi_{ABCD}\,\zeta^A\zeta^B\zeta^C\zeta^D = 0$$

[8] Robinson and Trautman (1962) also doubted the original claim by Newman *et al.* since they knew that the linearized equations had many solutions.

[9] Robinson and Trautman also had a fairly natural complex tetrad approach.

is a homogeneous quartic equation with four complex roots, $\zeta_i^A$. The related real null vectors, $Z_i^{\alpha\dot{\alpha}} = \zeta_i^\alpha \zeta_i^{\dot{\alpha}}$, are the four PNVs of Petrov. The spinor $\zeta^\alpha = \delta_0^\alpha$ gives a PNV if $\Psi_0 = 0$. It is a repeated root and therefore it is the principal null vector of an AS spacetime if $\Psi_1 = 0$ as well.

The Kerr (1963) letter presented the main results of my calculations but gave few details.[10] The methods that I used to solve the equations for AS spaces are essentially those used in "Exact Solutions" by Stephani *et al.* (2003), culminating in their Equation (27.27). I will try to use the same notation as in that book since it is almost identical to the one I used in 1963, but I may get some of the signs wrong. Beware!

Suppose that $(\mathbf{e}_a) = (\boldsymbol{m}, \bar{\boldsymbol{m}}, \boldsymbol{l}, \boldsymbol{k})$ is a null tetrad, i.e. a set of four null vectors where the last two are real and the first two are complex conjugates. The corresponding dual forms are $(\omega^a) = (\bar{m}, m, -k, -l)$ and the metric is[11]

$$ds^2 = 2(m\bar{m} - kl) = 2(\omega^1\omega^2 - \omega^3\omega^4). \tag{2.3}$$

The vector $\boldsymbol{k}$ is a PNV and so its direction is uniquely defined, but the other directions are not. The form of the metric tensor in (2.3) is invariant under a combination of a null rotation $(B)$ about $\boldsymbol{k}$, a rotation $(C)$ in the $\boldsymbol{m} \wedge \bar{\boldsymbol{m}}$ plane and a Lorentz transformation $(A)$ in the $\boldsymbol{l} \wedge \boldsymbol{k}$ plane,

$$\boldsymbol{k}' = \boldsymbol{k}, \qquad \boldsymbol{m}' = \boldsymbol{m} + B\boldsymbol{k}, \qquad \boldsymbol{l}' = \boldsymbol{l} + B\bar{\boldsymbol{m}} + \bar{B}\boldsymbol{m} + B\bar{B}\boldsymbol{k}, \tag{2.4a}$$

$$\boldsymbol{k}' = \boldsymbol{k}, \qquad \boldsymbol{m}' = e^{iC}\boldsymbol{m}, \qquad \boldsymbol{l}' = \boldsymbol{l}, \tag{2.4b}$$

$$\boldsymbol{k}' = A\boldsymbol{k}, \qquad \boldsymbol{m}' = \boldsymbol{m}, \qquad \boldsymbol{l}' = A^{-1}\boldsymbol{l}. \tag{2.4c}$$

The most important connection form is

$$\boldsymbol{\Gamma}_{41} = \Gamma_{41a}\omega^a = m^\alpha\, k_{\alpha;\beta}\, dx^\beta$$

The optical scalars of Ray Sachs (Sachs, 1961, 1962) for $\boldsymbol{k}$ are just the components of this form with respect to the $\omega^a$.

$$\sigma = \Gamma_{411} = \text{shear};$$
$$\rho = \Gamma_{412} = \text{complex divergence};$$
$$\kappa = \Gamma_{414} = \text{geodesy}.$$

[10] I spent many years trying to write up this research but, unfortunately, I could never decide whether to use spinors or a complex tetrad, and thus it did not get written up until Kerr and Debney (1970). George Debney also collaborated with Alfred Schild and myself on the Kerr–Schild metrics in Debney *et al.* (1969).

[11] I personally hate the minus sign in this expression and did not use it in 1963, but it seems to have become standard. By the time I finish this article I am sure that I will wish I had stuck with all positive signs!

The fourth component, $\Gamma_{413}$, is not invariant under a null rotation about $\boldsymbol{k}$,

$$\Gamma'_{413} = \Gamma_{413} + B\rho,$$

and has no real geometric significance. It can be set to zero using an appropriate null rotation. Also, since $\boldsymbol{k}$ is geodesic and shearfree for AS metrics, both $\kappa$ and $\sigma$ are zero and therefore

$$\boldsymbol{\Gamma}_{41} = \rho\omega^2. \tag{2.5}$$

If we use the simplest field equations,

$$R_{44} = 2R_{4142} = 0, \quad R_{41} = R_{4112} - R_{4134} = 0, \quad R_{11} = 2R_{4113} = 0,$$

a total of five real equations, and the fact that the metric is AS,

$$\Psi_0 = -2R_{4141} = 0, \quad 2\Psi_1 = -R_{4112} - R_{4134} = 0,$$

then the most important of the second Cartan equations simplifies to

$$\mathrm{d}\boldsymbol{\Gamma}_{41} - \boldsymbol{\Gamma}_{41} \wedge (\boldsymbol{\Gamma}_{12} + \boldsymbol{\Gamma}_{34}) = R_{41ab}\omega^a \wedge \omega^b = R_{4123}\omega^2 \wedge \omega^3. \tag{2.6}$$

Taking the wedge product of (2.6) with $\boldsymbol{\Gamma}_{41}$ and using (2.5),

$$\boldsymbol{\Gamma}_{41} \wedge d\boldsymbol{\Gamma}_{41} = 0. \tag{2.7}$$

This was the key step in my study of these metrics but this result was not found in quite such a simple way. At first, I stumbled around using individual component equations rather than differential forms to look for a useful coordinate system. It was only after I had found this that I realized that using differential forms from the start would have short-circuited several days analysis.

Equation (2.7) is the integrability condition for the existence of complex functions, $\zeta$ and $\Pi$, such that

$$\boldsymbol{\Gamma}_{41} = \mathrm{d}\bar{\zeta}/\Pi, \quad \boldsymbol{\Gamma}_{42} = \mathrm{d}\zeta/\bar{\Pi}.$$

The two functions $\zeta$ and its complex conjugate, $\bar{\zeta}$, will be used as (complex) coordinates. They are not quite unique since $\zeta$ can always be replaced by an arbitrary analytic function $\Phi(\zeta)$.

Using the transformations in (2.4b) and (2.4c),

$$\boldsymbol{\Gamma}_{4'1'} = Ae^{iC}\boldsymbol{\Gamma}_{41} = Ae^{iC}\mathrm{d}\bar{\zeta}/\Pi, \quad \Rightarrow \quad \Pi' = A^{-1}e^{-iC}\Pi.$$

$\Pi'$ can be set to 1 by choosing $Ae^{iC} = \Pi$, and that is what I did in 1963, but it is also common to just use the $C$-transformation to convert it to a real function $P$,

$$\mathbf{\Gamma}_{41} = \rho\omega^2 = \mathrm{d}\bar{\zeta}/P. \tag{2.8}$$

This is the derivation for two of the coordinates used in 1963. Since $\omega^1_\alpha k^\alpha = 0 \rightarrow \boldsymbol{k}(\zeta) = 0$, these functions, $\zeta$, $\bar{\zeta}$, are constant along the PNV.

The other two coordinates were very standard and were used by most people considering similar problems at that time. The simplest field equation is

$$R_{44} = 0 \quad \Rightarrow \quad \boldsymbol{k}\rho = \rho_{|4} = \rho^2,$$

so that the real part of $-\rho^{-1}$ is an affine parameter along the rays. This was the obvious choice for the third coordinate, $r$,

$$\rho^{-1} = -(r + i\Sigma).$$

There was no clear choice for the fourth coordinate, so $u$ was chosen so that $l^\alpha u_{,\alpha} = 1,\ k^\alpha u_{,\alpha} = 0$, a pair of consistent equations.

Given these four coordinates, the basis forms are

$$\begin{aligned}
\omega^1 = m_\alpha \mathrm{d}x^\alpha &= -\mathrm{d}\zeta/P\bar{\rho} = (r - i\Sigma)\mathrm{d}\zeta/P,\\
\omega^2 = \bar{m}_\alpha \mathrm{d}x^\alpha &= -\mathrm{d}\bar{\zeta}/P\rho = (r + i\Sigma)\mathrm{d}\bar{\zeta}/P,\\
\omega^3 = k_\alpha \mathrm{d}x^\alpha &= \mathrm{d}u + L\mathrm{d}\zeta + \bar{L}\mathrm{d}\bar{\zeta},\\
\omega^4 = l_\alpha \mathrm{d}x^\alpha &= \mathrm{d}r + W\mathrm{d}\zeta + \bar{W}\mathrm{d}\bar{\zeta} + H\omega^3.
\end{aligned}$$

where $L$ is independent of $R$, and the coefficients $\Sigma$, $W$ and $H$ have still to be determined.

When all this was substituted into the first Cartan equation, (2.42), and (2.6), the simplest component of the second Cartan equation, (2.43), $\Sigma$ and $W$ were calculated as functions of $L$ and its derivatives,[12]

$$\begin{aligned}
2i\Sigma &= P^2(\bar{\partial}L - \partial\bar{L}), \quad \partial = \partial_\zeta - L\partial_u,\\
W &= -(r + i\Sigma)L_{,u} + i\partial\Sigma.
\end{aligned}$$

The remaining field equations, the "hard" ones, were more complicated, but still fairly straightforward to calculate. Two gave $H$ as a function of a real "mass"

[12] In Kerr (1963) $\Omega$, $D$ and $\Delta$ were used instead of $L$, $\partial$ and $\Sigma$, but the results were the same, *mutatis mutandis*.

function $m(u, \zeta, \bar{\zeta})$ and certain functions of the higher derivatives of $P$ and $L$,[13]

$$H = \tfrac{1}{2}K - r(\ln P)_{,u} - \frac{mr + M\Sigma}{r^2 + \Sigma^2}.$$
$$M = \Sigma K + P^2 \mathrm{Re}[\partial\bar{\partial}\Sigma - 2\bar{L}_{,u}\partial\Sigma - \Sigma\partial_u\partial\bar{L}],$$
$$K = 2P^{-2}\mathrm{Re}[\partial(\bar{\partial}\ln P - \bar{L}_{,u})],$$

Finally, the first derivatives of the mass function, $m$, are given by the rest of the field equations, $R_{31} = 0$ and $R_{33} = 0$,

$$\partial(m + iM) = 3(m + iM)L_{,u}, \tag{2.9a}$$
$$\bar{\partial}(m - iM) = 3(m - iM)\bar{L}_{,u}, \tag{2.9b}$$
$$[P^{-3}(m + iM)]_{,u} = P[\partial + 2(\partial\ln P - L_{,u})]\partial I, \tag{2.9c}$$

where

$$I = \bar{\partial}(\bar{\partial}\ln P - \bar{L}_{,u}) + (\bar{\partial}\ln P - \bar{L}_{,u})^2. \tag{2.10}$$

As was said in Kerr (1963), there are two natural choices that can be made to restrict the coordinates and simplify the final results. One is to rescale $r$ so that $P = 1$ and $L$ is complex, the other is to take $L$ to be pure imaginary with $P \neq 1$. I chose to do the first since this gives the most concise form for $M$ and the remaining field equations. It also gives the smallest group of permissible coordinate transformations, simplifying the task of finding all possible Killing vectors. The results for this gauge are

$$M = \mathrm{Im}(\bar{\partial}\bar{\partial}\partial L), \tag{2.11a}$$
$$\partial(m + iM) = 3(m + iM)L_{,u}, \tag{2.11b}$$
$$\bar{\partial}(m - iM) = 3(m - iM)\bar{L}_{,u}, \tag{2.11c}$$
$$\partial_u[m - \mathrm{Re}(\bar{\partial}\bar{\partial}\partial L)] = |\partial_u\partial L|^2. \tag{2.11d}$$

Since all derivatives of $m$ are given, the commutators were calculated to see whether the field equations were completely integrable. This gives $m$ as a function of the higher derivatives of $L$ unless both $\Sigma_{,u}$ and $L_{,uu}$ are zero. As stated in Kerr (1963), if these are both zero then there is a coordinate system in which $P$ and $L$ are independent of $u$ and $m = cu + A(\zeta, \bar{\zeta})$, where $c$ is a real constant. If it is zero

[13] This expression for $M$ was first published by Robinson *et al.* (1969). The corresponding expression in Kerr (1963) is for the gauge when $P = 1$. The same is true for Equation (2.9c).

then the metric is independent of $u$ and therefore stationary. The field equations are

$$\begin{aligned} &\nabla[\nabla(\ln P)] = c, && \nabla = P^2\partial^2/\partial\zeta\partial\bar{\zeta}, && (2.12a)\\ &M = 2\Sigma\nabla(\ln P) + \nabla\Sigma, && m = cu + A(\zeta,\bar{\zeta}), && (2.12b)\\ &cL = (A + iM)_{\zeta}, \quad \Rightarrow && \nabla M = c\Sigma. && (2.12c)\end{aligned}$$

We shall call these metrics quasi-stationary.

In Kerr (1963) I stated that the solutions of these equations include the Kerr metric (for which $c = 0$). This is true but it is not how this solution was found. Furthermore, in spite of what many believe, its construction did not use the Kerr–Schild ansatz.

## 2.3 Symmetries in algebraically special spaces

The field equations, (2.9) or (2.11), are so complicated that some extra assumptions were needed to reduce them to a more manageable form. My next step in the hunt for physically interesting solutions was to assume that the metric is stationary. Fortunately, I had some tricks that allowed me to find all possible Killing vectors without actually solving Killing's equation.

The key observation is that any Killing vector generates a 1-parameter group which must be a subgroup of the group $\mathcal{C}$ of coordinate transformations that preserve all imposed coordinate conditions.

Suppose that $\{x^{\star a}, \omega^{\star}_a\}$ is another set of coordinates and tetrad vectors that satisfy the conditions imposed in the previous section. If we restrict our coordinates to those that satisfy $P = 1$ then $\mathcal{C}$ is the group of transformations $x \to x^{\star}$ for which

$$\begin{aligned} &\zeta^{\star} = \Phi(\zeta), && \omega^{1\star} = (|\Phi_{\zeta}|/\Phi_{\zeta})\omega^1,\\ &u^{\star} = |\Phi_{\zeta}|(u + S(\zeta,\bar{\zeta})), && \omega^{3\star} = |\Phi_{\zeta}|^{-1}\omega^3,\\ &r^{\star} = |\Phi_{\zeta}|^{-1}r, && \omega^{4\star} = |\Phi_{\zeta}|\omega^4,\end{aligned}$$

and the basic metric functions, $L^{\star}$ and $m^{\star}$, are given by

$$L^{\star} = (|\Phi_{\zeta}|/\Phi_{\zeta})[L - S_{\zeta} - \tfrac{1}{2}(\Phi_{\zeta\zeta}/\Phi_{\zeta})(u + S(\zeta,\bar{\zeta}))], \qquad (2.13a)$$

$$m^{\star} = |\Phi_{\zeta}|^{-3}m. \qquad (2.13b)$$

Let $\mathcal{S}$ be the identity component of the group of symmetries of a space. If we interpret these as coordinate transformations, rather than point transformations, then it is the set of transformations $x \to x^{\star}$ for which

$$g^{\star}_{\alpha\beta}(x^{\star}) = g_{\alpha\beta}(x^{\star}).$$

It can be shown that $\mathcal{S}$ is precisely the subgroup of $\mathcal{C}$ for which

$$m^\star(x^\star) = m(x^\star), \quad L^\star(x^\star) = L(x^\star).$$ [14]

Suppose that $x \to x^\star(x, t)$ is a 1-parameter group of motions,

$$\begin{aligned} \zeta^\star &= \psi(\zeta; t), \\ u^\star &= |\psi_\zeta|(u + T(\zeta, \bar{\zeta}; t)), \\ r^\star &= |\psi_\zeta|^{-1} r. \end{aligned}$$

Since $x^\star(x; 0) = x$, the initial values of $\psi$ and $T$ are

$$\psi(\zeta; 0) = \zeta, \quad T(\zeta, \bar{\zeta}; 0) = 0.$$

The corresponding infinitesimal transformation, $\mathbf{K} = K^\mu \partial/\partial x^\mu$ is

$$K^\mu = \left[\frac{\partial x^{\star\mu}}{\partial t}\right]_{t=0}.$$

If we define

$$\alpha(\zeta) = \left[\frac{\partial \psi}{\partial t}\right]_{t=0}, \quad V(\zeta, \bar{\zeta}) = \left[\frac{\partial T}{\partial t}\right]_{t=0},$$

then the infinitesimal transformation is

$$\mathbf{K} = \alpha \partial_\zeta + \bar{\alpha} \partial_{\bar{\zeta}} + \mathrm{Re}(\alpha_\zeta)(u \partial_u - r \partial_r) + V \partial_u. \tag{2.14}$$

Replacing $\Phi(\zeta)$ with $\psi(\zeta; t)$ in (2.13), differentiating this w.r.t. $t$, and using the initial values for $\psi$ and $T$, $\mathbf{K}$ is a Killing vector if and only if

$$\begin{gathered} V_\zeta + \tfrac{1}{2} \alpha_{\zeta\zeta} r + \mathbf{K} L + \tfrac{1}{2}(\alpha_\zeta - \bar{\alpha}_{\bar{\zeta}}) L = 0, \\ \mathbf{K} m + 3 \mathrm{Re}(\alpha_\zeta) m = 0. \end{gathered}$$

The transformation rules for $\mathbf{K}$ under an element ($\Phi$, $S$) of $\mathcal{C}$ are

$$\alpha^\star = \Phi_\zeta \alpha, \quad V^\star = |\Phi_\zeta| [V - \mathrm{Re}(\alpha_\zeta) S + \mathbf{K} S].$$

Since $\alpha$ is itself analytic, if $\alpha \neq 0$ for a particular Killing vector then, $\Phi$ can be chosen so that $\alpha^\star = 1$,[15] and then $S$ so that $V^\star = 0$. If $\alpha = 0$ then so is $\alpha^\star$, and $\mathbf{K}$ is

[14] Note that this implies that all derivatives of these functions are also invariant, and so $g_{\alpha\beta}$ itself is invariant.
[15] Or any other analytic function of $\zeta$ that one chooses.

already simple without the $(\Phi, S)$ transformation being used. There are therefore two canonical types for **K**,

$$\text{type 1}: \underset{1}{\mathbf{K}} = V\partial_u, \quad \text{or} \quad \text{type 2}: \underset{2}{\mathbf{K}} = \partial_\zeta + \partial_{\bar{\zeta}}. \tag{2.15}$$

These are asymptotically timelike and spacelike, respectively.

## 2.4 Stationary solutions

The obvious and easiest way to simplify the field equations was to assume that the metric was stationary. The type 2 Killing vectors are asymptotically spacelike and so I assumed that $\mathfrak{C}$ had a type 1 Killing vector $\underset{1}{\mathbf{K}} = V\partial_u$. The coordinates used in the last section assumed that $P = 1$. If we transform to the more general coordinates where $P \neq 1$, using an A-transformation (2.4c) with associated change in the $(r, u)$ variables, we get

$$\begin{aligned} \boldsymbol{k}' = A\boldsymbol{k}, \quad \boldsymbol{l}' = A^{-1}\boldsymbol{l}, r' = A^{-1}r, \quad u' = Au, \\ \underset{1}{\mathbf{K}} = V\partial_u = VA\partial_{u'} = \partial_{u'} \quad \text{if } VA = 1. \end{aligned}$$

The metric can therefore be assumed independent of $u$, but $P$ may not be constant. The basic functions, $L$, $P$ and $m$ are functions of $(\zeta, \bar{\zeta})$ alone, and the metric simplifies to

$$ds^2 = ds_o^2 + 2mr/(r^2 + \Sigma^2)k^2, \tag{2.16}$$

where the "base" metric is

$$(ds_0)^2 = 2(r^2 + \Sigma^2)P^{-2}\, d\zeta\, d\bar{\zeta} - 2l_0\, k, \tag{2.17a}$$

$$l_0 = dr + i(\Sigma_{,\zeta} d\zeta - \Sigma_{,\bar{\zeta}} d\bar{\zeta}) + \left[\tfrac{1}{2}K - \frac{M\Sigma}{(r^2 + \Sigma^2)}\right]k. \tag{2.17b}$$

Although the base metric is flat for Schwarzschild it is not so in general. $\Sigma$, $K$ and $M$ are all functions of the derivatives of $L$ and $P$,

$$\begin{aligned} \Sigma &= P^2 \text{Im}(L_{\bar{\zeta}}), \qquad K = 2\nabla^2 \ln P, \\ M &= \Sigma K + \nabla^2\Sigma, \quad \nabla^2 = P^2\partial_\zeta\partial_{\bar{\zeta}}, \end{aligned} \tag{2.18}$$

The mass function, $m$, and $M$ are conjugate harmonic functions,

$$m_\zeta = -iM_\zeta, \quad m_{\bar{\zeta}} = +iM_{\bar{\zeta}}, \tag{2.19}$$

and the remaining field equations are

$$\nabla^2 K = 0, \quad \nabla^2 M = 0. \tag{2.20}$$

If $m$ is a particular solution of these equations then so is $m + m_0$ where $m_0$ is an arbitrary constant. The most general situation where the metric splits in this way is when $P$, $L$ and $M$ are all independent of $u$ but $m = cu + A(\zeta, \bar{\zeta})$. The field equations for these are given in (2.12) and in Kerr (1963). We can state this as a theorem.

**Theorem 2.2** *If* $\mathrm{d}s_0^2$ *is any stationary (diverging) algebraically special metric, or more generally a solution of (2.12), then so is*

$$\mathrm{d}s_0^2 + \frac{2m_0 r}{r^2 + \Sigma^2} k^2,$$

*where* $m_0$ *is an arbitrary constant. These are the most general diverging algebraically special spaces that split in this way.*

These are "generalized Kerr–Schild" metrics with base spaces $\mathrm{d}s_0^2$ that are not necessarily flat.

The field equations for stationary AS metrics are certainly simpler than the original ones, (2.9), but they are still PDEs, not ODEs, and their complete solution is unknown.

## 2.5 Axial symmetry

We are getting close to Kerr. At this point I assumed that the metric was axially symmetric as well as stationary. I should have revisited the Killing equations to look for any Killing vector (KV) that commutes with the stationary one, $\partial_u$. However, I knew that it could not also be type 1[16] and therefore it must be type 2. It seemed fairly clear that it could be transformed to the canonical form $i(\partial_\zeta - \partial_{\bar{\zeta}})$ $(= \partial_y$ where $\zeta = x + iy)$ or equivalently $i(\zeta\partial_\zeta - \bar{\zeta}\partial_{\bar{\zeta}})$ $(= \partial_\phi$ in polar coordinates where $\zeta = \mathrm{Re}^{i\phi})$. I was getting quite eager at this point so I decided to just assume such a KV and see what turned up.[17]

From the first equation in (2.20), the curvature $\nabla^2(\ln P)$ of the 2-metric $P^{-2}\,\mathrm{d}\zeta\,\mathrm{d}\bar{\zeta}$ is a harmonic function,

$$\nabla^2 \ln P = P^2(\ln P)_{,\zeta\bar{\zeta}} = F(\zeta) + \bar{F}(\bar{\zeta}),$$

[16] If it were it would be parallel to the stationary KV and therefore a constant multiple of it.
[17] All possible symmetry groups were found for diverging AS spaces in George C. Debney's Ph.D. thesis. My 1963 expectations were confirmed there.

where $F$ is analytic. There is only one known solution of this equation for $F$ not a constant,

$$P^2 = P_0(\zeta + \bar{\zeta})^3, \quad \nabla^2 \ln P = -\frac{3}{2} P_0(\zeta + \bar{\zeta}), \tag{2.21}$$

where $P_0$ is an arbitrary constant. The mass function $m$ is then constant and the last field equation, $\nabla^2 M = 0$, can be solved for $L$. The final metric is given in Kerr and Debney (1970), Equation (6.14), but it is not worth writing out here since it is not asymptotically flat.

If $\mathfrak{E}$ is to be the metric for a localized physical source then the null congruence should be asymptotically the same as Schwarzschild. $F(\zeta)$ must be regular everywhere, including at infinity, and must therefore be constant,

$$\tfrac{1}{2}K = PP_{,\zeta\bar{\zeta}} - P_{,\zeta}P_{,\bar{\zeta}} = R_0 = \pm P_0^2, \quad \text{(say)}. \tag{2.22}$$

As was shown in Kerr and Debney (1970), the appropriate Killing equations for a $\underset{2}{\mathbf{K}}$ that commutes with $\underset{1}{\mathbf{K}}$ are

$$\begin{aligned} &\underset{2}{\mathbf{K}} = \alpha\partial_\zeta + \bar{\alpha}\partial_{\bar{\zeta}}, && \alpha = \alpha(\zeta), \\ &\underset{2}{\mathbf{K}}L = -\alpha_\zeta L, && \underset{2}{\mathbf{K}}\Sigma = 0, \\ &\underset{2}{\mathbf{K}}P = \mathrm{Re}(\alpha_\zeta)P, && \underset{2}{\mathbf{K}}m = 0. \end{aligned} \tag{2.23}$$

I do not remember the choice I made for the canonical form for $\underset{2}{\mathbf{K}}$ in 1963, but it was probably $\partial_y$. The choice in Kerr and Debney (1969) was

$$\alpha = i\zeta, \quad \Rightarrow \quad \underset{2}{\mathbf{K}} = i(\zeta\partial_\zeta - \bar{\zeta}\partial_{\bar{\zeta}}),$$

and that will be assumed here. For any function $f(\zeta, \bar{\zeta})$,

$$\underset{2}{\mathbf{K}}f = 0 \quad \Rightarrow \quad f(\zeta, \bar{\zeta}) = g(Z), \quad \text{where} \quad Z = \zeta\bar{\zeta}.$$

Now $\mathrm{Re}(\alpha_{,\zeta}) = 0$, and therefore

$$\underset{2}{\mathbf{K}}P = 0, \quad \Rightarrow \quad P = P(Z),$$

and therefore $K$ is given by

$$\tfrac{1}{2}K = P^2(\ln P)_{,\zeta\bar{\zeta}} = PP_{,\zeta\bar{\zeta}} - P_{,\zeta}P_{,\bar{\zeta}} = Z_0 \quad \Rightarrow \quad P = Z + Z_0,$$

after a $\Phi(\zeta)$-coordinate transformation. Note that the form of the metric is invariant under the transformation

$$\begin{aligned} r = A_0 r^*, \quad u = A_0^{-1} u^*, \quad \zeta = A_0 \zeta^*, \\ Z_0 = A_0^{-2} Z_0^*, \quad m_0 = A_0^{-3} m_0^*, \end{aligned} \tag{2.24}$$

where $A_0$ is a constant, and therefore $Z_0$ is a disposable constant. We will choose it later.

The general solution of (2.23) for $L$ and $\Sigma$ is

$$L = i\bar{\zeta} P^{-2} B(Z), \quad \Sigma = ZB' - (1 - Z_0 P^{-1})B,$$

where $B' = dB/dZ$. The complex "mass", $m + iM$, is an analytic function of $\zeta$ from (2.19), and is also a function of $Z$ from (2.23). It must therefore be a constant,

$$m + iM = \mu_0 = m_0 + iM_0.$$

Substituting this into (2.18), the equation for $\Sigma$,

$$\begin{aligned} \Sigma K + \nabla^2 \Sigma = M = M_0 \quad \longrightarrow \\ P^2[Z\Sigma'' + \Sigma'] + 2Z_0 \Sigma = M_0. \end{aligned}$$

The complete solution to this is

$$\Sigma = C_0 + \frac{Z - Z_0}{Z + Z_0}[-a + C_2 \ln Z],$$

where $\{C_0, a, C_2\}$ are arbitrary constants. This gives a four-parameter metric when these known functions are substituted into (2.16) and (2.17). However, if $C_2$ is non-zero then the final metric is singular at $r = 0$. It was therefore omitted in Kerr (1963). The "imaginary mass" is then $M = 2Z_0 C_0$ and so $C_0$ is a multiple of the NUT parameter. It was known in 1963 that the metric cannot be asymptotically flat if this is non-zero and so it was also omitted. The only constants retained were $m_0$, $a$ and $Z_0$. When $a$ is zero and $Z_0$ is positive the metric is that of Schwarzschild. It was not clear that the metric would be physically interesting when $a \neq 0$, but if it had not been so then this whole exercise would have been futile.

The curvature of the 2-metric, $2P^{-2}\,d\zeta\,d\bar{\zeta}$, had to have the same sign as Schwarzschild if the metric was to be asymptotically flat, and so $Z_0 = +P_0^2$. The basic functions in the metric then became

$$\begin{aligned} Z_0 = P_0^2, \quad P = \zeta\bar{\zeta} + P_0^2, \quad m = m_0, \quad M = 0, \\ L = ia\bar{\zeta} P^{-2}, \quad \Sigma = -a\frac{\zeta\bar{\zeta} - Z_0}{\zeta\bar{\zeta} + Z_0}. \end{aligned}$$

The metric was originally published in spherical polar coordinates. The relationship between these and the $(\zeta, \bar{\zeta})$ coordinates is

$$\zeta = P_0 \cot\tfrac{\theta}{2}\, e^{i\phi}.$$

At this point we choose $A_0$ in the transformation (2.24) so that

$$2P_0^2 = 1, \quad \Rightarrow \quad k = \mathrm{d}u + a\sin^2\theta\, \mathrm{d}\phi$$

Recalling the split of (2.16) and (2.17),

$$\mathrm{d}s^2 = \mathrm{d}s_0^2 + 2mr/(r^2 + a^2\cos^2\theta)k^2 \tag{2.25}$$

where $m = m_0$, a constant, and

$$\begin{aligned} \mathrm{d}s_0^2 = {} & (r^2 + a^2\cos^2\theta)(\mathrm{d}\theta^2 + \sin^2\theta\, \mathrm{d}\phi^2) \\ & - (2\, dr + \mathrm{d}u - a\sin^2\theta\, \mathrm{d}\phi)(\mathrm{d}u + a\sin^2\theta\, \mathrm{d}\phi). \end{aligned} \tag{2.26}$$

This is the original form of Kerr (1963), except that $u$ has been replaced by $-u$ to agree with current conventions, and $a$ has been replaced with its negative.[18]

Having found this fairly simple metric, I was desperate to see whether it was rotating. Fortunately, I knew that the curvature of the base metric, $\mathrm{d}s_0^2$, was zero, and so it was only necessary to find coordinates where this was manifestly Minkowskian. These were

$$(r + ia)e^{i\phi}\sin\theta = x + iy, \quad r\cos\theta = z, \quad r + u = -t.$$

This gives the Kerr–Schild form of the metric,

$$\begin{aligned} \mathrm{d}s^2 = {} & \mathrm{d}x^2 + \mathrm{d}y^2 + \mathrm{d}z^2 - \mathrm{d}t^2 + \frac{2mr^3}{r^4 + a^2z^2}\Big[\mathrm{d}t + \frac{z}{r}\mathrm{d}z \\ & + \frac{r}{r^2 + a^2}(x\,\mathrm{d}x + y\,\mathrm{d}y) - \frac{a}{r^2 + a^2}(x\,\mathrm{d}y - y\,\mathrm{d}x)\Big]^2. \end{aligned} \tag{2.27}$$

where the surfaces of constant $r$ are confocal ellipsoids of revolution about the $z$-axis,

$$\frac{x^2 + y^2}{r^2 + a^2} + \frac{z^2}{r^2} = 1. \tag{2.28}$$

Asymptotically, of course, $r$ is just the distance from the origin in the Minkowskian coordinates, and the metric is clearly asymptotically flat.

[18] We will see why later.

### *Angular momentum*

After the metric had been put into its Kerr–Schild form I went to Alfred Schild and told him I was about to calculate the angular momentum of the central body. He was just as excited as I was and so he joined me in my office while I computed. We were both heavy smokers at that time, so you can imagine what the atmosphere was like, Alfred puffing away at his pipe in an old arm chair, and myself chain-smoking cigarettes at my desk.

The Kerr–Schild form is particularly suitable for calculating the physical parameters of the solution. My PhD thesis at Cambridge was entitled "Equations of Motion in General Relativity". It had been claimed previously in the literature that it was only necessary to satisfy the momentum equations for singular particles to be able to integrate the EIH quasi-static approximation equations at each order. One thing shown in my thesis was the physically obvious fact that the angular momentum equations were equally important. Some of this was published in Kerr (1958) and (1960). Because of this previous work I was well aware how to calculate the angular momentum in this new metric.

It was first expanded in powers of $R^{-1}$, where $R = x^2 + y^2 + z^2$ is the usual Euclidean distance,

$$\begin{aligned} \mathrm{d}s^2 &= \mathrm{d}x^2 + \mathrm{d}y^2 + \mathrm{d}z^2 - \mathrm{d}t^2 + \frac{2m}{R}(\mathrm{d}t + \mathrm{d}R)^2 \\ &\quad - \frac{4ma}{R^3}(x\,\mathrm{d}y - y\,\mathrm{d}x)(\mathrm{d}t + \mathrm{d}R) + O(R^{-3}) \end{aligned} \tag{2.29}$$

Now, if $x^\mu \to x^\mu + a^\mu$ is an infinitesimal coordinate transformation, then $\mathrm{d}s^2 \to \mathrm{d}s^2 + 2\,\mathrm{d}a_\mu\,\mathrm{d}x^\mu$. If we choose

$$\begin{aligned} a_\mu \mathrm{d}x^\mu &= -\frac{am}{R^2}(x\,\mathrm{d}y - y\,\mathrm{d}x) \quad \Rightarrow \\ 2\,\mathrm{d}a_\mu\,\mathrm{d}x^\mu &= -4m\frac{4am}{R^3}(x\,\mathrm{d}y - y\,\mathrm{d}x)\,\mathrm{d}R, \end{aligned}$$

then the approximation in (2.29) simplifies to

$$\begin{aligned} \mathrm{d}s^2 &= \mathrm{d}x^2 + \mathrm{d}y^2 + \mathrm{d}z^2 - \mathrm{d}t^2 + \frac{2m}{R}(\mathrm{d}t + \mathrm{d}R)^2 \\ &\quad - \frac{4ma}{R^3}(x\,\mathrm{d}y - y\,\mathrm{d}x)\,\mathrm{d}t + O(R^{-3}). \end{aligned} \tag{2.30}$$

The leading terms in the linear approximation for the gravitational field around a rotating body were well known (for instance, see Papapetrou (1974) or Kerr (1960)). The contribution from the angular momentum vector, $\mathbf{J}$, is

$$4R^{-3}\,\epsilon_{ijk} J^i x^j\,\mathrm{d}x^k\,\mathrm{d}t.$$

A comparison of the last two equations showed that the physical parameters were[19]

$$\text{Mass} = m, \qquad \mathbf{J} = (0, 0, ma).$$

When I turned to Alfred Schild, who was still sitting in the armchair smoking away, and said “Its rotating!” he was even more excited than I was. I do not remember how we celebrated, but celebrate we did!

Robert Boyer subsequently calculated the angular momentum by comparing the known Lense–Thirring results for frame dragging around a rotating object in linearized relativity with the frame dragging for a circular orbit in a Kerr metric. This was a very obtuse way of calculating the angular momentum since the approximation (2.30) was the basis for the calculations by Lense and Thirring, but it did show that the sign was wrong in the original paper!

## 2.6 Singularities and topology

The first Texas Symposium on Relativistic Astrophysics was held in Dallas December 16–18, 1963, just a few months after the discovery of the rotating solution. It was organized by a combined group of Relativists and Astrophysicists and its purpose was to try to find an explanation for the newly discovered quasars. The source 3C 273B had been observed in March and was thought to be about a million million times brighter than the sun.

It had been long known that a spherically symmetric body could collapse inside an event horizon to become what was to be later called a black hole by John Wheeler. However, the Schwarzschild solution was non-rotating and it was not known what would happen if rotation was present. I presented a paper called “Gravitational collapse and rotation” in which I outlined the Kerr solution and said that the topological and physical properties of the event horizon may change radically when rotation is taken into account. It was not known at that time that Kerr was the only possible stationary solution for such a rotating black hole and so I discussed it as an example of such an object.

Although this was not pointed out in the original letter, Kerr (1963), the geometry of Kerr is even more complicated than the Kruskal extension of Schwarzschild. The metric is everywhere non-singular, except on the ring

$$z = 0, \qquad x^2 + y^2 = a^2.$$

[19] Unfortunately, I was rather hurried when performing this calculation and got the sign wrong. This is why the sign of the parameter $a$ in Kerr (1963) is different to that in all other publications, including this one. This way of calculating **J** was explained at the First Texas Symposium at the end of 1963, see Kerr (1965), but I did not check the sign at that time.

The Weyl scalar, $R_{abcd}R^{abcd} \to \infty$ near these points and so they are true singularities, not just coordinate ones. Furthermore, this ring behaves like a branch point in the complex plane. If one travels on a closed curve that threads the ring the initial and final metrics are different: $r$ changes sign. Equation (2.28) has one non-negative root for $r^2$, and therefore two real roots, $r_\pm$, for $r$. These coincide where $r^2 = 0$, i.e. on the disc $D$ bounded by the ring singularity

$$D: \quad z = 0, \quad x^2 + y^2 \leq a^2.$$

The disc can be taken as a branch cut for the analytic function $r$. We have to take two spaces, $E_1$ and $E_2$ with the topology of $R^4$ less the disc $D$. The points above $D$ in $E_1$ are joined to the points below $D$ in $E_2$ and vice versa. In $E_1$, $r > 0$ and the mass is positive at infinity; in $E_2$, $r < 0$ and the mass is negative. The metric is then everywhere analytic except on the ring.

It was trivially obvious to everyone that if the parameter $a$ is very much less than $m$ then the Schwarzschild event horizon at $r = 2m$ will be modified slightly but cannot disappear. For instance, the light cones at $r = m$ in Kerr all point inwards for small $a$. Before I went to the meeting I had calculated the behaviour of the time-like geodesics up and down the axis of rotation and found that horizons occurred at the points on the axis in $E_1$ where

$$r^2 - 2mr + a^2 = 0, \quad r = |z|.$$

but that there are no horizons in $E_2$ where the mass is negative. In effect, the ring singularity is "naked" in that sheet.

I made a rather hurried calculation of the two event horizons in $E_1$ before I went to the Dallas Symposium and claimed incorrectly there, Kerr (1965), that the equations for them were the two roots of

$$r^4 - 2mr^3 + a^2z^2 = 0,$$

whereas $z^2$ should be replaced by $r^2$ in this and the true equation is

$$r^2 - 2mr + a^2 = 0.$$

I attempted to calculate this using inappropriate coordinates and assuming that the equation would be: "$\psi(r, z)$ is null for some function of both $r$ and $z$". I did not realize that this function depended only on $r$.

The Kerr–Schild coordinates are a generalization of the Eddington–Finkelstein coordinates for Schwarzschild. For the latter future-pointing radial geodesics are well behaved but not those travelling to the past. Kruskal coordinates were designed to handle both. Similarly for Kerr, the coordinates given here only handle ingoing curves. This metric is known to be type D and therefore it has another set of Debever–Penrose vectors and an associated coordinate system for which the outgoing geodesics are well behaved, but not the ingoing ones.

This metric consists of three blocks, outside the outer event horizon, between the two horizons and within the inner horizon (at least for $m < a$, which is probably true for all existing black holes). Just as Kruskal extends Schwarzschild by adding extra blocks, Boyer and Lindquist (1967) and Carter (1968) independently showed that the maximal extension of Kerr has a similar proliferation of blocks. However, the Kruskal extension has no application to a real black hole formed by the collapse of a spherically symmetric body and the same is true for Kerr. In fact, even what I call $E_2$, the sheet where the mass is negative, is probably irrelevant for the final state of a collapsing rotating object.

Ever since this metric was first discovered people have tried to fit an interior solution. One morning during the summer of 1964 Ray Sachs and myself decided that we would try to do so. Since the original form is useless and the Kerr–Schild form was clearly inappropriate we started by transforming to the canonical coordinates for stationary axisymmetric solutions.

In Papapetrou (1966) there is a very elegant treatment of stationary axisymmetric Einstein spaces. He shows that if there is a real non-singular axis of rotation then the coordinates can be chosen so that there is only one off-diagonal component of the metric. We call such a metric quasi-diagonalizeable. All cross terms between $\{dr, d\theta\}$ and $\{dt, d\phi\}$ can be eliminated by transformations of the type

$$dt' = dt + A\,dr + B\,d\theta, \quad d\phi' = d\phi + C\,dr + D\,d\theta.$$

where the coefficients can be found algebraically. Papapetrou proved that $dt'$ and $d\phi'$ are perfect differentials if the axis is regular.[20]

Ray and I calculated the coefficients $A \ldots D$ and transformed the metric to the Boyer–Lindquist form,

$$\begin{aligned} dt &\to dt + \frac{2mr}{\Delta}\,dr \\ d\phi &\to -d\phi + \frac{a}{\Delta}\,dr \\ \Delta &= r^2 - 2mr + a^2, \\ \Sigma &= r^2 + a^2\cos^2\theta, \end{aligned}$$

[20] It is shown in Weir and Kerr (1977) that if the metric is also algebraically special then it is quasi-diagonalizeable precisely when it is type D. These metrics are the NUT parameter generalization of Kerr.

where, as before, $u = -(t + r)$. The right-hand sides of the first two equations are clearly perfect differentials as the Papapetrou analysis showed. The full Boyer–Lindquist form of the metric is

$$\begin{aligned} ds^2 = \frac{\Sigma}{\Delta}\,dr^2 - \frac{\Delta}{\Sigma}[dt - a\sin^2\theta\,d\phi]^2 \\ + \Sigma\,d\theta^2 + \frac{\sin^2\theta}{\Sigma}[(r^2 + a^2)\,d\phi - a\,dt]^2, \end{aligned} \tag{2.31}$$

after some tedious analysis that used to be easy but now seems to require an algebraic package such as Maple.

Having derived this canonical form, we studied the metric for at least ten minutes and then decided that we had no idea how to introduce a reasonable source into a metric of this form, and probably would never have. Presumably those who have tried to solve this problem in the last 43 years have had similar reactions. Soon after this failed attempt Robert Boyer came to Austin. He said to me that he had found a new quasi-diagonalized form of the metric. I said "Yes. It is the one with the polynomial $r^2 - 2mr + a^2$" but for some reason he refused to believe that we had also found this form. Since it did not seem a "big deal" at that time I did not pursue the matter further, but our relations were hardly cordial after that.

One of the main advantages of this form is that the event horizons can be easily calculated since the inverse metric is simple. If $f(r, \theta) = 0$ is a null surface then

$$\Delta(r) f_{,rr} + f_{,\theta\theta} = 0,$$

and therefore $\Delta \le 0$. The two event horizons are the surfaces $r = r_{\pm}$ where the parameters $r_{\pm}$ are the roots of $\Delta = 0$,

$$\Delta = r^2 - 2mr + a^2 = (r - r_+)(r - r_-).$$

If $a < m$ there are two distinct horizons between which all time-like lines point inwards; if $a = m$ there is only one event horizon; and for larger $a$ the singularity is bare! Presumably, any collapsing star can only form a black hole if the angular momentum is small enough: $a < m$. This seems to be saying that the body cannot rotate faster than light, if the final picture is that the mass is located on the ring radius $a$. However, it should be remembered that this radius is purely a coordinate radius, and that there is no way that the final stage of such a collapse is that all the mass is located at the singularity.

The reason for the last statement is that if the mass were to end on the ring then there would be no way to avoid the second asymptotically flat sheet where the mass appears negative. I do not believe that the star opens up like this along the axis of rotation. If we remember that the metric is discontinuous across the disc bounded by the singular ring then it is quite possible that a well-behaved finite source could be put between $z = 0_{\pm}$, $|R| < a$.[21] This would correspond to the surface of the final body being $r = 0$ in Boyer–Lindquist coordinates, say, but where the interior corresponds to $r < 0$. The actual surface may be more complicated than this but I am quite sure that this is closer to the final situation than that the matter all collapses onto the ring.

What I believe to be more likely is that the inner event horizon never actually forms. As the body continues to collapse inside its event horizon it spins faster and faster so that the geometry in the region between its outer surface and the outer event horizon approaches that between the two event horizons for Kerr. The surface of the body surface will appear to be asymptotically null. The full metric may not be geodesically complete. Many theorems have been claimed stating that a singularity must exist if certain conditions are satisfied, but they all make assumptions that may not be true for collapse to a black hole. Furthermore, these assumptions are often (usually?) unstated or unrecognized, and the proofs are dependent on other claims/theorems that may not be correct.

These are only two of a very large range of possibilities for the interior. What happens after the outer horizon forms is still a mystery after more than four decades. It is also the main reason why I said at the end of Kerr (1963) that "It would be desirable to calculate an interior solution . . . ". This statement has been taken by some to mean that I thought the metric only represented a real rotating star. This is untrue and is an insult to all those relativists of that era who had been looking for such a metric to see whether the event horizon of Schwarzschild would generalize to rotating singularities.

The metric was known to be type D with two distinct geodesic and shearfree congruences from the moment it was discovered. This means that if the other is used instead of $k$ then the metric must have the same form, i.e. it is invariant under a finite transformation that reverses "time" and possibly the axis of rotation in the appropriate coordinates. This also meant that there is an extension that is similar to the Kruskal–Szekeres extension of Schwarzschild. Both Boyer and Lindquist (1967) and a fellow Australasian, Brandon Carter (1968), solved the problem of constructing the maximal extensions of Kerr, and even that for charged Kerr. These are mathematically fascinating and the latter paper is a beautiful analysis of the problem, but the final result is of limited physical significance.

Brandon Carter's (1968) paper was one of the most significant papers on the Kerr metric during the mid sixties for another reason. He showed that there is an extra

[21] This has been done using $\delta$-functions, but I am thinking more of a non-singular source where the distance between the two sides of the disc is non-zero.

invariant for geodesic motion which is quadratic in the momentum components: $J = X_{ab}v^a v^b$ where $X_{ab}$ is a Killing tensor, $X_{(ab;c)} = 0$. This gave a total of four invariants with the two Killing vector invariants and $|\mathbf{v}|^2$ itself, enough to generate a complete first integral of the geodesic equations.

Another significant development was the "proof" that this is the only stationary metric with a simply connected bounded event horizon, i.e. the only possible black hole. Many contributed to this, including Steven Hawking (1972), Brandon Carter (1971) and another New Zealander, David Robinson (1975). The subsequent work in this area is discussed by David in an excellent article in this book and so I will not pursue this any further here.

## 2.7 Kerr–Schild metrics

One morning during the fall semester 1963, sometime before the First Texas Symposium, I tried generalizing the way that the field equations split for the Kerr metric by setting

$$ds^2 = ds_0^2 + \frac{2mr}{r^2 + \Sigma^2}k^2.$$

The base metric $ds_0^2$ was to be an algebraically special metric with $m = 0$. From an initial rough calculation this had to be flat. Also, it seemed that the coordinates could be manipulated so that $\partial L = L_\zeta - LL_u = 0$ and that the final metric depended on an arbitrary analytic function of the complex variable $\zeta$. At this point I lost interest since the metric had to be singular at the poles of the analytic function unless this function was quadratic and the metric was therefore Kerr.

Sometime after the Texas Symposium Jerzy Plebanski visited Austin. Alfred Schild gave one of his excellent parties for Jerzy during which I heard them talking about solutions of the Kerr–Schild type,[22] that is $ds^2 = ds_0^2 + hk^2$, where the first term is flat and $k$ is any null vector. I commented that there might be some algebraically special spaces with this structure depending on an arbitrary function of a complex variable but that this had not been checked.

At this point Alfred and I retired to his home office and calculated the simplest field equation, $R_{ab}k^a k^b = 0$. To our surprise this showed that the null vector had to be geodesic. We then calculated $k_{[a}R_{b]pq[c}k_{d]}k^p k^q$, found it to be zero and deduced that all metrics of this type had to be algebraically special. This meant that all such spaces with a diverging congruence might already be known. We checked my original calculations next day and found them to be correct.

[22] This name came later, of course.

As was stated in Theorem 2.2, $m$ is a unique function of $P$ and $L$ unless there is a canonical coordinate system where $m$ is linear in $u$ and $\{L, P\}$ are functions of $\{\zeta, \bar{\zeta}\}$ alone. If the base space is flat then $m_{,u} = c = 0$ and the metric is stationary. The way these Kerr–Schild metrics were found originally was by showing that in a coordinate system where $P = 1$ the canonical coordinates could be chosen so that $\partial L = 0$. Transforming from these coordinates to ones where $P \neq 1$ and $\partial_u$ is a Killing vector,

$$P_{,\zeta\zeta} = 0, \quad L = P^{-2}\bar{\phi}(\bar{\zeta}), \tag{2.32}$$

where $\phi(\zeta)$ is analytic. From the first of these $P$ is a real bilinear function of $\zeta$ and therefore of $\bar{\zeta}$,

$$P = p\zeta\bar{\zeta} + q\zeta + \bar{q}\bar{\zeta} + c.$$

This can be simplified to one of three canonical forms, $P = 1, 1 \pm \zeta\bar{\zeta}$ by a linear transformation on $\zeta$. We will assume henceforth that

$$P = 1 + \zeta\bar{\zeta}.$$

The only problem was that this analysis depended on results for algebraically special metrics and these had not been published and would not be for several years. We had to derive the same results by a more direct method. The metric was written as

$$\mathrm{d}s^2 = \mathrm{d}x^2 + \mathrm{d}y^2 + \mathrm{d}z^2 - \mathrm{d}t^2 + hk^2, \tag{2.33a}$$

$$k = (\mathrm{d}u + \bar{Y}\,\mathrm{d}\zeta + Y\,\mathrm{d}\bar{\zeta} + Y\bar{Y}\,\mathrm{d}v)/(1 + Y\bar{Y}), \tag{2.33b}$$

where $Y$ is the old coordinate $\zeta$ used in (2.32),[23] and[24]

$$u = z + t, \quad v = t - z, \quad \zeta = x + iy.$$

The tetrad used to calculate the field equations was defined naturally from the identity

$$\mathrm{d}s^2 = (\mathrm{d}\zeta + Y\,\mathrm{d}v)(\mathrm{d}\bar{\zeta} + \bar{Y}\,\mathrm{d}v) - (\mathrm{d}v + hk)k.$$

Each of these spaces has a symmetry which is also a translational symmetry for the base Minkowski space, $\mathrm{d}s_0^2$. The most interesting situation is

[23] $Y$ is the ratio of the two components of the spinor corresponding to $\boldsymbol{k}$.

[24] Note that certain factors of $\sqrt{2}$ have been omitted to simplify the results. This does lead to a factor 2 appearing in (2.35).

when this is time-like and so it will be assumed that the metric is independent of $t = \frac{1}{2}(u + v)$.

If $\phi(Y)$ is the same analytic function as in (2.32) then $Y$ is determined as a function of the coordinates by

$$Y^2\bar{\zeta} + 2zY - \zeta + \phi(Y) = 0. \tag{2.34}$$

and the coefficient of $k^2$ in (2.33) is

$$h = 2m\mathrm{Re}(2Y_\zeta), \tag{2.35}$$

where $m$ is a real constant. Differentiating (2.34) with respect to $\zeta$ gives

$$Y_\zeta = (2Y\bar{\zeta} + 2z + \phi')^{-1}. \tag{2.36}$$

The Weyl spinor invariant is given by

$$\Psi_2 = c_0 m Y_\zeta^3,$$

where $c_0$ is some power of 2, and the metric is therefore singular precisely where Y is a repeated root of its defining equation, (2.34).

If the $k$-lines are projected onto the Euclidean 3-space $t = 0$ with $\{x, y, z\}$ as coordinates so that $\mathrm{d}s_E^2 = \mathrm{d}x^2 + \mathrm{d}y^2 + \mathrm{d}z^2$, then the perpendicular from the origin meets the projected k-line at the point

$$F_0 : \quad \zeta = \frac{\phi - Y^2\bar{\phi}}{P^2}, \quad z = -\frac{\bar{Y}\phi + Y\bar{\phi}}{P^2},$$

and the distance of the line from the origin is

$$D = \frac{|\phi|}{1 + Y\bar{Y}},$$

a remarkably simple result. This was used by Kerr and Wilson (1978) to prove that unless $\phi$ is quadratic the singularities are unbounded and the spaces are not asymptotically flat. The reason why I did not initially take the general Kerr–Schild metric seriously was that this was what I expected.

Another point that is easily calculated is $Z_0$ where the line meets the plane $z = 0$,

$$Z_0 : \quad \zeta = \frac{\phi + Y^2\bar{\phi}}{1 - (Y\bar{Y})^2}, \quad z = 0,$$

The original metric of this type is Kerr where

$$\phi(Y) = -2iaY, \qquad D = \frac{2|a||Y|}{1+|Y|^2} \leq |a|.$$

If $\phi(Y)$ is any other quadratic function then it can be transformed to the same value by using an appropriate Euclidean rotation and translation about the t-axis. The points $F_0$ and $Z_0$ are the same for Kerr, so that $F_0$ lies in the z-plane and the line cuts this plane at a point inside the singular ring provided $|Y| \neq 1$. The lines where $|Y| = 1$ are the tangents to the singular ring lying entirely in the plane $z = 0$ outside the ring. When $a \to 0$ the metric becomes Schwarzschild and all the $Y$-lines pass through the origin.

When $\phi(Y) = -2iaY$ (2.34) becomes

$$Y^2\bar{\zeta} + 2(z - ia)Y - \zeta = 0.$$

There are two roots, $Y_1$ and $Y_2$ of this equation,

$$Y_1 = \frac{r\zeta}{(z+r)(r-ia)}, \qquad 2Y_{1,\zeta} = +\frac{r^3 + iarz}{r^4 + a^2z^2}$$
$$Y_2 = \frac{r\zeta}{(z-r)(r+ia)}, \qquad 2Y_{2,\zeta} = -\frac{r^3 + iarz}{r^4 + a^2z^2}$$

where $r$ is a real root of (2.28). This is a quadratic equation for $r^2$ with only one non-negative root and therefore two real roots differing only by sign, $\pm r$. When these are interchanged, $r \leftrightarrow -r$, the corresponding values for $Y$ are also swapped, $Y_1 \leftrightarrow Y_2$.

When $Y_2$ is substituted into the metric then the same solution is returned except that the mass has changed sign. This is the other sheet where $r$ has become negative. It is usually assumed that $Y$ is the first of these roots, $Y_1$. The coefficient $h$ of $k^2$ in the metric, (2.33), is then

$$h = 2m\,\mathrm{Re}(2Y_\zeta) = \frac{2mr^3}{r^4 + a^2z^2}.$$

This gives the metric in its KS form, (2.27).

The results were published in two places, Kerr and Schild (1965a,b). The first of these was a talk that Alfred gave at the Galileo Centennial in Italy, the second was an invited talk that I gave, but Alfred wrote, at the Symposium on Applied Mathematics of the American Mathematical Society, April 25, 1964. The manuscript had to be presented before the conference so that the participants had some chance of understanding results from distant fields. We stated on page 205 that

Together with their graduate student, Mr. George Debney, the authors have examined solutions of the nonvacuum Einstein–Maxwell equations where the metric has the form (2.1).[25] Most of the results mentioned above apply to this more general case. This work is continuing.

## 2.8 Charged Kerr

What was this quote referring to? When we had finished with the Kerr–Schild metrics, we looked at the same problem with a non-zero electromagnetic field. The first stumbling block was that $R_{ab}k^a k^b = 0$ no longer implied that the $k$-lines are geodesic. The equations were quite intractable without this and so it had to be added as an additional assumption. It then followed that the principal null vectors were shearfree, so that the metrics had to be algebraically special. The general forms of the gravitational and electromagnetic fields were calculated from the easier field equations. The E–M field proved to depend on two functions called $A$ and $\gamma$ in Debney, Kerr and Schild (1969).

When $\gamma = 0$ the "difficult" equations are linear and similar to those for the purely gravitational case. They were readily solved giving a charged generalization of the original Kerr–Schild metrics. The congruences are the same as for the uncharged metrics, but the coefficient of $k^2$ is

$$h = 2m\,\mathrm{Re}(2Y_{,\varsigma}) - |\psi|^2|2Y_{,\varsigma}|^2. \tag{2.37}$$

where $\psi(Y)$ is an extra analytic function generating the electromagnetic field. This is best expressed through a potential,

$$f = \tfrac{1}{2}F_{\mu\nu}\,\mathrm{d}x^\mu\,\mathrm{d}x^\nu = -\mathrm{d}\alpha,$$
$$\alpha = -P(\psi Z + \bar{\psi}\bar{Z})k - \tfrac{1}{2}(\chi\,\mathrm{d}\bar{Y} + \bar{\chi}\,\mathrm{d}Y),$$

where

$$\chi = \int P^{-2}\psi(Y)\,\mathrm{d}Y,$$

$\bar{Y}$ being kept constant in this integration.

The most important member of this class is charged Kerr. For this,

$$h = \frac{2mr^3 - |\psi(Y)|^2 r^2}{r^4 + a^2 z^2}. \tag{2.38}$$

[25] Equation (2.33) in this paper. It refers to the usual Kerr–Schild ansatz.

Asymptotically, $r = R$, $k = \mathrm{d}t - \mathrm{d}R$, a radial null-vector and $Y = \tan(\frac{1}{2}\theta)e^{i\phi}$. If the analytic function $\psi(Y)$ is non-constant then it must be singular somewhere on the unit sphere and so the gravitational and electromagnetic fields will be also. The only physically significant charged Kerr–Schild is therefore when $\psi$ is a complex constant, $e + ib$. The imaginary part, $b$, can be ignored as it gives a magnetic monopole, and so we are left with $\psi = e$, the electric charge,

$$\begin{aligned} \mathrm{d}s^2 = \mathrm{d}x^2 + \mathrm{d}y^2 + \mathrm{d}z^2 - \mathrm{d}t^2 + \frac{2mr^3 - e^2r^2}{r^4 + a^2z^2}\Big[\mathrm{d}t + \frac{z}{r}\mathrm{d}z \\ + \frac{r}{r^2 + a^2}(x\,\mathrm{d}x + y\,\mathrm{d}y) - \frac{a}{r^2 + a^2}(x\,\mathrm{d}x - y\,\mathrm{d}y)\Big]^2, \end{aligned} \tag{2.39}$$

The electromagnetic potential is

$$\alpha = \frac{er^3}{r^4 + a^2z^2}\left[\mathrm{d}t - \frac{a(x\,\mathrm{d}y - y\,\mathrm{d}x)}{r^2 + a^2}\right],$$

where a pure gradient has been dropped. The electromagnetic field is

$$(F_{xt} - iF_{yz}, F_{yt} - iF_{zx}, F_{zt} - iF_{xy}) = \frac{er^3}{(r^2 + iaz)^3}(x, y, z + ia).$$

In the asymptotic region this field reduces to an electric field,

$$\mathbf{E} = \frac{e}{R^3}(x, y, z),$$

and a magnetic field,

$$\mathbf{H} = \frac{ea}{R^5}(3xz, 3yz, 3z^2 - R^2).$$

This is the electromagnetic field of a body with charge $e$ and magnetic moment $(0, 0, ea)$. The gyromagnetic ratio is therefore $ma/ea = m/e$, the same as that for the Dirac electron. This was first noticed by Brandon Carter and was something that fascinated Alfred Schild.

This was the stage we had got to before March 1964. We were unable to solve the equations where the function $\gamma$ was non-zero so we enlisted the help of our graduate student, George Debney. Eventually we realized that we were unable to solve the more general equations and so we suggested to George that he drop this investigation. He then tackled the problem of finding all possible groups of symmetries in diverging algebraically special spaces. He succeeded very well with this, solving many of the ensuing field equations for the associated metrics. This work formed the basis for his PhD thesis and was eventually published in Kerr and Debney (1970).

In Janis and Newman (1965) and Newman and Janis (1965) the authors defined and calculated multipole moments for the Kerr metric, using the Kerr–Schild coordinates as given in Kerr (1963). They then claimed that this metric is that of a ring of mass rotating about its axis of symmetry. Unfortunately, this cannot be so because the metric is multivalued on its symmetry axis and is consequently discontinuous there. The only way that this can be avoided is by assuming that the space contains matter on the axis near the centre. As was acknowledged in a footnote to the second paper, this was pointed out to the authors by the referee and myself before the paper was published, but they still persisted with their claim.

## 2.9 Newman's construction of the Kerr–Newman metric

Newman knew that the Schwarzschild, Reissner–Nordström and Kerr metrics all have the same simple form in Eddington–Finkelstein or Kerr–Schild coordinates (see Eq. 2.33). Schwarzschild and its charged generalization have the same null congruence, $k = \mathrm{d}r - \mathrm{d}t$; only the coefficient $h$ is different,

$$h = \frac{2m}{r} \quad \longrightarrow \quad h = \frac{2m}{r} - \frac{e^2}{r^2}.$$

For these the complex divergence of the underlying null congruence is

$$\rho = \bar{\rho} = 1/r.$$

Newman hoped to find a charged metric with the same congruence as Kerr but with $h$ generalized to something like the Reissner–Nordström form with $e^2/r^2$ replaced by $e^2\rho^2$. This does not quite work since $\rho$ is complex for Kerr so he had to replace $\rho^2$ with something real.

There are many real rational functions of $\rho$ and $\bar{\rho}$ that reduce to $\rho^2$ when $\rho$ is real, so he wrote down several possibilities and distributed them to his graduate students. Each was checked to see whether it was a solution of the Einstein–Maxwell equations. The simplest, $\rho^2 \to \rho\bar{\rho}$, worked! The appropriate electromagnetic field was then calculated, a non-trivial problem.

The reason that this approach was successful has nothing to do with "complexifying the Schwarzschild and Reissner–Nordström metrics" by some complex coordinate transformation, as stated in the original papers. It works because all these metrics are of Kerr–Schild form and the *general* Kerr–Schild metric can be charged by replacing the uncharged $h$ with its appropriate charged version,

$h = 2m\,\mathrm{Re}(2Y_\zeta) - |\psi|^2|2Y_{,\zeta}|^2 \longrightarrow 2m\,\mathrm{Re}(\rho) - e^2\rho\bar{\rho}$, without changing the congruence.

The charged solution was given in Newman *et al.* (1965). They claimed that the metric can be generated by a classical charged rotating ring. As in the previous paper Newman and Janis (1965), it was then admitted in a footnote that the reason why this cannot be true had already been explained to them.

## 2.10 Appendix: standard notation

Let $\{\mathbf{e}_a\}$ and $\{\omega^a\}$ be dual bases for tangent vectors and linear 1-forms, respectively, i.e. $\omega^a(\mathbf{e}_b) = \delta^a_b$. Also let $g_{ab}$ be the components of the metric tensor,

$$\mathrm{d}s^2 = g_{ab}\omega^a\omega^a, \qquad g_{ab} = \mathbf{e}_a \cdot \mathbf{e}_b.$$

The components of the connection in this frame are the Ricci rotation coefficients,

$$\Gamma^a_{bc} = -\omega^a{}_{\mu;\nu}e_b{}^\mu e_c{}^\nu, \qquad \Gamma_{abc} = g_{as}\Gamma^s_{bc},$$

The commutator coefficients $D^a_{bc} = -D^a_{cb}$ are defined by

$$[\mathbf{e}_b, \mathbf{e}_c] = D^a_{bc}\mathbf{e}_a, \quad \text{where} \quad [\mathbf{u}, \mathbf{v}](\mathrm{f}) = \mathbf{u}(\mathbf{v}(\mathrm{f})) - \mathbf{v}(\mathbf{u}(\mathrm{f})).$$

or equivalently by

$$\mathrm{d}\omega^a = D^a_{bc}\omega^b \wedge \omega^c. \tag{2.40}$$

Since the connection is symmetric, $D^a_{bc} = -2\Gamma^a_{[bc]}$, and since it is metrical

$$\Gamma_{abc} = \tfrac{1}{2}(g_{ab|c} + g_{ac|b} - g_{bc|a} + D_{bac} + D_{cab} - D_{abc}),$$
$$\Gamma_{abc} = g_{am}\Gamma^m_{bc}, \qquad D_{abc} = g_{am}D^m_{bc}.$$

If it is assumed that the $g_{ab}$ are constant, then the connection components are determined solely by the commutator coefficients and therefore by the exterior derivatives of the tetrad vectors,

$$\Gamma_{abc} = \tfrac{1}{2}(D_{bac} + D_{cab} - D_{abc}).$$

The components of the curvature tensor are

$$\Theta^a_{bcd} \equiv \Gamma^a_{bd|c} - \Gamma^a_{bc|d} + \Gamma^e_{bd}\Gamma^a_{ec} - \Gamma^e_{bc}\Gamma^a_{ed} - D^e_{cd}\Gamma^a_{be}. \tag{2.41}$$

We must distinguish between the expressions on the right, the $\Theta^a_{bcd}$, and the curvature components, $R^a_{bcd}$, which the N–P formalism treat as extra variables, their $(\Psi_i)$.

A crucial factor in the discovery of the spinning black hole solutions was the use of differential forms and the Cartan equations. The connection 1-forms $\mathbf{\Gamma}^a_b$ are defined as

$$\mathbf{\Gamma}^a_b = \Gamma^a_{bc}\omega^c.$$

These are skew-symmetric when $g_{ab|c} = 0$,

$$\mathbf{\Gamma}_{ba} = -\mathbf{\Gamma}_{ab}, \qquad \mathbf{\Gamma}_{ab} = g_{ac}\mathbf{\Gamma}^c_a.$$

The first Cartan equation follows from (2.40),

$$\mathrm{d}\omega^a + \mathbf{\Gamma}^a_b\,\omega^b = 0. \tag{2.42}$$

The curvature 2-forms are defined from the second Cartan equations,

$$\mathbf{\Theta}^a_b \equiv \mathrm{d}\mathbf{\Gamma}^a_b + \mathbf{\Gamma}^a_c \wedge \mathbf{\Gamma}^c_b = \tfrac{1}{2}R^a_{bcd}\omega^c\omega^d. \tag{2.43}$$

The exterior derivative of (2.42) gives

$$\mathbf{\Theta}^a_b \wedge \omega^b = 0 \quad \Rightarrow \quad \Theta^a_{[bcd]} = 0,$$

which is just the triple identity for the Riemann tensor,

$$R^a_{[bcd]} = 0. \tag{2.44}$$

Similarly, from the exterior derivative of (2.43),

$$\mathrm{d}\mathbf{\Theta}^a_b - \mathbf{\Theta}^a_f \wedge \mathbf{\Gamma}^f_b + \mathbf{\Gamma}^a_f \wedge \mathbf{\Theta}^f_b = 0,$$

that is

$$\mathbf{\Theta}^a_{b[cd;e]} \equiv 0, \quad \rightarrow \quad R^a_{b[cd;e]} = 0.$$

This equation says nothing about the Riemann tensor, $R^a_{bcd}$ directly. It says that certain combinations of the derivatives of the expressions on the right hand side of (2.41) are linear combinations of these same expressions.

$$\Theta_{ab[cd|e]} + D^s_{[cd}\Theta_{e]sab} - \Gamma^s_{a[c}\Theta_{de]sb} - \Gamma^s_{b[c}\Theta_{de]as} \equiv 0. \tag{2.45}$$

These are the true Bianchi identities. As a consequence, if the components of the Riemann tensor are thought of as variables, along with the components of the metric and the base forms, then these variables have to satisfy

$$R_{ab[cd|e]} = -2R^{a}_{be[c}\Gamma^{e}_{df]}. \tag{2.46}$$

## References


Bondi, H., van der Burg, M. G. J. & Metzner, A. W. K. (1962), "Gravitational waves in general relativity. 7. Waves from axisymmetric isolated systems", *Proc. Roy. Soc. Lond.* **A 269**, 21–52.

Boyer, R. H. & Lindquist, R. W. (1967), "Maximal analytic extension of the Kerr metric", *J. Math. Phys.* **8**, 265–280.

Carter, B. (1968), "Global structure of the Kerr family of gravitational fields", *Phys. Rev.* **174**, 1559–1571.

Carter, B. (1971), "Axisymmetric black hole has only two degrees of freedom", *Phys. Rev. Lett.* **26**, 331–333.

Debney, G. C., Kerr, R. P. & Schild, A. (1969), "Solutions of the Einstein and Einstein–Maxwell equations", *J. Math. Phys.* **10**, 1842–1854.

Goldberg, J. N. & Kerr, R. P. (1961), "Some applications of the infinitesimal-holonomy group to the Petrov classification of Einstein spaces", *J. Math. Phys.* **2**, 327.

Goldberg, J. N. & Sachs, R. K. (1962), "A theorem on Petrov types", *Acta Phys. Polon., Suppl.* **22**, 13.

Hawking, S. W. (1972), "Black holes in general relativity", *Commun. Math. Phys.* **25**, 152–166.

Janis, A. I. & Newman, E. T. (1965), "Structure of gravitational sources", *J. Math. Phys.* **6**, 902–914.

Kerr, R. P. (1958), "The Lorentz-covariant approximation method in general relativity I", *Nuovo Cimento* **13**, 470–491.

Kerr, R. P. (1960), "On the quasi-static approximation method in general relativity", *Nuovo Cimento* **16**, 26–60.

Kerr, R. P. (1963), "Gravitational field of a spinning mass as an example of algebraically special metrics", *Phys. Rev. Lett.* **11**, 237–238.

Kerr, R. P. (1965), "Gravitational collapse and rotation", in *Quasi-stellar Sources and Gravitational Collapse, including the Proceedings of the First Texas Symposium on Relativistic Astrophysics*, ed. I. Robinson, A. Schild & E. L. Schücking (University of Chicago Press, Chicago), pp. 99–109.

Kerr, R. P. & Debney, G. C. (1970), "Einstein spaces with symmetry groups", *J. Math. Phys.* **11**, 2807–2817.

Kerr, R. P. & Goldberg, J. N. (1961), "Einstein spaces with four-parameter holonomy groups", *J. Math. Phys.* **2**, 332.

Kerr, R. P. & Schild, A. (1965a), "A new class of vacuum solutions of the Einstein field equations", in *Atti del Congresso Sulla Relativita Generale: Galileo Centenario.*

Kerr, R. P. & Schild, A. (1965b), "Some algebraically degenerate solutions of Einstein's gravitational field equations, *Proc. Symp. Appl. Math.* **17**, ed. R. Finn (American Mathematical Society, Providence, Rhode Island), pp. 199–209.

Kerr, R. P. & Wilson, W. B. (1978), "Singularities in the Kerr–Schild Metrics", *Gen. Rel. Grav.* **10**, 273–281.

Newman, E. & Penrose, R. (1962), "An approach to gravitational radiation by a method of spin coefficients", *J. Math. Phys.* **3**, 566–768; err. **4** (1963) 988.

Newman, E., Tamburino, L. & Unti, T. (1963), "Empty-space generalization of the Schwarzschild metric", *J. Math. Phys.* **4**, 915–923.

Newman, E. T. & Janis, A. I. (1965), "Note on the Kerr spinning-particle metric", *J. Math. Phys.* **6**, 915–917.

Newman, E. T., Couch, E., Chinnapared, K., Exton, A., Prakash, A. & Torrence, R. (1965), "Metric of a rotating, charged mass", *J. Math. Phys.* **6**, 918–919.

Papapetrou, A. (1966), "Champs gravitationnels stationnaires à symétric axial", *Ann. Inst. H. Poincaré*, **4**, 83.

Papapetrou, A. (1974), *Lectures on General Relativity* (Reidel, Dordrecht), p. 116.

Petrov, A. Z. (1954), "The classification of spaces defining gravitational fields", in *Scientific Proceedings of Kazan State University* **114**, 55;
Translation by J. Jezierski & M. A. H. MacCallum, with introduction by M. A. H. MacCallum, *Gen. Rel. Grav.* **32**, 1661 (2000).

Pirani, F. A. E. (1957), "Invariant formulation of gravitational radiation theory", *Phys. Rev.* **105**, 1089–1099.

Robinson, D. C. (1975), "Uniqueness of the Kerr black hole", *Phys. Rev. Lett.* **34**, 905–906.

Robinson, I. & Trautman, A. (1962), "Some spherical gravitational waves in general relativity", *Proc. Roy. Soc. Lond.* **A 265**, 463–473.

Robinson, I. & Trautman, A. (1964), "Exact degenerate solutions of Einstein's equations, in *Proceedings on Theory of Gravitation, Conference in Warszawa and Jablonna, 1962*, ed. L. Infeld (Gauthier-Villars, Paris and PWN, Warszawa), pp. 107–114.

Sachs, R. K. (1961), "Gravitational waves in general relativity, VI. Outgoing radiation condition", *Proc. Roy. Soc. Lond.* **A 264**, 309.

Sachs, R. K. (1962), "Gravitational waves in general relativity, VIII. Waves in asymptotically flat space-time", *Proc. Roy. Soc. Lond.* **A 270**, 103.

Stephani, H., Kramer, D., MacCallum, M. A. H., Hoenselaers, C. & Herlt, E. (2003), *Exact Solutions to Einstein's Field Equations*, 2nd edn (Cambridge University Press, Cambridge).

Weir, G. J. & Kerr, R. P. (1977), "Diverging type – D metrics", *Proc. Roy. Soc. Lond.* **A 355**, 31–52.

# 3

# Roy Kerr and twistor theory

Roger Penrose

## 3.1 Introduction: Austin Texas, 1963

It was a day in early December 1963, in the aftermath of the Kennedy assassination, at a time when I was on a year-long visiting appointment at the University of Texas in Austin. Alfred Schild had acquired an impressive assembly of general relativists to work there at that time. The office to the right of mine was that of Engelbert Schücking, and in the office to the left was Roy Kerr. A little farther down on the left was Ray Sachs's office. Jerry Kristian was around somewhere too, but I think that his office was in a different area. There were frequent contacts with the group at Dallas where were Ivor Robinson, Wolfgang Rindler, and Pista (István) Ozsváth. But what was special about that particular day? It was a remarkable and unexpected encounter that I had with Roy Kerr, which has had a strong influence on my thinking ever since.

For some weeks earlier, various ideas had been buzzing around my head, partly stimulated by remarks from Engelbert on the need for a complex-manifold way of looking at spacetime. I had formed a view that there ought to be some kind of complex geometry which achieved, for the entire Minkowski space, what the division into a southern and northern hemisphere did for the Riemann sphere. The holomorphic extension of a function defined on the equator of that sphere, into one hemisphere or the other, is what most neatly expresses its splitting into positive and negative frequency parts. This splitting, when taken with respect to the time $t$, has profound significance for quantum field theory, a point which had earlier been deeply impressed upon me by Engelbert.

I think (if my memory serves me correctly) that it had been on the very previous day when a realization had come to me, while being driven from San Antonio back

The Kerr Spacetime: Rotating Black Holes in General Relativity. Ed. David L. Wiltshire, Matt Visser and Susan M. Scott. Published by Cambridge University Press. 

to Austin[1] by Pista (with a follow-up a little later on the same day at home in Austin itself), that a certain complex projective 3-space (a $\mathbb{CP}^3$) that I would later refer to as "twistor space" must be the space that I had been searching for. Like the two-real-dimensional Riemann sphere with its one-real-dimensional equator, this six-real-dimensional (twistor) space $\mathbb{PT}$, as I now label it, had a five-real-dimensional "equator" $\mathbb{PN}$ which divided it into an upper ("northern") part $\mathbb{PT}^+$ and a lower ("southern") part $\mathbb{PT}^-$, which I had hopes would play the required frequency-splitting role – although, as it turned out, the correct understanding of how this actually worked would not finally come to me[2] until about a decade and a half later!

The key realization leading me to twistor space concerned a certain type of shear-free congruence of twisting rays (i.e. null geodesics)[3] in Minkowski space $\mathbb{M}$ which Ivor Robinson had described to me – and which I later named a "Robinson congruence", in Ivor's honour. Some days earlier, I had studied the geometry of Ivor's congruences, and had persuaded myself that they could be described geometrically in terms of the twisted geometry of Clifford parallels in $S^3$, where the $S^3$ would be a constant-time slice through $\mathbb{M}$, compactified by a point $i^0$ at spatial infinity. The tangents to the (oriented) Clifford(–Hopf) circles on this $S^3$ provided the spatial projections of the null tangent directions to the rays of the Robinson congruence. I had come to the plausible conclusion, based on Ivor's original construction, that the various Robinson congruences in $\mathbb{M}$ ought to provide the points, in a natural way, of some complex manifold, but I had not previously made the effort to work out the dimensionality or nature of this space. Moreover, I was also aware that there would be a *real* submanifold consisting of *special* Robinson congruences, defined by the fact that the rays of such a congruence would be those rays meeting a given ray in $\mathbb{M}$. Accordingly, each point of this real submanifold would represent a single ray in Minkowski space (possibly at infinity).

I had originally somehow imagined that this space of special Robinson congruences might be of moderately high co-dimension. But on the trip back from San Antonio, I decided to think through what the various dimensions actually were, and was somewhat surprised to find the co-dimension 1, and that the family of Robinson congruences would constitute a complex 3-manifold, divided into two halves (corresponding to right-handed and left-handed Clifford parallels, respectively) by the real 5-submanifold of special Robinson congruences. It had suddenly struck me that this must be just the very space that I had been searching for!

Upon returning to where I lived in Austin, I immediately set to work to sort out this geometry, adapting Ivor's construction to the 2-spinor formalism that I

[1] This was on Sunday, 1st December 1963, according to Zsuzsi Ozsváth.
[2] Penrose (1979, 1987).
[3] In this article I consistently use the term "ray" for a null geodesic.

had long been familiar with. It did not take me long to realize that my complex (Robinson-congruence-)space must be a complex projective 3-space ($\mathbb{CP}^3$), and that it was indeed divided into two parts by a five-real-dimensional hypersurface each of whose points could be taken to represent a ray in Minkowski 4-space $\mathbb{M}$ (perhaps at infinity).

When I arrived in my office in the morning – on what I recall to have been the very next day – coming armed with these geometrical and 2-spinor descriptions,[4] I was distinctly surprised to overhear, in the next office, Roy Kerr explaining to Ray Sachs his way of generating *all* the (analytic) shear-free ray congruences in $\mathbb{M}$. What really struck me was that Roy was, in effect, using just the *same* "twistorial" complex coordinate descriptions as I had been led to in my own search for a complex space, but where my basic motivations were seemingly completely different from Roy's. But in one important respect, my own picture was a special case of Roy's – in the sense that a Robinson congruence is a special case of a general shear-free ray congruence. Roy's construction was to represent shear-free ray congruences in terms of the vanishing of a general holomorphic function of the three complex coordinates that he had been led to. In this special case, when the ray congruence is a Robinson congruence, we simply use a *linear* function instead of a general holomorphic one.

## 3.2 Twistors and Robinson congruences

Let us first see the basis of Ivor Robinson's construction. I shall express the central idea using the 2-spinor formalism, although that was not the explicit procedure that Ivor used. Suppose that we have some congruence of rays $\gamma$, with null tangent vectors which are the future-null vectors $\ell^a = \gamma^A \bar{\gamma}^{A'}$ (in 2-spinor notation[5]), so that $\bar{\gamma}^{A'} \gamma^A \gamma^B \nabla_{AA'}\gamma_B = 0$ expresses the geodeticity of the rays.[6] The *shear* of this congruence is measured by a complex quantity $\sigma$, defined by

$$\gamma^A \gamma^B \nabla_{AA'}\gamma_B = \sigma \, \bar{\gamma}_{A'}, \tag{3.1}$$

so the condition for the congruence to be *shear-free* is

$$\gamma^A \gamma^B \nabla_{AA'}\gamma_B = 0, \tag{3.2}$$

where I am taking "shear-freeness" to include "geodeticity".

[4] However, I had presumably used conventions with regard to the up/down and primed/unprimed placing of the 2-spinor indices that would have been in accord with the notation of Penrose (1967), which are less logical than those which I adopted later (e.g. Penrose and Rindler 1986, and as given here).

[5] For the 2-spinor and abstract-index notation used here, see Penrose and Rindler (1984).

[6] This equation is $\ell^a \nabla_a \gamma_B \propto \gamma_B$ (transvected with $\gamma^B$), telling us that the null "flagpole" direction of the spinor is parallel-propagated in the tangent direction $\ell^a = \bar{\gamma}^{A'} \gamma^A$ to $\gamma$.

In flat space-time $\mathbb{M}$, we can consider $\nabla_{AA'}$ to be $\partial/\partial x^{AA'}$, where $x^{AA'}$ $(= x^a)$ is the (abstract-indexed) 2-spinor version of the position vector of the point $x$ of $\mathbb{M}$ under consideration. It is clear that if we apply a *translation* to $\mathbb{M}$, which amounts to adding a real constant vector $K^a$ to $x^a$, then the shear-free condition will be unaffected. It is also reasonably evident that this also applies (locally in $\mathbb{M}$) if $K^a$ is *complex* (since the formal expressions are identical, the shear-free condition $\gamma^A \gamma^B \nabla_{AA'} \gamma_B = 0$ being *holomorphic* in $\gamma^A$, i.e. does not involve $\bar{\gamma}^{A'}$) provided that the field $\gamma^A$ is analytic, in $\mathbb{M}$, and that $K^a$ is small enough that the complex translation does not carry us outside the local complexification region (in the complexification $\mathbb{CM}$ of $\mathbb{M}$) where the holomorphically extended $\gamma^A$ remains non-singular. Such a complex translation would lead us to a field of *complex* null vectors in the real Minkowski space $\mathbb{M}$, these therefore having the form[7]

$$\gamma^A \beta^{A'}, \tag{3.3}$$

where $\beta^{A'}$ is now (generally) distinct from $\bar{\gamma}^{A'}$. The "trick" here is simply to replace this $\beta^{A'}$ by $\bar{\gamma}^{A'}$ to obtain a new field of *real* null vectors. The above "shear-free" (and geodetic) condition remains true, since it does not involve $\beta^{A'}$ (or $\bar{\gamma}^{A'}$).

To obtain a Robinson congruence, all we need do is to apply this procedure to the very simple shear-free ray congruence which consists of all the rays $Z$ meeting a fixed ray $Y$ – a *special* Robinson congruence. Another way to think of a special Robinson congruence, which renders its shear-free character evident, is as the generators of the (1-parameter) family of light cones whose vertices lie on $Y$. This is a particular instance of the more general situation of the generators of a family of light cones whose vertices lie on some given curve $\mathscr{Y}$, the general situation of a *non-rotating* shear-free ray congruence in $\mathbb{M}$.

For an explicit description of this procedure for finding a Robinson congruence in terms of formulae, we need an equation expressing the condition for rays $Y$ and $Z$ to meet. To specify a ray, say $Z$, we can take the position vector $r^a$ of a point $R$ on the ray together with a *null* tangent vector $p^a$, defining its direction, Both vectors of the pair $(r^a,\ p^a)$ are to be real – and let us take the null $p^a$ to be *future* pointing. The ray $Y$ could be specified correspondingly by the pair $(s^a,\ q^a)$; where $s^a$ and $q^a$ are both real, with $q^a$ future-null. The condition for $X$ and $Y$ to intersect is therefore

$$r^a + b\,p^a = s^a + c\,q^a, \qquad \text{for some } b, c \in \mathbb{R}. \tag{3.4}$$

[7] See Penrose and Rindler (1984), p. 126.

Writing this in 2-spinor form (i.e. replacing $a$ by $AA'$) we find, by transvecting through by $\bar{\kappa}_A \pi_{A'}$, where

$$p^{AA'} = \bar{\pi}^A \pi^{A'}, \qquad q^{AA'} = \bar{\kappa}^A \kappa^{A'}, \tag{3.5}$$

(future-null vectors always having this form) that this condition for $X$ and $Y$ to intersect can be re-expressed, without mention of $b$ or $c$, as

$$\bar{\kappa}_A r^{AA'} \pi_{A'} = \bar{\kappa}_A s^{AA'} \pi_{A'}. \tag{3.6}$$

We shall find it convenient to re-express this equation in terms of the quantities

$$\omega^A = i\, r^{AA'} \pi_{A'} \quad \text{and} \quad \eta_A = i\, s^{AA'} \kappa_{A'}, \tag{3.7}$$

where the *twistors*

$$Z^\alpha = (\omega^A, \pi_{A'}), \qquad Y^\alpha = (\eta^A, \kappa_{A'}), \tag{3.8}$$

together with their complex conjugates

$$\bar{Z}_\alpha = (\bar{\pi}_A, \bar{\omega}^{A'}), \qquad \bar{Y}_\alpha = (\bar{\kappa}_A, \bar{\eta}^{A'}), \tag{3.9}$$

which are *dual* twistors (note the reversal of the order of spinor parts), enable us to express the above condition for $X$ and $Y$ to meet as

$$0 = i\, \bar{\kappa}_A r^{AA'} \pi_{A'} - i\, \bar{\kappa}_A s^{AA'} \pi_{A'} = \bar{\kappa}_A \omega^A + \bar{\eta}^{A'} \pi_{A'} \tag{3.10}$$

i.e.

$$\bar{Y}_\alpha Z^\alpha = 0, \tag{3.11}$$

which I shall sometimes write without indices as

$$\bar{\mathbf{Y}} \cdot \mathbf{Z} = 0. \tag{3.12}$$

Equivalently, we can use the complex conjugate of this equation to represent the meeting of the rays $Y$ and $Z$, namely $\bar{\mathbf{Z}} \cdot \mathbf{Y} = 0$, i.e. $\bar{Z}_\alpha Y^\alpha = 0$.

We note that since the ray $Z$ obviously meets itself, we must have the condition

$$\bar{\mathbf{Z}} \cdot \mathbf{Z} = 0, \tag{3.13}$$

(i.e. $\bar{Z}_\alpha Z^\alpha = 0$) holding, and this applies, correspondingly, for all other rays in $\mathbb{M}$. In particular, we must have $\bar{\mathbf{Y}} \cdot \mathbf{Y} = 0$, and under these circumstances (and taking $\mathbf{Y}$

as fixed) the equation $\bar{\mathbf{Y}} \cdot \mathbf{Z} = 0$ (together with $\bar{\mathbf{Z}} \cdot \mathbf{Z} = 0$) represents the condition, on $\mathbf{Z}$, that the ray $Z$ belongs to the special Robinson congruence of rays meeting the particular ray $Y$. However, we also obtain a (shear-free) ray congruence if we allow $\bar{\mathbf{Y}} \cdot \mathbf{Y} \neq 0$, and it is not hard to see that this gives us the more general situation of a (*non*-special) Robinson congruence, according to the procedure outlined above, at the beginning of this section.

One advantage of using the twistor description of a ray, as opposed to simply using the pair $(r^a,\ p^a)$ to represent the ray $Z$ is that whereas we can replace $r^a$ according to

$$r^a \mapsto r^a + w\, p^a, \quad \text{for any } w \in \mathbb{R}, \tag{3.14}$$

and the ray $Z$ is unaltered, the twistor description in terms of the pair $(\omega^A,\ \pi_{A'}) = \mathbf{Z}$ remains *unchanged* by this replacement. We can also allow for a rescaling of $p^a$ by a positive factor

$$p^a \mapsto k\, p^a, \quad \text{for any } 0 < k \in \mathbb{R}, \tag{3.15}$$

(with $r^a$ unchanged). In *twistor* terms, this can be represented as

$$\mathbf{Z} \mapsto \lambda\, \mathbf{Z}, \quad \text{for any } 0 \neq \lambda \in \mathbb{C}, \tag{3.16}$$

where $\bar{\lambda}\,\lambda = k$, as this corresponds to the simultaneous rescalings $\omega^A \mapsto \lambda\, \omega^A$, $\pi_{A'} \mapsto \lambda\, \pi_{A'}$. Thus, the ray $Z$ determines, and is determined by, a *projective* null twistor, denoted $\mathbb{P}\mathbf{Z}$ (and the same notation will be used when $\mathbf{Z}$ is not null).

The twistors $\mathbf{Z}$ constitute a complex vector 4-space $\mathbb{T}$, called (non-projective) *twistor space*. The space of *null* twistors $\mathbf{Z}$ (satisfying $\bar{\mathbf{Z}} \cdot \mathbf{Z} = 0$, where we may observe that the Hermitian form $\bar{\mathbf{Z}} \cdot \mathbf{Z}$ has signature $+ + - -$), is a seven-real-dimensional subspace $\mathbb{N}$, of $\mathbb{T}$, which divides $\mathbb{T}$ into the space $\mathbb{T}^+$, of *positive* twistors (for which $\bar{\mathbf{Z}} \cdot \mathbf{Z} > 0$), and the space $\mathbb{T}^-$, of *negative* twistors (for which $\bar{\mathbf{Z}} \cdot \mathbf{Z} < 0$). The *projective* versions of these spaces (the spaces of their one-complex-dimensional vector subspaces) are, respectively, $\mathbb{PT}$, $\mathbb{PN}$, $\mathbb{PT}^+$, and $\mathbb{PT}^-$. We see from the above that the *projective null twistors* (elements of the space $\mathbb{PN}$) – for which we provisionally assume $\pi_{A'} \neq 0$ – precisely represent the *rays* in $\mathbb{M}$. If we wish to include these exceptional twistors for which $\pi_{A'} = 0$, i.e. of the form $(\omega^A, 0)$, then we must include "rays at infinity" for $\mathbb{M}$. This requires that we form the *conformal compactification* $\mathbb{M}^\#$ of $\mathbb{M}$ (where $\mathbb{M}^\#$ can be viewed as a real 4-quadric of signature $+ + - - - -$ in $\mathbb{RP}^5$; see Section 3.3).[8] The conformal

[8] See Penrose and Rindler (1986), Sections 9.1–3; in fact, one clear route to the definition and geometry of $\mathbb{M}^\#$ is through the study of this null-twistor correspondence.

space $\mathbb{M}^{\#}$ is $\mathbb{M}$ with an additional "light cone at infinity", and it is the family of generators of this cone $\mathscr{I}$ which are described by the (non-zero) twistors of the form $(\omega^A, 0)$. These twistors constitute a two-complex-dimensional subspace of $\mathbb{T}$, which I shall refer to as $\mathbf{I}$; projectively, this is a line $\mathbb{P}\mathbf{I}$ in $\mathbb{PN}$.

More generally, any point $P$ whatever, in compactified Minkowski space $\mathbb{M}^{\#}$, is represented by a projective line $\mathbf{P}$ in $\mathbb{PT}$ which lies entirely within $\mathbb{PN}$. The converse is also true: every projective line $\mathbf{P}$ in $\mathbb{PN}$ represents a point $P$ in $\mathbb{M}^{\#}$. The non-compact space $\mathbb{PN}-\mathbb{P}\mathbf{I}$ correspondingly represents the rays in $\mathbb{M}$, rather than $\mathbb{M}^{\#}$, and the projective lines entirely in $\mathbb{PN}-\mathbb{P}\mathbf{I}$ represent the points of $\mathbb{M}$.

Being a complex projective line, $\mathbf{P}$ is a *Riemann sphere* (a Riemann surface of topology $S^2$), and this $S^2$ is a family of points in $\mathbb{PN}$, each of which represents a ray in $\mathbb{M}^{\#}$ through $P$. These rays are the generators of the *light cone* of $P$ so we can therefore identify the Riemann sphere $\mathbf{P}$ as the *celestial sphere* of $P$. Celestial spheres have a natural (Lorentz-invariant) *conformal* structure, and indeed this is the kind of structure that a Riemann sphere also possesses.

We now have the interpretation of the points of $\mathbb{PN}$ as rays in $\mathbb{M}^{\#}$, and it is natural to ask for an interpretation, also, of the *general* points of $\mathbb{PT}$. Here we see the seminal role, for twistor theory, of the Robinson congruences. Let us consider a general point $\mathbb{P}\mathbf{Z}$ of $\mathbb{PT}$. Then rather than attempting to assign a meaning to $\mathbb{P}\mathbf{Z}$ directly, we can consider the *dual* twistor $\bar{\mathbf{Z}}$ instead. The family of rays $X$ whose twistor descriptions are given by null twistors $\mathbf{X}$ (so $\bar{\mathbf{X}} \cdot \mathbf{X} = 0$) for which

$$\bar{\mathbf{Z}} \cdot \mathbf{X} = 0 \tag{3.17}$$

is, by considerations given above, a *Robinson congruence*. Moreover, knowledge of this Robinson congruence completely fixes the projective dual twistor $\mathbb{P}\bar{\mathbf{Z}}$ so it also fixes the projective twistor $\mathbb{P}\mathbf{Z}$.

We note that, in a certain sense, the Robinson congruence is more directly associated with the *dual* twistor $W_\alpha = \bar{Z}_\alpha$, than with the twistor $Z^\alpha$ itself. (We can write such a dual twistor in abstract form, without indices, as $\mathbf{W}$, or alternatively adopt the abstract-index form $W_\alpha$ for the same entity.) The dual twistor $\mathbf{W}$ is an element of the dual twistor space $\mathbb{T}^*$, where $\mathbb{T}^*$ is simultaneously the dual and the complex conjugate of the complex vector 4-space $\mathbb{T}$. As with $\mathbb{T}$, there will be a *projective* version $\mathbb{PT}^*$ of this space, the projective dual twistor corresponding to the element $\mathbf{W} \in \mathbb{T}^*$ being written $\mathbb{P}\mathbf{W}$. In twistor-geometric terms, any (non-zero) projective dual twistor $\mathbb{P}\mathbf{W}$ will have an interpretation as a complex projective *plane* in $\mathbb{PT}$. We can consider the intersection of this plane (a four-real-dimensional space) with the 5-space $\mathbb{PN}$. This three-real-dimensional intersection represents a 3-real-parameter family of rays in $\mathbb{M}^{\#}$, in other words, a *congruence* of rays. This is the Robinson congruence associated with the dual twistor $\mathbf{W}$. If we write

$$W_\alpha = (\lambda_A, \mu^{A'}), \tag{3.18}$$

then the complex conjugate of $W_\alpha$ is the twistor

$$\bar{W}^\alpha = (\bar{\mu}^A, \bar{\lambda}_{A'}), \tag{3.19}$$

and its scalar product with the twistor $Z^\alpha = (\bar{\omega}^A,\ \pi_{A'})$ is the quantity

$$\mathbf{W} \cdot \mathbf{Z} = W_\alpha\, Z^\alpha = \lambda_A\, \omega^A + \mu^{A'}\, \pi_{A'}. \tag{3.20}$$

### 3.3 Other interpretations of twistor space

In fact, there are various ways, other than using Robinson congruences, of providing an interpretation of a (projective) twistor $\mathbf{Z}$ (or dual twistor $\mathbf{W}$). The most direct is to formulate twistor space in terms of *complexified* Minkowski space $\mathbb{CM}$. Up to this point, our interpretations have been basically in terms of real space-time geometry. Even Ivor Robinson's complex translation was immediately re-interpreted in terms of real structures by replacing the complex null vector $\gamma^A\, \beta^{A'}$ by the real one $\ell^a = \gamma^A\, \bar{\gamma}^{A'}$. Yet, if we are happy to remain within the realm of complex geometry, the twistor correspondence appears as a much more immediate one. Indeed, it is essentially one that was well familiar to geometers of the late nineteenth and early twentieth centuries.

We may arrive at this complex correspondence simply by extending the relation between a twistor $\mathbf{Z}$, given by $\mathbf{Z} = (\omega^A,\ \pi_{A'})$ and a point $R$ of $\mathbb{M}$, with position vector $r^a = r^{AA'}$, which was given early in the previous section by

$$\omega^A = i\, r^{AA'}\, \pi_{A'}. \tag{3.21}$$

I refer to this as expressing *incidence* between $R$ and $\mathbf{Z}$. Provided that $r^a$ is a *real* vector (so that indeed $R \in \mathbb{M}$, rather than $\mathbb{CM}$), then incidence expresses the condition that the point $R$ lie on the *ray* determined by $\mathbf{Z}$, and for this incidence equation to be solvable for a real $r^a$, we require $\bar{\mathbf{Z}} \cdot \mathbf{Z} = 0$. But if we now allow $r^a$ to be a *complex* vector (i.e. we allow $r^{AB'}$ to be non-Hermitian), so now $R \in \mathbb{CM}$, and also permit $\bar{\mathbf{Z}} \cdot \mathbf{Z} \neq 0$, then we find that the solutions of the incidence equation $\omega^A = i\, r^{AA'}\, \pi_{A'}$, for fixed $(\omega^A, \pi_{A'})$ do not give us simply the complexification of a null geodesic, but something of *two* complex dimensions (and therefore of *four* real ones). This complex plane is referred to as an *alpha*-plane, and it has the property that the holomorphic metric of $\mathbb{CM}$ (as naturally extended from the real metric of $\mathbb{M}$) *vanishes identically* when restricted to such a plane.

In fact, there are two types of complex plane in $\mathbb{CM}^{\#}$ with this property (i.e. that are *totally null* in this sense), the other being called a *beta*-plane. The complex tangent vectors to an alpha-plane and to a beta-plane are all null, and therefore these tangent vectors all have the 2-spinor form

$$\gamma^{A}\pi^{A'}. \tag{3.22}$$

Moreover, these tangent vectors are all orthogonal to each other. The difference between an alpha- and a beta-plane is that for an alpha-plane the tangent vectors have the above form where $\pi^{A'}$ is held fixed and $\gamma^{A}$ varies (so orthogonality follows from $\pi^{A'}\pi_{A'} = 0$), whereas for a beta-plane it is $\gamma^{A}$ which we hold fixed and it is $\pi^{A'}$ which varies (orthogonality following from $\gamma^{A}\gamma_{A} = 0$). An alpha-plane is referred to as *self-dual* and a beta-plane as *anti-self-dual*.

As has just been asserted, in effect, it is the alpha-planes in $\mathbb{CM}^{\#}$ that correspond to the *points* in $\mathbb{PT}$. Dually, the beta-planes of $\mathbb{CM}^{\#}$ correspond to the complex *planes* in $\mathbb{PT}$. The equation of incidence between a dual twistor $W^{\alpha} = (\lambda_{A}, \mu^{A'})$ and a point $x$ of $\mathbb{M}$, or of $\mathbb{CM}$, is (via complex conjugation of the incidence relation $\omega^{A} = i\, r^{AA'}\pi_{A'}$)

$$\mu^{A'} = -i\,\lambda_{A}\,x^{AA'}, \tag{3.23}$$

which asserts that the point $x$ lies on the beta-plane defined by $\mathbf{W}$. In addition, as we have seen, the *points* in $\mathbb{CM}^{\#}$ correspond to the complex projective *lines* in $\mathbb{PT}$. Such a line may be thought of as the join of two points in $\mathbb{PT}$, corresponding to a point of $\mathbb{CM}^{\#}$ arising as the intersection of two alpha-planes. Or it may be thought of as the intersection of two planes in $\mathbb{PT}$, corresponding to a point of $\mathbb{CM}^{\#}$ arising as the intersection of two beta-planes.

In fact, this correspondence between $\mathbb{PT}$ and $\mathbb{CM}^{\#}$ is a realization of one that has been familiar to geometers for well over a century, known as the *Klein* (or Grassmann) correspondence, where the lines of projective 3-space are represented by the points of a quadric hypersurface in projective 5-space. The most direct way to see how $\mathbb{M}^{\#}$ can be realized as a quadric in this way is to think of the real pseudo-Euclidean 6-space of metric signature $(++----)$ with coordinates $U, T, X, Y, Z, W$, the metric form being

$$\mathrm{d}s^2 = \mathrm{d}U^2 + \mathrm{d}T^2 - \mathrm{d}X^2 - \mathrm{d}Y^2 - \mathrm{d}Z^2 - \mathrm{d}W^2, \tag{3.24}$$

where we are concerned with the "null cone" $\mathscr{K}$ of the origin

$$U^2 + T^2 - X^2 - Y^2 - Z^2 - W^2 = 0. \tag{3.25}$$

The intersection of $\mathscr{K}$ with the null hyperplane $U - W = 1$ has the standard Minkowski metric, and may be regarded as a realization of $\mathbb{M}$, though imbedded in a non-standard "paraboloidal" way. To see how $\mathbb{M}$ is compactified as $\mathbb{M}^{\#}$, we think of the *projective* version of this 6-space, namely the $\mathbb{RP}^5$ of straight lines through the origin. The family of those which lie on $\mathscr{K}$ gives a realization of $\mathbb{M}^{\#}$ as a quadric in this $\mathbb{RP}^5$. The light cone $\mathscr{I}$ at infinity is given by the subfamily of these lines which lie on the hyperplane $U - W = 0$, as these do not meet the paraboloidally imbedded $\mathbb{M}$.

In order to see how this quadric can be viewed as a "Klein quadric", whose points correspond to the entire family of lines in some projective 3-space, one must pass to a complex description. This is because the Klein quadric for lines in $\mathbb{RP}^3$ would be given by a quadratic form of signature $(+++---)$, rather than the $(++----)$ that we have here. That would correspond to a spacetime metric of signature $(++--)$,[9] rather than our standard Lorentzian $(+---)$. What is characteristic of the latter spacetime signature is the role of the subspace $\mathbb{PN}$, of $\mathbb{CP}^3$, which is given by the vanishing of a Hermitian-quadratic form of signature $(++--)$.

As a historical comment, it is worth mentioning that Sophus Lie had noted, already by 1872, that oriented spheres in complex Euclidean 3-space ($\mathbb{CE}^3$) could be put into correspondence with lines in $\mathbb{CP}^3$. To see that this is basically the twistor correspondence, we can think of an oriented sphere in ordinary *real* Euclidean 3-space ($\mathbb{E}^3$) as the wave-front of a flash of light, exploding outwards if the orientation is positive and imploding inwards if the orientation is negative, with light-speed $c$ in each case. In four-dimensional Minkowskian terms, these oriented spheres can be viewed as just a three-dimensional constant-time "slice" through the light cones of the various points of $\mathbb{M}$. Lie's correspondence really refers to the complexification of this, so that it represents points in $\mathbb{CM}$ as lines in $\mathbb{CP}^3$. In effect, this is indeed the twistor correspondence.

In the other direction, we may ask for a more clearly *physical* interpretation of a twistor $\mathbf{Z}$, than that which is given by either the rather formal and intangible complex-space descriptions above, or by the somewhat geometrically complex description in terms of Robinson congruences. This can be done by associating a twistor $Z^\alpha = (\omega^A, \pi_{A'})$ with a 4-momentum and 6-angular momentum, according to

$$p_a = \bar{\pi}_A \pi_{A'} \quad \text{and} \quad M^{ab} = i\,\omega^{(A}\,\bar{\pi}^{B)}\,\varepsilon^{A'B'} - i\,\varepsilon^{AB}\,\bar{\omega}^{(A'}\,\pi^{B')}. \tag{3.26}$$

This definition ensures the requirements, for the momentum and angular momentum of a massless particle, that $p_a$ be null and future pointing (assuming that

[9] Some authors have found it convenient to express things in terms of such a "pseudo-twistor theory", and then to perform a kind of "reverse Wick rotation" to obtain the quantities of physical interest. See Witten (2003).

$\pi_{A'} \neq 0$) and that the Pauli–Lubañski vector

$$S_a = \tfrac{1}{2}\, e_{abcd}\, p^b\, M^{cd} \tag{3.27}$$

should be proportional to the momentum $p_a$

$$S_a = s p_a, \tag{3.28}$$

where $s$ is the *helicity*. Conversely, any $p_a$ and $M^{ab}$ satisfying these standard requirements can be expressed in the above form, for some $(\omega^A, \pi_{A'})$, with $\pi_{A'} \neq 0$, the only freedom in $\omega^A$ and $\pi_{A'}$ being the simultaneous phase freedom

$$\omega^A \to e^{i\theta}\, \omega^A, \quad \pi_{A'} \to e^{i\theta}\, \pi_{A'}, \quad \text{where } \theta \in \mathbb{R}, \tag{3.29}$$

i.e. $\mathbf{Z} \to e^{i\theta}\mathbf{Z}$, and we find

$$s = \tfrac{1}{2}\, \bar{\mathbf{Z}} \cdot \mathbf{Z}. \tag{3.30}$$

In addition, we find that the behaviour of $(\omega^A, \pi_{A'})$, under change of origin is consistent with that for the pair $(p_a, M^{ab})$. Thus, if we translate our origin of coordinates in $\mathbb{M}$ by an amount $q^a$, so that our original position vector $r^a$ is now replaced by $r^a - q^a$, we find

$$\pi_{A'} \to \pi_{A'} \quad \text{and} \quad \omega^A \to \omega^A - i\, q^{AA'}\, \pi_A, \tag{3.31}$$

which is consistent with the standard

$$p_a \to p_a \quad \text{and} \quad M^{ab} \to M^{ab} - q^a\, p^b + q^b\, p^a. \tag{3.32}$$

### 3.4 Kerr's theorem for shear-free ray congruences

Now let us return to Roy Kerr's basic contribution to twistor theory, alluded to above. With the spinor position-vector description $x^{AA'}$ we have been using, we have coordinates for $\mathbb{M}$ (with $u$, $v$, real and $\zeta$ complex)

$$u = x^{00'}, \quad \zeta = x^{01'}, \quad \bar{\zeta} = x^{10'}, \quad v = x^{11'}, \tag{3.33}$$

the spin frame $(o^A, \iota^A)$ being constant throughout $\mathbb{M}$. The Minkowski metric now has the form

$$\mathrm{d}s^2 = 2\,\mathrm{d}u\,\mathrm{d}v - 2\,\mathrm{d}\zeta\,\mathrm{d}\bar{\zeta}. \tag{3.34}$$

Suppose that we have some shear-free congruence of rays $\gamma$, with null tangent vectors $\gamma^A \bar{\gamma}^{A'}$. Then we can express the condition that $\mathrm{d}x^{AA'}$ represents displacement along these rays by the relation

$$\gamma^A \, \mathrm{d}x^{AA'} = 0, \tag{3.35}$$

where we are assuming that $x^a$ is *real*. We can choose to scale $\gamma^A$ so that

$$\gamma^0 = -\gamma^1 = 1, \qquad \gamma_1 = \gamma^0 = \xi, \tag{3.36}$$

so the complex number $\xi$ (allowing $\xi = \infty$) defines the direction of the flagpole of $\gamma^A$. Then the preceding equation becomes

$$\mathrm{d}u + \xi \, \mathrm{d}\bar{\zeta} = 0 = \mathrm{d}\zeta + \xi \, \mathrm{d}v. \tag{3.37}$$

We now regard the "shear-free equation"

$$\gamma^A \gamma^B \nabla_{AA'} \gamma_B = 0 \tag{3.38}$$

as an equation on $\xi$, as a function of $u, \zeta, \bar{\zeta}, v$, and we find (since the spin-frame is constant) that the chain rule

$$\mathrm{d}\xi = \frac{\partial \xi}{\partial u} \, \mathrm{d}u + \frac{\partial \xi}{\partial \zeta} \, \mathrm{d}\zeta + \frac{\partial \xi}{\partial \bar{\zeta}} \, \mathrm{d}\bar{\zeta} + \frac{\partial \xi}{\partial v} \, \mathrm{d}v, \tag{3.39}$$

leads to

$$\xi \frac{\partial \xi}{\partial u} = \frac{\partial \xi}{\partial \bar{\zeta}}, \qquad \xi \frac{\partial \xi}{\partial \zeta} = \frac{\partial \xi}{\partial v} \tag{3.40}$$

as the equations to be solved. Put

$$P = \frac{\partial \xi}{\partial u}, \qquad Q = \frac{\partial \xi}{\partial \zeta}, \tag{3.41}$$

so that

$$\xi P = \frac{\partial \xi}{\partial \bar{\zeta}}, \qquad \xi Q = \frac{\partial \xi}{\partial v}; \tag{3.42}$$

and we have

$$\mathrm{d}\xi = P \, (\mathrm{d}u + \xi \, \mathrm{d}\bar{\zeta}) + Q \, (\mathrm{d}\zeta + \xi \, \mathrm{d}v), \tag{3.43}$$

so it follows that $\mathrm{d}\xi = 0$ whenever $\mathrm{d}u + \xi \, \mathrm{d}\bar{\zeta} = 0 = \mathrm{d}\zeta + \xi \, \mathrm{d}v$. Hence $\xi$ is constant over the regions where $u + \xi \, \bar{\zeta}$ and $\zeta + v$ are both constant. This will hold when there is a functional relation

$$F(\xi, u + \xi\,\bar{\zeta}, \zeta + \xi\,v) = 0, \tag{3.44}$$

where – the arguments being complex variables – we require $F$ to be a *holomorphic* function. This was Roy Kerr's ingenious method of solving the problem of finding general shear-free ray congruences in Minkowski space.

In fact, this is indeed the complete solution for congruences which are real-*analytic* in the real spacetime $\mathbb{M}$. In the context of entirely real variables, this kind of procedure would have provided a complete solution, analytic or not, just as it does for analytic congruences in the present context. But there are some subtleties in the present situation, where we are concerned with mixtures of real and complex variables.[10] It was noted in Section 3.2 that a (non-rotating) shear-free ray congruence is formed by the generators of a family of light cones whose vertices lie on a given curve $\mathscr{Y}$. If this curve is not analytic, then the congruence will be non-analytic. But the function $F$, above, is necessarily analytic, since it is holomorphic, and any ray congruence obtainable by this method is necessarily real-analytic.

Now, what is the relation between Roy Kerr's variables $\xi$, $u + \xi\,\bar{\zeta}$, $\zeta + \xi\,v$ and twistor variables? Recall the basic twistor relation $\omega^A = i\,x^{AA'}\,\pi_{A'}$, expressing incidence between the point $x$ of $\mathbb{M}$ (or $\mathbb{CM}$), with position vector $x^{AA'}$ and the twistor $\mathbf{Z} = (\omega^A,\ \pi_{A'})$. In the present context, it is more relevant to consider the dual (or complex conjugate) of this, namely the relation (3.23)

$$\mu^{A'} = -i\,\lambda_A\,x^{AA'}, \tag{3.45}$$

which expresses incidence between the point $x$ of $\mathbb{M}$ (or $\mathbb{CM}$) and the *dual* twistor $\mathbf{W} = (\lambda_A,\ \mu^{A'})$. In components, this is

$$\mu^{0'} = -i\lambda_0\,x^{00'} - i\lambda_1\,x^{10'}, \qquad \mu^{1'} = -i\lambda_0\,x^{01'} - i\lambda_1\,x^{11'}, \tag{3.46}$$

i.e.

$$\mu^{0'} = -i\lambda_0\,u - i\lambda_1\bar{\zeta}, \qquad \mu^{1'} = -i\lambda_0\,\zeta - i\lambda_1\,v. \tag{3.47}$$

If we take $\lambda_A$ to be proportional to $\gamma_A$, whose flagpole points along the direction, at the point $x$, of the ray (incident with $x$) of the congruence under consideration, then we have $\xi = \lambda_1/\lambda_0$, so the above gives us

$$\lambda_0 : \lambda_1 : \mu^{0'} : \mu^{1'} = 1 : \xi : -i(u + \xi\,\bar{\zeta}) : -i(\zeta + \xi\,v), \tag{3.48}$$

[10] There are close connections with the much-studied "Hans Lewy problem"; see Lewy (1956, 1957), Andreotti and Hill (1972); also Penrose (1983), Penrose and Rindler (1986).

so we see that the Kerr variables that we encountered above are (apart from the two factors of $-i$) just the ratios of the components of a (dual) twistor $\mathbf{W} = (\lambda_A, \mu^A)$.

### 3.5 Twistor geometry of Kerr's theorem and his black-hole congruence

What is the geometrical significance of this for twistor theory? The equation of incidence $\mu^{A'} = -i\lambda_A x^{AA'}$, in complex terms, expresses the condition that the point $x$ lie on the beta-plane determined by $\mathbf{W}$. Hence, Kerr's construction tells us that the rays of an (analytic) shear-free ray congruence in $\mathbb{M}$ arise as the intersections of $\mathbb{M}$ with the beta-planes of a two-complex-parameter holomorphic family in $\mathbb{CM}$. As a whole, beta-planes form a three-complex-parameter family in $\mathbb{CM}$. The single holomorphic equation $F(\xi, u + \xi\,\bar{\zeta}, \zeta + \xi\, v) = 0$ reduces this family to a family of two complex parameters. Taking intersections with $\mathbb{M}$, the equation $F = 0$ correspondingly reduces the entire five-real-parameter family of rays in $\mathbb{M}$ to the three-real-parameter family of rays in the shear-free congruence. (Most beta-planes do not meet $\mathbb{M}$ at all; only a five-real-parameter family does, giving the entire system of rays in $\mathbb{M}$.) In terms of twistor geometry, the equation $F = 0$ (which we can regard as a homogeneous equation $F(\mathbf{W}) = 0$, in the components of a dual twistor $\mathbf{W}$) determines a *holomorphic 2-surface* $\mathcal{S}$ in $\mathbb{CP}^3$, where this $\mathbb{CP}^3$ is $\mathbb{PT}^*$. The shear-free ray congruence in $\mathbb{M}$ itself arises as the three-real-dimensional intersection $\mathcal{S} \cap \mathbb{PN}^*$, where we are interpreting the elements of $\mathbb{PN}^*$ as giving us rays in $\mathbb{PN}^*$ in just the same way as elements of $\mathbb{PN}$ do, as was described in Section 3.2.

Of course, we could equally well have phrased things in terms of (projective) *twistor* space $\mathbb{PT}$ rather than dual twistor space $\mathbb{PT}^*$. The difference is totally unimportant here; dual twistor space has arisen in the present discussion, rather than twistor space merely because it had seemed more natural to use the unprimed spinor $\lambda^A$ in the discussion of shear-free congruences, rather than the primed one $\bar{\lambda}^{A'}$. This is immaterial. Phrasing Kerr's construction in terms of twistor space rather than dual twistor space would amount to thinking of the (analytic) shear-free congruence as arising at the intersections with $\mathbb{M}$ of a two-complex-parameter holomorphic family of *alpha*-planes in $\mathbb{CM}$, rather than beta-planes (i.e. arising as an intersection $\bar{\mathcal{S}} \cap \mathbb{PN}$, where $\bar{\mathcal{S}} \subset \mathbb{PT}$ is the complex conjugate of $\mathcal{S} \subset \mathbb{PT}^*$). In fact, it is both, because for a *real* ray congruence the alpha-planes that arise are simply the complex conjugates of the beta-planes, and each real ray arises as the intersection of an alpha-plane with its complex-conjugate beta-plane (where such an intersection exists).

Let us now consider the special kind of surface $\mathcal{S}$ that arises in the particular case of the shear-free congruence in $\mathbb{M}$ which (via the now well-known

Kerr–Schild construction[11]) led to the famous Kerr metric describing a general rotating stationary vacuum black hole. For this, $\mathcal{S}$ is a quadric surface $\mathcal{Q}$ in $\mathbb{PT}^*$, i.e. given by a homogeneous *quadratic* expression

$$\mathbf{WW} : \mathbf{Q} = 0, \quad \text{i.e.} \quad W_\alpha \, W_\beta \, Q^{\alpha\beta} = 0, \tag{3.49}$$

where we may assume the symmetry

$$Q^{\alpha\beta} = Q^{\beta\alpha}. \tag{3.50}$$

Surfaces of this kind are the next simplest after the planes which give rise to Robinson congruences, and it will be appropriate to give them some special consideration here.

The general quadric surface $\mathcal{Q}$ is ruled by two one-complex-parameter families of lines in $\mathbb{CP}^3$ ($= \mathbb{PT}^*$ here) – where in fact we have, both topologically and holomorphically, $\mathcal{Q} \simeq \mathbb{CP}^1 \times \mathbb{CP}^1$. Each of these families represents a *curve* in $\mathbb{CM}^\#$ which is a (complex) Minkowskian circle.[12] These two circles are related to each other by the remarkable fact that every point of either one of the circles is null-separated from every point of the other (so a complex ray connects any given pair of points, one taken from each of the two circles).

In the case of the $\mathcal{Q}$ which gives rise to a (non-spinning) Schwarzschild black hole, one of these families of lines is "real" in the sense that it corresponds, in $\mathbb{CM}^\#$, to the complexification of a *real* Minkowskian circle in $\mathbb{M}^\#$. Here this Minkowskian "circle" is actually a timelike straight line: the "source" line in $\mathbb{M}$ describing the Schwarzschild particle. The other circle is purely imaginary and happens to lie entirely at (complex) conformal infinity $\mathbb{C}\mathscr{I}$ (see Section 3.3). This case is characterized by two conditions on $Q^{\beta\gamma}$. The first condition ($I_{\alpha\beta}$ having $\varepsilon^{A'B'}$ as its only non-zero spinor part) is

$$I_{\alpha\beta} \, Q^{\beta\gamma} \, I_{\gamma\eta} = 0, \tag{3.51}$$

which ensures that one of the circles is a straight line in $\mathbb{CM}$ (since it contains the point $I$ at infinity which is the vertex of $\mathscr{I}$, and then the other circle lies entirely in $\mathbb{C}\mathscr{I}$). The second condition is that the inverse $Q^{(-1)}_{\alpha\beta}$ of $Q^{\alpha\beta}$, as defined by

$$Q^{\alpha\beta} \, Q^{(-1)}_{\beta\gamma} = \delta^\alpha_\gamma, \tag{3.52}$$

[11] Kerr and Schild (1965).

[12] Minkowskian circles are either (non-null) straight lines in $\mathbb{CM}$ or the curves obtained from these by conformal motions of $\mathbb{CM}^\#$. The real timelike ones are the world-lines of uniformly accelerating particles. The real spacelike ones are ordinary instantaneous circles (or straight lines) in some Lorentz frame. In a general curved (pseudo-)Riemannian space these generalize to the notion of a "conformal circle"; see Yano (1957), p. 158.

is proportional to $\bar{Q}_{\alpha\beta}$, i.e.

$$Q^{(-1)}_{\alpha\beta} = k\ \bar{Q}_{\alpha\beta}. \tag{3.53}$$

From this last equation, we deduce

$$k = 4\{\bar{Q}_{\alpha\beta}\ Q^{\alpha\beta}\}^{-1}, \tag{3.54}$$

and the timelike nature of the source line is determined by

$$k < 0. \tag{3.55}$$

In the case of a "tachyonic" particle we would have $k > 0$, the "source" circle being a spacelike straight line; in this case the other Minkowskian circle would also be real and spacelike (but at infinity). We also obtain the shear-free ray congruences that provide the weak-field limit of the accelerating Levi–Civita "C-metrics", with an *accelerating* source line – again a Minkowskian circle – timelike or spacelike according as $k < 0$ or $k > 0$. In these accelerating cases, the other condition $I_{\alpha\beta}\ Q^{\beta\gamma}\ I_{\gamma\eta} = 0$ fails, however.

In the case from which the Kerr black hole with non-zero spin arises (by the Kerr–Schild construction), we may think of the source world-line as being timelike and straight, but displaced into the complex, by a complex translation. We find that the inverse of $Q^{\alpha\beta}$ is not now proportional to its complex conjugate, but that (with a suitable phase factor chosen for $Q^{\alpha\beta}$) the condition (with $I^{\gamma\beta}$ having $\varepsilon^{GB}$ as its only non-zero part, so $I^{\gamma\beta} I_{\gamma\eta} = 0$)

$$\bar{Q}_{\alpha\beta}\ I^{\gamma\beta} = Q^{\gamma\beta}\ I_{\alpha\beta} \tag{3.56}$$

must hold (or that the two sides of this equation are merely proportional if a general phase factor is chosen), which has the condition $I_{\alpha\beta}\ Q^{\beta\gamma}\ I_{\gamma\eta} = 0$ as a consequence. In this case the shear-free ray congruence is *stationary*, as is required for the Kerr–Schild construction (which, of course, must also hold in the special case of the Schwarzschild congruence).

Since (with non-zero spin) the inverse $Q^{(-1)}_{\alpha\beta}$ is no longer proportional to the complex conjugate $\bar{Q}_{\alpha\beta}$, we have two distinct quadric surfaces in twistor space $\mathbb{PT}$, each defined from the original quadric $\mathcal{Q}$ in $\mathbb{PT}^*$, namely $\mathcal{Q}^*$ and $\bar{\mathcal{Q}}$, defined respectively by

$$Q^{(-1)}_{\alpha\beta}\ Z^\alpha\ Z^\beta = 0 \quad \text{and} \quad \bar{Q}_{\alpha\beta}\ Z^\alpha\ Z^\beta = 0. \tag{3.57}$$

It is the second of these $\bar{\mathcal{Q}}$ which, via $\bar{\mathcal{Q}} \cap \mathbb{PN}$, simply gives the original shear-free ray congruence again (as was defined by $Q_{\alpha\beta}\ W^\alpha\ W^\beta = 0$, via $\mathcal{Q} \cap \mathbb{PN}^*$). This is a consequence of the fact that this shear-free congruence is *real*, i.e. the same as its complex conjugate. The equation $\mathcal{Q}^{(-1)}_{\alpha\beta} Z^{\alpha} Z^{\beta} = 0$ defining $\mathcal{Q}^*$, on the other hand, gives us a *different* shear-free ray congruence, via $\mathcal{Q}^* \cap \mathbb{PN}$, which we can refer to as the *reciprocal* ray congruence to the original one. This reciprocal congruence is, however, very simply related to the original one here; it corresponds to a black hole with the same mass-centre world-line, but with opposite spin.

### 3.6 String histories and reciprocal surfaces

This example of reciprocal shear-free ray congruences is actually an instance of a general phenomenon; but in the general case, the relation between a shear-free ray congruence and its reciprocal is considerably more subtle, and is most easily visualized in terms of another geometrical object in $\mathbb{M}$ that bears a corresponding relation to each of them, namely a *string history*. By a "string history" I mean a *real, oriented, timelike 2-surface* $\mathcal{H}$ in $\mathbb{M}$ (or $\mathbb{M}^{\#}$), which I shall normally assume to be both analytic and suitably "generic" – which here essentially means *not* ruled by a 1-parameter family of rays. Thus we may think of $\mathcal{H}$ as a string (which may be either a closed loop or an open string segment) which is moving with less than light speed in some arbitrary (but analytic) way. The role of the string history $\mathcal{H}$, for our shear-free ray congruence, is that it is the envelope of that subfamily of rays of the congruence which is defined by the property that the *rotation of the congruence vanishes* along these rays. I shall call these rays the *non-rotating* rays of the congruence. These non-rotating rays are precisely the rays which are in contact with $\mathcal{H}$, in a direction consistent with its given orientation. In more prosaic terms, we may think of a "laser searchlight" at each point of the string, pointing in one direction along the string. As the string evolves, the light rays of the searchlights (extended also into the past) provide us with the family of non-rotating rays of the congruence. It is remarkable that this two-real-parameter family of rays, defined directly by the string history, actually determines the entire shear-free ray congruence.[13]

The timelike 2-surface $\mathcal{H}$ is ruled by *two* families of null non-geodesic curves (since its tangent 2-planes, being timelike, contain two distinct null directions, these being tangents to the curves). It is the choice of one of these families of null curves, in preference to the other, that determines $\mathcal{H}$'s orientation. In terms of our laser searchlights, the choice is in which of the two directions along the string the searchlights are to point. The reversal of this choice (i.e. the reversal of $\mathcal{H}$'s orientation) entails passing from the congruence to its reciprocal congruence. In the case of the ray congruence that gives rise to the Kerr

[13] See Penrose (1997a,b); also Newman (2004).

black hole, the string history is the "Kerr ring", as it evolves in time (moving uniformly in a straight line), this being where the curvature singularity occurs in the full black-hole solution. Reversing its orientation simply reverses the spin.

To see why, in the analytic case, the string history $\mathcal{H}$ itself (somewhat remarkably) actually *determines* the shear-free ray congruence, we revert to the complex geometry of $\mathbb{CM}^{\#}$ and $\mathbb{PT}^{*}$. Since $\mathcal{H}$ is analytic, we can extend it holomorphically into $\mathbb{CM}$ (or $\mathbb{CM}^{\#}$) as a complex 2-surface $\mathbb{C}\mathcal{H}$ (at least by a certain amount, until singularities arise in the extension). Now $\mathcal{H}$ is timelike, so it is non-null, and the orthogonal complement of the tangent plane at each of its points is also non-null and therefore contains two (complex) null directions, each orthogonal to $\mathbb{C}\mathcal{H}$. Extending each of these null directions to a complex null geodesic (a complex ray), we generate two complex null hypersurfaces $\mathcal{N}$ and $\mathcal{N}'$. Their intersection is $\mathbb{C}\mathcal{H}$ (locally, at least): $\mathcal{N} \cap \mathcal{N}' = \mathbb{C}\mathcal{H}$ and they are each orthogonal to $\mathbb{C}\mathcal{H}$. This is a complexification of the familiar situation in Lorentzian 4-geometry, whereby a spacelike surface is always locally the intersection of two null hypersurfaces, each orthogonal to it. Now, had $\mathcal{H}$ indeed been spacelike, then $\mathcal{N}$ and $\mathcal{N}'$ would have each been (complexifications of) distinct *real* null hypersurfaces. But here $\mathcal{H}$ is timelike, and the only other way of it arising as a real 2-surface within this complex intersection is for $\mathcal{N}$ and $\mathcal{N}'$ to be complex conjugates of each other (that is, the coefficients in their equations are respective complex conjugates; they are still both holomorphic loci). Since the *real* points of $\mathcal{N}$ occur at its intersection with its complex conjugate (i.e. with $\mathcal{N}'$), we find that $\mathcal{H}$ arises as the family of real points of this uniquely determined complex null hypersurface $\mathcal{N}$ (where we may regard the distinction between $\mathcal{N}$ and $\mathcal{N}'$ arising from the orientation of $\mathcal{H}$). Thus we have the striking fact that any oriented timelike surface $\mathcal{H}$ in $\mathbb{M}^{\#}$ is (locally at least) *equivalent* to a complex null hypersurface $\mathcal{N}$ in $\mathbb{CM}^{\#}$, where $\mathcal{N}$ is, in fact, completely general as a complex null hypersurface.

Let us next see why $\mathcal{N}$ is equivalent to our original shear-free null congruence (whence the entire null congruence must actually be determined by its subfamily of non-rotating rays!). First, how do we represent, twistorially, a complex null hypersurface $\mathcal{N}$ in $\mathbb{CM}^{\#}$? Indeed, how do we represent a complex ray? A complex ray in $\mathbb{CM}^{\#}$ is uniquely the intersection of an alpha-plane $Z$ with a beta-plane $W$ (which happen to intersect, so the corresponding twistor $\mathbf{Z}$ and dual twistor $\mathbf{W}$ satisfy $\mathbf{W} \cdot \mathbf{Z} = 0$). In $\mathbb{PT}$, this geometry is a plane $\mathbb{P}\mathbf{W}$ with a point $\mathbb{P}\mathbf{Z}$ on it; in $\mathbb{PT}^{*}$ the configuration is the same, but now $\mathbb{P}\mathbf{W}$ represents the point and $\mathbb{P}\mathbf{Z}$ the plane through it. When the null hypersurface $\mathcal{N}$ is swept out by its complex two-parameter family of complex rays, the points $\mathbb{P}\mathbf{Z}$ in $\mathbb{PT}$ sweep out a complex 2-surface $\mathcal{S}^{*}$ in $\mathbb{PT}$ and it turns out (because $\mathcal{N}$ is a null hypersurface and not just an ordinary ruled surface ruled by a complex two-parameter family of complex rays)[14] that the tangent plane at each non-singular point of $\mathcal{S}^*$ is precisely the plane $\mathbb{P}\mathbf{W}$ which is associated with the point $\mathbb{P}\mathbf{Z}$ according to the preceding prescription. Accordingly $\mathcal{S}^*$ is also the envelope of the 2-parameter family of planes $\mathbb{P}\mathbf{W}$, each contact point being the corresponding $\mathbb{P}\mathbf{Z}$. In $\mathbb{PT}^*$, the situation is just the same, but with the roles of $\mathbb{P}\mathbf{Z}$ and $\mathbb{P}\mathbf{W}$ reversed, so that we have another surface $\mathcal{S}$, now in the dual twistor space $\mathbb{PT}^*$, which is both the locus of points $\mathbb{P}\mathbf{W}$ and the envelope of planes $\mathbb{P}\mathbf{Z}$, where the contact point of each plane $\mathbb{P}\mathbf{Z}$ is the corresponding point $\mathbb{P}\mathbf{W}$.

The two surfaces $\mathcal{S}$ and $\mathcal{S}^*$ are *reciprocals* (or duals) of one another, in the sense that if we regard one as a point locus and pass to the dual projective space, then we get the other from its plane envelope (modulo singularities). In the case of a non-singular quadric surface, the reciprocal is again a non-singular quadric surface. This is exactly the situation encountered above (in Section 3.5) with $\mathcal{Q}$ and $\mathcal{Q}^*$, given by the relation between $W_\alpha\, W_\beta\, Q^{\alpha\beta} = 0$ and $Z^\alpha\, Z^\beta\, Q^{(-1)}_{\alpha\beta} = 0$. But for surfaces of higher order, the situation is likely to be much more complicated. For example, the reciprocal of a general *cubic* surface is a particular kind of surface of order 12 (that is to say, if $\mathcal{S}$ is given by $W_\alpha\, W_\beta\, W_\gamma\, C^{\alpha\beta\gamma} = 0$, with $C^{\alpha\beta\gamma}$ general and symmetric, then $\mathcal{S}^*$ is given by an equation $Z^\alpha\, Z^\beta \ldots Z^\zeta\, C^{(-1)}_{\alpha\beta\ldots\zeta} = 0$, where $C^{(-1)}_{\alpha\beta\ldots\zeta} = 0$ has 12 indices). Accordingly, the original congruence and its reciprocal would differ greatly from each other, in this case.

Recall that the two null surfaces $\mathcal{N}$ and $\mathcal{N}'$ which intersect in $\mathbb{P}\mathcal{H}$ are complex conjugates of one another. The family of beta-planes through generators of $\mathcal{N}$ provides us with the locus of points in $\mathbb{PT}^*$ that is the surface $\mathcal{S}$. Accordingly, by complex conjugation, the family of alpha-planes through generators of $\mathcal{N}'$ provides us with the locus of points in $\mathbb{PT}$ that is the complex-conjugate surface $\bar{\mathcal{S}}$. But if we were to think of $\mathcal{N}$ in this way, namely as providing a family of *alpha*-planes through the complex rays which generate it, then we get a picture in $\mathbb{PT}$ of a locus of points which is the *reciprocal* surface to $\mathcal{S}$ in $\mathbb{PT}^*$, namely the surface $\mathcal{S}^*$ in $\mathbb{PT}$. In general, $\mathcal{S}^*$ and $\bar{\mathcal{S}}$ are different (though they coincide in the case when $\mathcal{H}$ degenerates to a timelike curve and the entire congruence is non-rotating). Likewise, we can think of the corresponding picture in $\mathbb{PT}^*$, where we now think of $\mathcal{N}'$ as providing a family of *beta*-planes, giving us a locus $\bar{\mathcal{S}}^*$ in $\mathbb{PT}^*$, which is generally distinct from $\mathcal{S}$.

In the above, we have seen that the complex surface $\mathcal{S}$ in $\mathbb{PT}^*$ (or, equivalently, the surface $\bar{\mathcal{S}}$ in $\mathbb{PT}$) can be interpreted in various different ways in $\mathbb{M}$ and in $\mathbb{CM}$. We have Roy Kerr's original description of general shear-free ray congruences in $\mathbb{M}$, which amounts to considering the spacetime translation of $\mathcal{S} \cap \mathbb{PN}^*$. Thinking of $\mathcal{S}$ as a *point* locus in $\mathbb{PT}^*$ gives us the family of beta-planes that this corresponds

[14] So each null generator is displaced orthogonally from its neighbours; see Penrose (1997a,b) for a fuller discussion of these issues.

to in $\mathbb{CM}$. But we can also regard $\mathcal{S}$ as an envelope of *planes* in $\mathbb{PT}^*$ which, when we pass to a description in $\mathbb{PT}$ gives us the locus of points which, via Kerr's $\mathcal{S}^* \cap \mathbb{PN}$ gives us the *reciprocal* congruence. Intermediate between these two is to think of $\mathcal{S}$ as an envelope of the *lines* in $\mathbb{PT}^*$ which touch $\mathcal{S}$. Each of these lines gives us a point in $\mathbb{CM}^\#$ and the locus of these points is the null hypersurface $\mathcal{N}$ whose family of real points provides us with the string history $\mathcal{H}$. Each generator of $\mathcal{N}$ is represented as that family of lines passing through a point $\mathbb{P}\mathbf{W}$ on $\mathcal{S}$ which lie in the tangent plane $\mathbb{P}\mathbf{Z}$ to $\mathcal{S}$ at that point. $\mathbb{P}\mathbf{W}$ and $\mathbb{P}\mathbf{Z}$ represent, respectively, the alpha-plane and the beta-plane through that generator.

One can even carry this geometry further[15] by considering the family of lines in $\mathbb{PT}^*$ which have higher-order (i.e. triple-point or more) contact with $\mathcal{S}$. These form a two-complex parameter family, and have the interpretation as a (generic) complex null 2-surface $\mathcal{C}$ in $\mathbb{CM}^\#$. The relation of $\mathcal{C}$ to our previous geometry is that it is the locus of caustic points of $\mathcal{N}$; moreover $\mathcal{N}$ can be reconstructed from $\mathcal{C}$, as the complex null hypersurface generated by complex rays touching $\mathcal{C}$.

Finally, it may be remarked that many of these geometrical considerations can be applied also in curved spacetime. In particular, some of these ideas apply in a rather restricted way within those particular spacetimes which contain a shear-free ray congruence (e.g. algebraically special vacuums such as the Kerr black hole).[16] But in general spacetimes, the ideas can be used by applying the concept of "shear-freeness" only at the rays' intersections with some particular hypersurface (usually either spacelike or null). This could be an ordinary finite hypersurface or one taken at conformal infinity, leading, respectively, to the concept of "hypersurface twistors" or "asymptotic twistors", respectively.[17] Asymptotic twistor space for an asymptotically flat spacetime has a close relation to (and, as a historical fact, received a considerable input from) the notion of "$\mathcal{H}$-space" initiated by E.T. Newman.[18] In some more recent work, $\mathcal{H}$-space has been related to asymptotically shear-fee ray congruences and the notion, referred to towards the end of Section 3.5 above, of complex-displaced world-lines.[19]

## Acknowledgments

I am grateful to E. T. Newman for discussions, and to NSF for support under PHY 009-0091.

[15] See Penrose (1997a,b).

[16] This has relevance to Trautman's notion of "optical geometry" (Trautman 1985, Robinson and Trautman 1985) and has been studied, partly in relation to the analyticity issues by Lewandowski and Nurowski (1990), Lewandowski *et al.* (1990), and by Tafel *et al.* (1991).

[17] See Penrose and MacCallum (1972), Penrose (1975), Penrose and Rindler (1986).

[18] See Newman (1976), Penrose (1992).

[19] Kozameh and Newman (2005).

## References


Andreotti, A. & Hill, C. D. (1972), "E. E. Levi convexity and the Hans Lewy problem. Part I: Reduction to vanishing theorems; Part II Vanishing theorems", *Ann. Scuola Norm. Sup. Pisa Cl. Sci.* **26**, 747–806.

Kerr, R. P. & Schild, A. (1965), "Some algebraically degenerate solutions of Einstein's gravitational field equations", *Proc. Symp. Appl. Math.* **17**, 199–209.

Kozameh, C. & Newman, E. T. (2005), "The large footprints of $\mathcal{H}$-space on asymptotically flat space-times", *Class. Quantum Grav.* **22**, 4659–4665.

Lewandowski, J. & Nurowski, P. (1990), "Einstein's equations and realizability of CR structures", *Class. Quantum Grav.* **7**, L241–L246.

Lewandowski, J., Nurowski, P. & Tafel, J. (1990), "Einstein's equations and realizability of CR structures", *Class. Quantum Grav.* **7**, 309–328.

Lewy, H. (1956), "On the local character of a solution of an atypical linear differential equation in three variables and a related theorem for regular functions of two complex variables", *Ann. Math.* **64**, 514–522.

Lewy, H. (1957), "An example of a smooth linear partial differential equation without solution", *Ann. Math.* **66**, 155–158.

Lie, S. (1872), "Über Complexe, unbesondere Linien- und Kugelcomplexe mit Anwendung auf die Theorie partieller Differentialgleichungen", *Math. Ann.* **5**, 145–265.

Newman, E. T. (1976), "Heaven and its properties", *Gen. Rel. Grav.* **7**, 107–111.

Newman, E. T. (2004), "Maxwell fields and shear-free null geodesic congruences", *Class. Quantum Grav.* **21**, 1–25.

Penrose, R. (1967), "Twistor algebra", *J. Math. Phys.* **8**, 345–366.

Penrose, R. (1975), "Twistor theory: its aims and achievements", in *Quantum Gravity, an Oxford Symposium*, ed. C. J. Isham, R. Penrose & D. W. Sciama (Oxford University Press, Oxford).

Penrose, R. (1979), "On the twistor description of massless fields", in *Complex Manifold Techniques in Theoretical Physics*, ed. D. E. Lerner & P. D. Sommers (Pitman, San Francisco), pp. 55–91.

Penrose, R. (1983), "Physical space-time and non-realizable CR-structures", *Proc. Symp. Pure Math.* **39**, 401–422.

Penrose, R. (1987), "On the origins of twistor theory", in *Gravitation and Geometry: a volume in honour of I. Robinson*, ed. W. Rindler & A. Trautman (Bibliopolis, Naples).

Penrose, R. (1992), "$\mathcal{H}$-space and twistors", in *Recent Advances in General Relativity* (Einstein Studies, vol. 4), ed. A. I. Janis & J. R. Porter (Birkhäuser, Boston), pp. 6–25.

Penrose, R. (1997a), "Twistor geometry of light rays", *Class. Quantum Grav.* **14**, A299–A323.

Penrose, R. (1997b), "Five things you can do with a surface in $\mathbb{PT}$", *Twistor Newsletter* **42**, 1–5.

Penrose, R. & MacCallum, M. A. H. (1972), "Twistor theory: an approach to the quantization of fields and space-time", *Phys. Rep.* **6**, 241–315.

Penrose, R. & Rindler, W. (1984), *Spinors and Space-Time, vol. 1: Two-Spinor Calculus and Relativistic Fields* (Cambridge University Press, Cambridge).

Penrose, R. & Rindler, W. (1986), *Spinors and Space-Time, vol. 2: Spinor and Twistor Methods in Space-Time Geometry* (Cambridge University Press, Cambridge).

Robinson, I. & Trautman, A. (1985), "Integrable optical geometry", *Lett. Math. Phys.* **10**, 179–182.

Tafel, J., Nurowski, J. & Lewandowski, P. (1991), "Pure radiation fields of the Einstein equations", *Class. Quantum Grav.* **8**, L83–L88.
Trautman, A. (1985), "Optical structures in relativistic theories", *Astérisque, numéro hors série*, 401–420.
Witten, E. (2003), "Perturbative gauge theory as a string theory in twistor space", *Commun. Math. Phys.* **252**, 189–258.
Yano, K. (1957), *The Theory of Lie Derivatives and its Applications* (North-Holland, Amsterdam).

# 4

# Global and local problems solved by the Kerr metric

Brandon Carter

## 4.1 Introduction

One of the most salient historical milestones in the development of theoretical astrophysics was the publication by Roy Kerr in 1963 of a letter [1] which concluded with the memorable assertion that:

> Among the solutions ... there is one which is stationary ... and also axisymmetric. Like the Schwarzschild metric, which it contains, it is type D ... $m$ is a real constant ... The metric is
>
> $$\begin{aligned} ds^2 = {} & (r^2 + a^2\cos^2\theta)(d\theta^2 + \sin^2\theta\, d\phi^2) + 2(du + a\sin^2\theta\, d\phi)(dr + a\sin^2\theta\, d\phi) \\ & - \left(1 - \frac{2mr}{r^2 + a^2\cos^2\theta}\right)(du + a\sin^2\theta\, d\phi)^2 , \end{aligned} \tag{4.1}$$
>
> where $a$ is a real constant. This may be transformed to an asymptotically flat coordinate system ... we find that $m$ is the Schwarzschild mass and $ma$ the angular momentum ... It would be desirable to calculate an interior solution ...

The purpose of this article to provide a succinct overview of the many amazing features that were subsequently found to characterise this outstandingly remarkable solution of Einstein's equations. This overview will be concerned mainly with purely mathematical aspects, but what makes this subject so important is the substantial if not yet rigorously conclusive evidence that (whereas the hope of finding a well behaved interior solution has not been fulfilled) the Kerr solution will arise naturally as the generic outcome of gravitational collapse to what is now known as a "black hole", as characterised by the feature first recognised in his metric whenever $a^2 < m^2$ by Kerr himself [2], namely the "horizon" consisting of

The Kerr Spacetime: Rotating Black Holes in General Relativity. Ed. David L. Wiltshire, Matt Visser and Susan M. Scott. Published by Cambridge University Press. 

a regular null hypersurface behind which the badly behaved interior regions are hidden.

## 4.2 Transformation to Kerr–Schild form

Already in his original Physical Review Letter of 1963 [1] Roy Kerr drew attention to the existence of a coordinate transformation whereby, as discussed in a subsequent article he wrote in collaboration with Alfred Schild [3] his solution $\mathrm{d}s^2 = g_{\mu\nu}\,\mathrm{d}x^\mu\,\mathrm{d}x^\nu$ can be seen to have a metric tensor that is decomposable as a sum

$$g_{\mu\nu} = \eta_{\mu\nu} + 2(m/U)n_\mu n_\nu \tag{4.2}$$

in which the first term is just a **flat** background metric $\eta_{\mu\nu}$ and the second is proportional to the tensor product with itself of a covector $n_\mu$, that is **null** in the sense that

$$g_{\mu\nu}n^\mu n^\nu = 0\,. \tag{4.3}$$

In terms of the original coordinates, the null covector can be taken to be given by the expression

$$n_\mu\,\mathrm{d}x^\mu = \mathrm{d}u + a\sin^2\theta\,\mathrm{d}\phi\,, \tag{4.4}$$

with the corresponding proportionality factor given by

$$U = (r^2 + a^2\cos^2\theta)/r\,. \tag{4.5}$$

The flatness of the background contribution $\eta_{\mu\nu}$ becomes manifest when it is written in terms of the new Kerr–Schild coordinate system as obtained by setting

$$\bar{t} = u - r, \qquad \bar{z} = a\cos\theta, \qquad \bar{x} + i\bar{y} = (r - ia)\mathrm{e}^{i\phi}\sin\theta, \tag{4.6}$$

which provides an expression of the familiar Minkowski form

$$\eta_{\mu\nu}\,\mathrm{d}x^\mu\,\mathrm{d}x^\nu = \mathrm{d}\bar{x}^2 + \mathrm{d}\bar{y}^2 + \mathrm{d}\bar{z}^2 - \mathrm{d}\bar{t}^2, \tag{4.7}$$

while the corresponding expression for the null covector takes the form

$$n_\mu\,\mathrm{d}x^\mu = \mathrm{d}\bar{t} + \frac{\bar{z}\,\mathrm{d}\bar{z}}{r} + \frac{(r\bar{x} - a\bar{y})\,\mathrm{d}\bar{x} + (r\bar{y} + a\bar{x})\,\mathrm{d}\bar{y}}{r^2 + a^2}\,. \tag{4.8}$$

As described in detail in his subsequent articles with Schild and Debney [3–5] (see his accompanying article in this volume) Kerr's original derivation [3] of his metric was based on the postulate that the (trace free) Weyl part of the curvature tensor – which is the same as the complete Riemann tensor in the pure vacuum case, i.e. when the Ricci trace part vanishes – should be algebraically special, meaning that it has less than the four independent null eigenvectors it would have in the generic case. Within this algebraically special category, the further requirement that it should be stationary axisymmetric and asymptotically flat guided Kerr to a metric of type D, meaning that – as in its non-rotating Schwarzschild limit – its Weyl tensor has only two independent degenerate (and in such a case necessarily shear free) null eigenvectors, of which, in the Kerr metric (either) one is identifiable as the Kerr–Schild vector (4.4). A rather different approach was used in 1965 by Newman and his collaborators [6] to obtain the astrophysically interesting generalisation from a pure Einstein vacuum solution to a solution of the source free **Einstein–Maxwell** vacuum equation that turned out [4, 7] to have exactly the same gyromagnetic ratio as the Dirac electron. That feature – which is also exhibited by the alternative generalisation to an **axion–dilaton** solution obtained more recently by Sen [8, 9] – has suggested various lines of research into the possibility of using the Kerr metric for classical modelling of the internal structure of spinning particles [10–12].

The geometric properties of the quotient manifold with respect to trajectories of the preferred asymptotically timelike Killing vector have been used to obtain other local characterisations of the Kerr metric by Perjés [13] and by Simon [14]. The latter has been followed up by Mars, who has shown [15, 16] how the Simon criterion is related to the Weyl tensor degeneracy property postulated by Kerr. Starting from the recognition that an essential feature is that the self dual part of the (necessarily antisymmetric) covariant derivative of the relevant Killing vector is an eigen-bivector of the self dual part of the Weyl tensor, Mars [17] has used work of Senovilla [18] to show that the Kerr solution is obtainable as a vacuum limit of the Wahlquist family [19] of perfect fluid solutions.

It should however be said that physical relevance of the non-vacuum members of the Wahlquist family is questionable. That might also have been said of the result of another kind of approach – based on the separability properties to be described below – that was used in 1968 by the present author [20, 21] to obtain the generalisation needed to allow for the possibility of a **cosmological constant**, something whose physical interest was not so obvious then as it has since become [22, 23].

Another kind of generalisation to which much attention has been given [24] is to spacetimes with more than four dimensions. As an ansatz for generalising the Kerr solution to higher dimensions, the Kerr–Schild decomposition (4.2) has been found to be particularly useful. The pure Einstein vacuum case was generalised in this way in 1986 by Myers and Perry [25]. By relaxing the requirement that the background metric $\eta_{\mu\nu}$ should be flat to the weaker requirement that it be of **de Sitter** type, I showed in 1972 [26] that the original generalisation [21] with cosmological term was also expressible in the form (4.2), and this generalised Kerr–Schild property was subsequently used to obtain further generalisations, first to five spacetime dimensions by Hawking and collaborators [27] in 1998, and most recently to arbitrarily high dimensions by Gibbons, Pope, and Page [28]. Some very recent work [29–31] on higher dimensional generalisations has focussed on separability and related hidden symmetry properties of the kind discussed below.

Interest in such generalisations is partly motivated by the vogue for speculative superstring and brane models in spacetimes of very high dimension (ten or even eleven). However in the remainder of this brief review I shall restrict attention to the case of ordinary four-dimensional spacetime in which, as will be discussed, there are strong (not just speculative) reasons for believing the Kerr metric to be directly applicable as a fairly accurate description of the inner geometry of astrophysical systems whose external features are directly observed.

### 4.3 Transformation to Boyer–Lindquist form

Although Kerr foresaw at the outset [1] that his solution would be astrophysically useful, its precise significance took several years to elucidate: on one hand there were pessimistic doubts about its dynamical stability, while on the other hand there were over-optimistic hopes that it might be used as an exterior for uncollapsed rotating star configurations. As a preliminary to the resolution of such issues, one of the first things to be done was to clarify the purely geometrical properties of the solution, of which the most important is the feature that, as first pointed out by Kerr himself [2], when $a^2 < m^2$ the singular interior region is hidden behind a well behaved null hypersurface of the kind that is now known as a "black hole horizon".

To start with, in its original version (4.1) it was already obvious from the ignorability of the coordinates $u$ and $\phi$ that the Kerr solution was characterised by a continuous 2-parameter isometry group generated by time displacements and axial rotations. It was predicted by a 1966 theorem of Papapetrou [32] that with any such stationary axisymmetric vacuum solution there would also be a discrete isometry involving simultaneous time and angle reversal, but this was not made manifest until the following year when Boyer and Lindquist [33] obtained a coordinate transformation of the appropriate form, namely

$$\mathrm{d}t = \mathrm{d}u - (r^2 + a^2)\Delta^{-1}\,\mathrm{d}r, \quad \mathrm{d}\varphi = -\mathrm{d}\phi + a\,\Delta^{-1}\,\mathrm{d}r, \tag{4.9}$$

in which the relevant discriminant is given by

$$\Delta = r^2 - 2mr + a^2\,. \tag{4.10}$$

This converts the expression for null covector (4.4) to the form

$$n_\mu \, \mathrm{d}x^\mu = \mathrm{d}t - a \sin^2\theta \, \mathrm{d}\varphi + \varrho^2 \Delta^{-1} \, \mathrm{d}r, \tag{4.11}$$

using the notation

$$\varrho = \sqrt{r^2 + a^2 \cos^2\theta}, \tag{4.12}$$

and it converts the expression for the metric itself to the form

$$\begin{aligned} \mathrm{d}s^2 = \varrho^2 \left( \frac{\mathrm{d}r^2}{\Delta} + \mathrm{d}\theta^2 \right) + (r^2 + a^2) \sin^2\theta \, \mathrm{d}\varphi^2 \\ + \frac{2mr}{\varrho^2} (\mathrm{d}t - a \sin^2\theta \, \mathrm{d}\varphi)^2 - \mathrm{d}t^2, \end{aligned} \tag{4.13}$$

in which the invariance under simultaneous reversal of the new ignorable time and angle coordinates $t$ and $\varphi$ is evident from the absence of cross terms of their differentials with the non-ignorable differentials, $\mathrm{d}r$ or $\mathrm{d}\theta$. However unless the rotation parameter is inordinately large, or to be more precise so long as $a^2 \leq m^2$ it can be seen that – as the price to be paid for the algebraic simplicity that has made it the most widely known expression for the Kerr solution – this Boyer–Lindquist version (4.13) suffers from the disadvantage of having a coordinate singularity on null "horizons" that occur where $\Delta$ vanishes.

## 4.4 Transformation to canonical separable form

As well as the ordinary "circular" symmetry generated by the continuous action of the time and angle Killing vectors, $k^\mu \partial/\partial x^\mu = \partial/\partial t$ and $h^\mu \partial/\partial x^\mu = \partial/\partial\varphi$, together with the discrete inversion symmetry under which their orientation is simultaneously reversed as described above, the Kerr metric also has a hidden symmetry that was first revealed via the introduction by the present author [7] of a certain **canonical** tetrad given by the expressions

$$\begin{aligned} \vartheta^{\hat{0}}_{\mu} \, \mathrm{d}x^\mu &= \sqrt{\Delta} \, \frac{\mathrm{d}t - a \sin^2\theta \, \mathrm{d}\varphi}{\varrho}, \qquad & \vartheta^{\hat{1}}_{\mu} \, \mathrm{d}x^\mu &= (\varrho/\sqrt{\Delta}) \, \mathrm{d}r, \\ \vartheta^{\hat{3}}_{\mu} \, \mathrm{d}x^\mu &= \sin\theta \, \frac{(r^2 + a^2) \, \mathrm{d}\varphi - a \, \mathrm{d}t}{\varrho}, \qquad & \vartheta^{\hat{2}}_{\mu} \, \mathrm{d}x^\mu &= \varrho \, \mathrm{d}\theta. \end{aligned} \tag{4.14}$$

The corresponding tetrad components $C^{\hat{a}}_{\ \hat{c}\hat{e}\hat{g}}$ of the Weyl tensor (which, because the Ricci tensor vanishes, is the same as the Riemann tensor) can be read out from the corresponding curvature 2-form with components $\Omega^{\hat{a}}_{\ \hat{c}\,\mu\nu} = C^{\hat{a}}_{\ \hat{c}\,\hat{e}\hat{g}} \vartheta^{\hat{e}}_{\ [\mu} \vartheta^{\hat{g}}_{\ \nu]}$ that are given (correcting sign errors in an earlier evaluation by myself [26]) in exterior product notation, $\Omega^{\hat{a}}_{\hat{c}} = \frac{1}{2} C^{\hat{a}}_{\hat{c}\,\hat{e}\hat{g}} \vartheta^{\hat{e}} \wedge \vartheta^{\hat{g}}$ by Marck [34] as

$$\begin{aligned}
\Omega^{\hat{0}}_{\hat{1}} &= 2I_{\mathrm{I}}\, \vartheta^{\hat{0}} \wedge \vartheta^{\hat{1}} - 2I_{\mathrm{II}}\, \vartheta^{\hat{2}} \wedge \vartheta^{\hat{3}}, &\qquad \Omega^{\hat{0}}_{\hat{3}} &= I_{\mathrm{I}}\, \vartheta^{\hat{0}} \wedge \vartheta^{\hat{3}} - I_{\mathrm{II}}\, \vartheta^{\hat{1}} \wedge \vartheta^{\hat{2}}, \\
\Omega^{\hat{3}}_{\hat{2}} &= 2I_{\mathrm{II}}\, \vartheta^{\hat{0}} \wedge \vartheta^{\hat{1}} + 2I_{\mathrm{I}}\, \vartheta^{\hat{2}} \wedge \vartheta^{\hat{3}}, &\qquad \Omega^{\hat{3}}_{\hat{1}} &= I_{\mathrm{II}}\, \vartheta^{\hat{0}} \wedge \vartheta^{\hat{2}} - I_{\mathrm{I}}\, \vartheta^{\hat{1}} \wedge \vartheta^{\hat{3}}, \\
\Omega^{\hat{1}}_{\hat{2}} &= -I_{\mathrm{I}}\, \vartheta^{\hat{1}} \wedge \vartheta^{\hat{2}} + I_{\mathrm{II}}\, \vartheta^{\hat{3}} \wedge \vartheta^{\hat{0}}, &\qquad \Omega^{\hat{0}}_{\hat{2}} &= -I_{\mathrm{I}}\, \vartheta^{\hat{0}} \wedge \vartheta^{\hat{2}} - I_{\mathrm{II}}\, \vartheta^{\hat{1}} \wedge \vartheta^{\hat{3}},
\end{aligned}$$

in terms of the (only) independent curvature invariants

$$I_{\mathrm{I}} = mr(r^2 - 3a^2\cos^2\theta)/\varrho^6, \qquad I_{\mathrm{II}} = ma\cos\theta(3r^2 - a^2\cos^2\theta)/\varrho^6.$$

In terms of the canonical tetrad (4.14), the Kerr–Schild null vector (4.4) is given by

$$\sqrt{\Delta}\, n_\mu = \varrho\big(\vartheta^{\hat{0}}_{\mu} + \vartheta^{\hat{1}}_{\mu}\big), \tag{4.15}$$

while the metric itself is expressible simply as a sum of four terms in the form

$$g_{\mu\nu} = \sum_{i=1}^{3} \vartheta^{\hat{i}}_{\mu} \vartheta^{\hat{i}}_{\nu} - \vartheta^{\hat{0}}_{\mu} \vartheta^{\hat{0}}_{\nu}. \tag{4.16}$$

This form turned out to be such that, when substituted into the expressions for the geodesic Hamilton–Jacobi equation [7] or the scalar wave equation [21] the corresponding four terms in the inverse metric $g^{\mu\nu}$ give contributions in which the dependence on the non-ignorable coordinates $r$ and $\theta$ separate out from each other. The ensuing possibility of obtaining an explicit analytic solution is particularly useful in the case of the null geodesics representing photon trajectories in astrophysically important applications, such as to line profiles of accretion discs [35]. Using a dot for differentiation with respect to an affine parameter, the first integrals of geodesic motion will be given by

$$\varrho^2 \dot{t} = (\mathcal{E}(r^2 + a^2)^2 - 2\Phi amr)/\Delta - \mathcal{E}a^2\sin^2\theta, \tag{4.17}$$

$$\varrho^2 \dot{\phi} = a(2\mathcal{E}m - \Phi a)/\Delta + \Phi/\sin^2\theta, \tag{4.18}$$

$$(\varrho^2 \dot{r})^2 = \Upsilon_r^2 - \Xi_r\,\Delta, \qquad (\varrho^2\dot{\theta})^2 = \Xi_\theta - \Upsilon_\theta^2/\sin^2\theta, \tag{4.19}$$

using the abbreviations

$$\Upsilon_r = \mathcal{E}(r^2 + a^2) - \Phi a, \qquad \Upsilon_\theta = \mathcal{E}a\sin^2\theta - \Phi, \tag{4.20}$$

$$\Xi_r = \mathcal{K} + m_o^2 r^2, \qquad \Xi_\theta = \mathcal{K} - m_o^2 a^2\cos^2\theta, \tag{4.21}$$

for single variable functions in which the parameters $\mathcal{E}$, $\Phi$, and $\mathcal{K}$ are constants of integration whose interpretation will be discussed below, while $m_o^2$ is a parameter interpretable as the square of the relevant particle mass in both the null case, for which it vanishes, and the timelike case for which the affine parameter can be made to agree with proper time by calibrating the units so as to obtain $m_o^2 = 1$. The "tachyonic" case of a spacelike geodesic with proper length parameterisation is obtainable by setting $m_o^2 = -1$.

The possibility of generalisation of such separability from the scalar case to higher spin was subsequently demonstrated by Unruh [36] for the massless neutrino equation, by Teukolsky [37] for the Maxwellian (spin 1) and gravitational perturbation (spin-2) cases, and finally by Chandrasekhar [38, 39] for the massive Dirac equation. It is to be remarked that thc latter authors made their algebra considerably more complicated than necessary by working, not with the canonical tetrad (4.14), but with a relatively boosted (Kinnersley type) tetrad whose use has the drawback of obscuring the symmetry between the incoming and outgoing families of principal null directions. A more efficient treatment was used for the integration of the equations of parallel transport by Marck [34, 40].

## 4.5 The Killing–Yano tensor

The amazing and unexpected cancellations involved in the successively more intricate separation schemes that have just been referred to motivated a program of investigation led by Roger Penrose whose aim [41] was to obtain a deeper understanding of the mechanism of the hidden symmetry involved, which turned out to hinge on the existence [42] in the Kerr metric of a certain Killing–Yano 2-form, which is given in terms of the canonical tetrad (4.14) by the expression

$$f_{\mu\nu} = 2a\cos\theta\, \vartheta^{\hat{1}}_{[\mu}\vartheta^{\hat{0}}_{\nu]} + 2r\, \vartheta^{\hat{2}}_{[\mu}\vartheta^{\hat{3}}_{\nu]}.$$

using square brackets to denote index antisymmetrisation.

The defining property of a Killing–Yano tensor is that it should be an antisymmetric solution of the equation

$$\nabla_\mu f_{\nu\rho} = \nabla_{[\mu} f_{\nu\rho]}. \tag{4.22}$$

It can easily be seen that any such antisymmetric tensor provides a symmetric tensor

$$K_{\mu\nu} = f_{\mu\rho} f^{\rho}_{\nu} \tag{4.23}$$

that will be a solution of the more widely known Killing tensor equation, namely

$$\nabla_{(\mu} K_{\nu\rho)} = 0, \tag{4.24}$$

and it can also be seen [43] that it will provide "secondary" and "primary" solutions

$$\tilde{k}^\mu = K^\mu_\nu k^\nu = a^2 k^\mu + a h^\mu \tag{4.25}$$

and

$$k^\mu = \frac{1}{6} \varepsilon^{\mu\nu\rho\sigma} \nabla_\nu f_{\rho\sigma} \tag{4.26}$$

of the ordinary Killing vector equation

$$\nabla_{(\mu} k_{\nu)} = 0. \tag{4.27}$$

For free motion of a particle of mass $m_o$ with an affinely propagated tangent vector $p^\nu$ evolving according to the geodesic equation

$$p^\nu \nabla_\nu p^\mu = 0, \tag{4.28}$$

with constant magnitude given by $p_\nu p^\nu = -m_o^2$, the time and angle Killing vectors will of course provide two more constants of motion

$$\mathcal{E} = -k^\nu p_\nu, \qquad \Phi = h^\nu p_\nu, \tag{4.29}$$

that will be interpretable in the usual way as representing energy and axial angular momentum respectively, while finally a "fourth" (hidden) constant of the motion will be provided by the Killing tensor according to the specification

$$\mathcal{K} = K^{\mu\nu} p_\mu p_\nu. \tag{4.30}$$

This extra constant was originally brought to light by the 1968 discovery of the associated separability of the geodesic Hamilton–Jacobi equation [7]. A deeper insight was obtained in 1973 by the Penrose group's discovery [42] that $K^{\mu\nu}$ is not merely a generic solution of the Killing equation (4.27) but more particularly that it is decomposable according to (4.23) as the square of a solution $f_{\mu\nu}$ of the much more restrictive Killing–Yano equation (4.22). This provides the corollary that the "fourth" constant (4.30) will be the square

$$\mathcal{K} = \mathcal{J}_\mu \mathcal{J}^\nu, \tag{4.31}$$

of a total angular momentum vector,

$$\mathcal{J}_\mu = f_{\mu\nu} p^\mu, \tag{4.32}$$

that is itself conserved in the sense of obeying the parallel propagation equation

$$p^\nu \nabla_\nu \mathcal{J}_\mu = 0.$$

These classical constants of motion for free particle motion have first quantised analogues underlying the concomitant separability [21] of the scalar wave equation, in the form of differential operators that **commute** with the d'Alembertian operator

$$\Box = \nabla^\nu \nabla_\nu \tag{4.33}$$

acting on the scalar field. The ordinary Killing property is enough by itself to ensure that the corresponding (self adjoint) energy and axial angular momentum operators, namely

$$\mathcal{E} = -ik^\nu \nabla_\nu, \qquad \Phi = ih^\nu \nabla_\nu,$$

will be conserved in the sense of commuting with the d'Alembertian,

$$[\mathcal{E}, \Box] = 0, \qquad [\Phi, \Box] = 0.$$

To show [44] that the (self adjoint) second-order differential operator corresponding to the "fourth" constant of the motion, namely the generalised squared total angular momentum operator

$$\mathcal{K} = \nabla_\mu K^{\mu\nu} \nabla_\nu,$$

will also satisfy such a conservation condition, i.e.

$$[\mathcal{K}, \Box] = 0,$$

the satisfaction of the Killing tensor equation (4.24) is not quite sufficient: it is also necessary to invoke a Killing–Yano integrability condition of the form $K^\rho{}_{[\mu} R_{\nu]\rho} = 0$, but it is obvious that this will hold automatically in the pure vacuum Kerr solution for which the Ricci tensor $R_{\mu\nu}$ vanishes by the Einstein equation.

Since the scalar separability property [21] associated with the foregoing commutation properties was shown to be extensible to a spin-$\frac{1}{2}$ field, first in the massless (neutrino) case governed by the Weyl equation [36] and later in the massive (electron) case governed by the full Dirac equation [38, 39] it was natural to look for corresponding operators that would commute with the Dirac operator

$$\mathcal{D} = \gamma^\mu \nabla_\mu$$

acting on spinors. As in the scalar case, the Killing property (4.24) alone is sufficient to ensure that the corresponding spinor energy and angular momentum operators [45], namely

$$\mathcal{E} = -ik^\nu \nabla_\nu + \tfrac{1}{4} i (\nabla_\mu k_\nu) \gamma^\mu \gamma^\nu, \qquad \Phi = ih^\nu \nabla_\nu + \tfrac{1}{4} i (\nabla_\mu h_\nu) \gamma^\mu \gamma^\nu,$$

will satisfy the expected commutation conditions

$$[\mathcal{E}, \mathcal{D}] = 0, \qquad [\Phi, \mathcal{D}] = 0.$$

The further specification [45]

$$\mathcal{J} = i\gamma^\mu (\gamma^5 f_\mu{}^\nu \nabla_\nu - k_\mu),$$

with $k^\mu$ given by (4.26), provides another operator that is also conserved in the sense of satisfying the commutation condition

$$[\mathcal{J}, \mathcal{D}] = 0,$$

as a consequence of the Killing–Yano condition (4.22). Just as the Dirac operator $\mathcal{D}$ was originally introduced as a sort of square root of the ordinary d'Alembertian operator $\Box$, similarly this new spinor angular momentum operator $\mathcal{J}$ is loosely interpretable as a sort of square root of the scalar squared angular momentum operator $\mathcal{K}$.

Due to the complications resulting from constraints in the second-order version or gauge dependence in the first-order versions, such a neat formulation in terms of commutating operators has not been obtained for Teukolsky's extension [37] of separability to the (massless) spin-1 (electromagnetic) and spin-2 (gravitational perturbation) equations. The separability of the latter is particularly important for the work initiated by Press and Teukolsky [46], and developed more recently by Whiting [47] in 1989, whose aim is to confirm that the Kerr background is sufficiently stable to be astrophysically relevant in the context of gravitational collapse as discussed below.

### 4.6 Solution of the parallel propagation problem

Another potentially important consequence of the Killing–Yano property was discovered by Marck [34, 40, 48], who showed that it provides an explicit analytic solution of the problem of parallel propagating a tetrad of basis vectors along a geodesic. This result is useful in contexts involving polarisation and tidal effects, but as far as I know it has so far been applied only to the spherical Schwarzschild case [49–51]. The solution for the generic Kerr case is not as well known as it deserves to be, perhaps because the original publications are not easily available on-line. It is therefore worthwhile to recapitulate the main result which is as follows.

For a timelike geodesic, it is obvious that, as the first element in such a basis, a timelike unit vector, $b_{\bar{0}}^{\mu}$ say, satisfying the parallel transport condition will be obtained by setting $p_\mu = m_o\, b_{\bar{0}}^{\mu}$, where $p^\mu$ is the geodesic tangent vector introduced in (4.28). For the solution given by (4.17) to (4.19) the canonical tetrad components of this basis vector will be given, using the notation (4.20) by

$$b_{\bar{0}}^{\hat{0}} = \Upsilon_r/\varrho\,\sqrt{\Delta}, \qquad b_{\bar{0}}^{\hat{1}} = \varrho\,\dot{r}/\sqrt{\Delta},$$
$$b_{\bar{0}}^{\hat{2}} = \varrho\,\dot{\theta}, \qquad b_{\bar{0}}^{\hat{3}} = -\Upsilon_\theta/\varrho\,\sin\theta.$$

As pointed out by Penrose [42] the total angular momentum vector (4.32) will also be parallel propagated, so by adjustment of its amplitude it will immediately provide a second orthogonal (and therefore spacelike) unit vector of the required kind, with canonical tetrad components given by

$$b_{\bar{2}}^{\hat{0}} = a\varrho\,\cos\theta\,\dot{r}/\sqrt{\mathcal{K}\Delta}, \qquad b_{\bar{2}}^{\hat{1}} = a\cos\theta\,\Upsilon_r/\varrho\sqrt{\mathcal{K}\Delta},$$
$$b_{\bar{2}}^{\hat{2}} = -r\Upsilon_\theta/\varrho\sqrt{\mathcal{K}}\sin\theta, \qquad b_{\bar{2}}^{\hat{3}} = -\varrho\, r\dot{\theta}/\sqrt{\mathcal{K}}.$$

Marck's next step [40] was to observe that the parallel transported dyad obtained in this way can be easily extended to a complete orthonormal tetrad by taking the other two basis vectors to have canonical components given in terms of the ratio

$$\alpha = \sqrt{\Xi_\theta/\Xi_r},$$

of the square roots of the quantities defined in (4.21) by

$$\tilde{b}_{\bar{1}}^{\hat{0}} = \alpha\varrho\, r\dot{r}/\sqrt{\mathcal{K}\Delta}, \qquad \tilde{b}_{\bar{1}}^{\hat{1}} = \alpha r\Upsilon_r/\varrho\sqrt{\mathcal{K}\Delta},$$
$$\tilde{b}_{\bar{1}}^{\hat{2}} = a\cot\theta\,\Upsilon_\theta/\alpha\varrho\sqrt{\mathcal{K}}, \qquad \tilde{b}_{\bar{1}}^{\hat{3}} = \varrho\, a\cos\theta\,\dot{\theta}/\alpha\sqrt{\mathcal{K}},$$

and

$$\tilde{b}_{\bar{3}}^{\hat{0}} = \alpha\Upsilon_r/\varrho\sqrt{\Delta}, \qquad \tilde{b}_{\bar{3}}^{\hat{1}} = \alpha\varrho\,\dot{r}/\sqrt{\Delta},$$
$$b_{\bar{3}}^{\hat{2}} = \varrho\,\dot{\theta}/\alpha, \qquad \tilde{b}_{\bar{3}}^{\hat{3}} = -\Upsilon_\theta/\alpha\varrho\,\sin\theta.$$

Unlike the dyad $b^{\mu}_{\bar{0}}$ and $b^{\mu}_{\bar{2}}$, the vectors $\tilde{b}^{\mu}_{\bar{1}}$ and $\tilde{b}^{\mu}_{\bar{3}}$ of this latter pair are not parallel transported, but Marck has shown [34] that orthonormal basis vectors $b^{\mu}_{\bar{1}}$ and $b^{\mu}_{\bar{3}}$ of the required parallel transported form will be given by a rotation of the form

$$b^{\mu}_{\bar{1}} = \tilde{b}^{\mu}_{\bar{1}} \cos\Psi - \tilde{b}^{\mu}_{\bar{3}} \sin\Psi, \qquad b^{\mu}_{\bar{3}} = \tilde{b}^{\mu}_{\bar{1}} \sin\Psi + \tilde{b}^{\mu}_{\bar{3}} \cos\Psi, \tag{4.34}$$

in which the rate of variation of the rotation angle $\Psi$ is given by

$$\frac{\varrho^2}{m_o\sqrt{\mathcal{K}}}\,\dot{\Psi} = \frac{\Upsilon_r}{\Xi_r} - \frac{a\Upsilon_\theta}{\Xi_\theta}.$$

It can be seen from (4.19) that this is soluble in the separated form

$$\Psi = \Psi_r + \Psi_\theta,$$

with

$$\Psi_r = \int^r \frac{\pm m_o\sqrt{\mathcal{K}}\,\Upsilon_r\,\mathrm{d}r}{\Xi_r\sqrt{\Upsilon_r^2 - \Xi_r\Delta}}, \qquad \Psi_\theta = \int^\theta \frac{\pm m_o\sqrt{\mathcal{K}}\,\Upsilon_\theta\,\mathrm{d}\theta}{\Xi_\theta\sqrt{\Xi_\theta - \Upsilon_\theta^2/\sin^2\theta}}, \tag{4.35}$$

in which the $\pm$ signs are independently chosen to be those of $\dot{r}$ and $\dot{\theta}$ respectively.

The case of a spacelike geodesic – with $b^{\mu}_{\bar{0}}$ spacelike but $\tilde{b}^{\mu}_{\bar{3}}$ and $b^{\mu}_{\bar{3}}$ timelike – is obtainable straightforwardly by analytic extrapolation of the formula by taking a purely imaginary value for $m_o$, so the rotation (4.34) is transformed to a Lorentzian boost. In the more delicate case of a null geodesic, it can be seen that the variation of $\Psi$ will simply be proportional to that of the affine parameter, but for this degenerate limit $m_o \to 0$ (which is astrophysically important for applications [52] involving polarised radiation) it is necessary to use a special treatment that has been given by Marck in a separate article [40].

## 4.7 The black hole property for $a^2 \leq m^2$

The remarkable properties described in the preceding section are of a purely local nature, but in so far as its astrophysical significance is concerned what matters is its global geometry, whose salient features were mainly elucidated in the period 1966–1968 [7, 33, 53, 54].

The unique importance of the Kerr solution depends on a feature whose mathematical properties in a generic context were first investigated by Boyer [53, 55], but that was originally discovered in his solution by Kerr himself [2], namely the property of describing what has since come to be known as a **black hole** whenever $a^2 \leq m^2$. In this case what has been whimsically referred to as "asymptopia" (loosely where the value of the radial coordinate $r$ becomes infinite) is simultaneously visible and accessible (by non-superluminal trajectories) only in a "domain of outer communications" – in other words the part outside the hole – constituting a non-singular region characterised by the inequality $\Delta > 0$ and bounded by past and future null outer and inner **horizons** on which $\Delta = 0$ where

$$r = m \pm c, \qquad c = \sqrt{m^2 - a^2}.$$

The part inside, where $\Delta < 0$ is commonly called the "black hole" region, but in stricter terminology this entire "hole" region should be described as the (overlapping) union of a **white hole** region, which is the part inaccessible by a future directed trajectory from asymptopia, and a strictly defined **black hole** region, which is the part from which asymptopia is inaccessible by a future directed trajectory.

The complete topology of the maximally extended Kerr solution has a rather complicated (time periodic) structure whose restriction to the symmetry axis [54] is as shown in the conformal projection diagram (plotting null rays at 45 degrees to the axes) in Fig. 4.1. The $r$ coordinate ranges all the way from infinite positive to infinite negative values, but the only parts that are thought to be astrophysically relevant – as the ultimate outcome of a gravitational collapse scenario – are the parts of the domain of outer communications where $r > m + c$, and the **black** parts of the hole region.

The stable part of the Kerr solution can not extend beyond the inner horizon where $r = m - c$, because as well as containing irremovable **ring singularities** where $r^2 + a^2\cos^2\theta \to 0$, it can be seen that the other parts of the domain of outer communications, where $r < m - c$, will be causally pathological, exhibiting "time machine" behaviour [7] due to the presence – as indicated by the double shading in Fig. 4.2 – of regions near the singularity where the circular trajectories generated by the axial Killing vector $h^\nu$ become timelike.

## 4.8 The black hole equilibrium problem

The overwhelming astrophysical importance of the Kerr solution derives from the presumption that the causally well behaved asymptotically flat part where $r > m + c$ provides a generic description of the ultimate state of the domain of outer communications surrounding a black hole formed by gravitational collapse. The assumption that no more general solution will be needed for the stationary external vacuum state that will ultimately be attained (after the decay, on an astronomically short timescale, of any dynamical mode that may have been excited) has been generally taken for granted in the astrophysical community since the publication

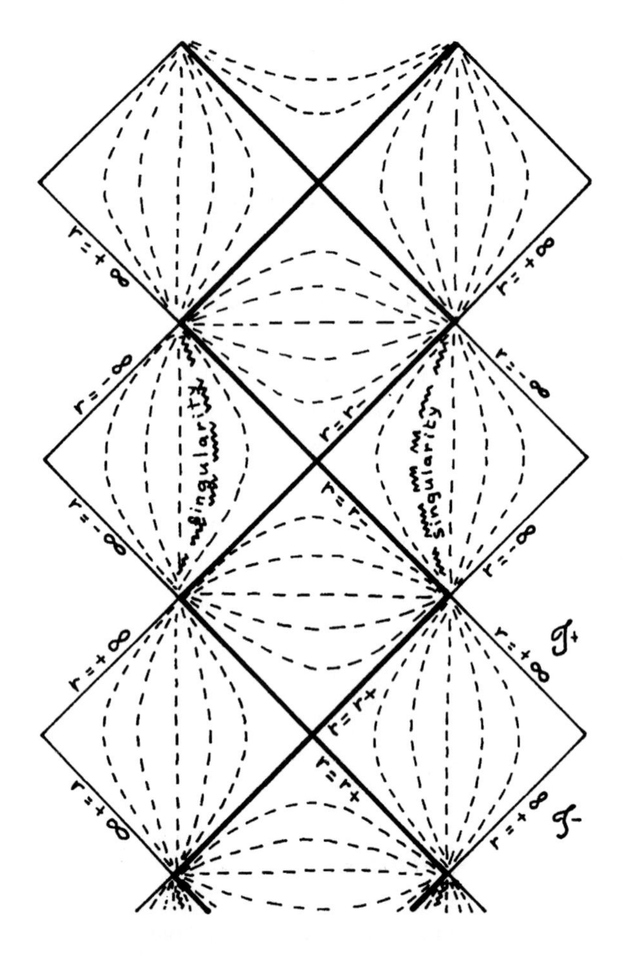


Fig. 4.1. Reproduction from 1972 Les Houches notes [26] of conformal projection of $\{r, t\}$ section (such as symmetry axis [54]) for fixed value of angular coordinates, which (except on the equator) passes smoothly through the ring singularity of the extended Kerr solution.

in 1971 of a **no-hair theorem** [56] – proving that no other such vacuum black hole equilibrium state is obtainable by continuous axisymmetric variation from the spherical Schwarzschild solution, which had already been shown by Israel [57] to be the only strictly static (non-rotating) possibility. (Israel originally thought this implied that the regular horizon in the Schwarzschild solution was an unstable feature, but after the demonstration to the contrary by Vishveshwara [58] it became clear that the formation of a regular horizon was a generic phenomenon.)

The extension of the no-hair theorem to a complete **uniqueness theorem** – to the effect that, apart from the Kerr solution, no other such well behaved black hole equilibrium solution of the vacuum Einstein equations exists at all – has been

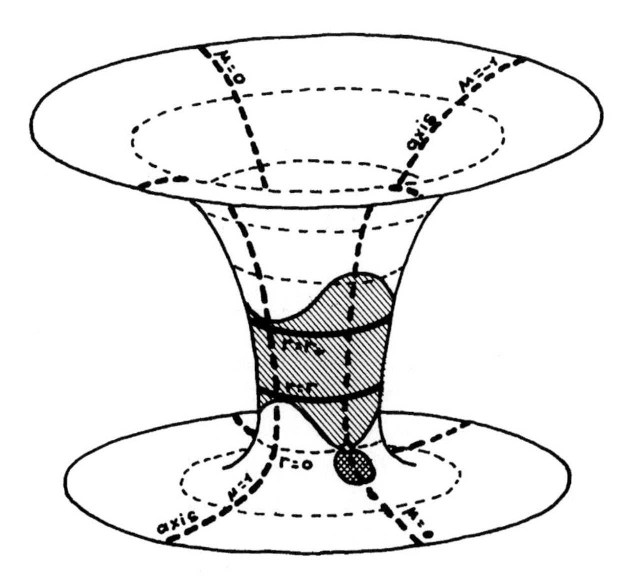


Fig. 4.2. Reproduction from 1972 Les Houches notes [26] of sketch of polar $\{r, \theta\}$ section (for fixed $u$ and $\phi$) extending from physical asymptotically flat region of positive radius through horizons at $r = r_+ = m + c$ and $r = r_- = m - c$ to unphysical asymptotically flat region of negative radius. The ring singularity intersects this section at two points of contact between the shaded parts representing the ergoregion, where the time translation Killing vector is spacelike, and the causality violating region, where the axial Killing vector is spacelike.

the objective of an extensive program of subsequent mathematical work, which is the subject of an accompanying article in this Festschrift proceedings by David Robinson. An important preliminary step was the work of Hawking [59] who used a plausible but [60] still not quite rigorously derived analyticity assumption to justify the supposition that such a state should be axisymmetric unless non-rotating, as in the static case covered by Israel's theorem. The demonstration of uniqueness for the rotating axisymmetric case was completed a few years later – first by Robinson [61] and afterwards using more elegant methods by subsequent work of which a newly archived account is available [62] – but the demonstration that the non-rotating case would necessarily be static, and hence covered by Israel's theorem, turned out to be rather more difficult than had been anticipated: this remaining lacuna was not dealt with adequately until the much more recent work due to Wald in collaboration with Sudarsky [63] and Chruściel [64, 65], as briefly described in a recent review by the present author [66].

In so far as the rotating case is concerned, the first essential step, based on preceding work of Papapetrou [32] is the demonstration [56, 62] that outside the horizon, whose null generator, $l^\mu = k^\mu + \Omega\, h^\mu$, has uniform angular velocity $\Omega$, the coordinates for the stationary axisymmetric asymptotically flat Einstein vacuum metric can be chosen in such a way that it takes the form

$$ds^2 = \varrho^2\, d\hat{s}^2 + X(d\varphi - \omega\, dt)^2 - (\lambda^2 - c^2)(1 - \mu^2)\, dt^2,$$

where $d\hat{s}^2$ is the metric of the two-dimensional space geometry given by

$$d\hat{s}^2 = d\lambda^2/(\lambda^2 - c^2) + d\mu^2/(1-\mu^2). \tag{4.36}$$

The domain of regularity of this ellipsoidal type coordinate system is bounded by the north and south polar branches of the symmetry axis where $\mu \to \pm 1$ and by the horizon where $\lambda \to c$, where $c$ is a positive scale parameter. Within this two-dimensional domain the Einstein equations reduce to a coupled partial differential system for just two independent unknowns, which can be taken to be the squared magnitude $X$ of the axial Killing vector and the rotational "frame dragging" coefficient $\omega$ which it is convenient to replace, as the next step by a corresponding potential $Y$ defined by the relation $X^2\, \partial\omega/\partial\lambda = (1-\mu^2)\, \partial Y/\partial\mu$. This provides a formulation analogous to the one introduced originally by Ernst [67] but with the crucially important difference that the roles of the time translation and rotation Killing vectors are interchanged, so as to take advantage of the fact that whereas the squared magnitude of the former changes sign on an "ergosurface" (within which the particle energy defined by (4.29) can be negative) on the other hand the squared magnitude $X$ of the latter must always be positive – as a necessary condition for absence from the exterior region under consideration of the kind of causal bad behaviour exhibited by the non-physical interior part of the Kerr solution. This requirement ensures the positive definiteness of the action given by

$$\int d\lambda\, d\mu\, (|\hat{\nabla} X|^2 + |\hat{\nabla} Y|^2)/X^2, \tag{4.37}$$

for the Einstein equations governing the unknowns $X$ and $Y$ in this formulation, which thereby takes the form of a well behaved elliptic boundary value problem, as specified by the conditions [56, 62] of regularity on the rigidly rotating horizon (characterised by a uniform angular velocity $\Omega$) where $\lambda = c$ by appropriate boundary conditions on the axis where $\mu = \pm 1$ and finally at large radius $\lambda \to \infty$ by an asymptotic flatness condition expressed in terms of angular momentum $ma$.

The original 1971 "no-hair" theorem [56] was obtained for this system by equating a divergence with zero volume integral (by the boundary conditions) to a certain positive definite function of the infinitesimal difference between nearby solutions. This relation, and its 1975 extension by Robinson [61] to the finite difference between any distinct solutions – so as to obtain a complete uniqueness theorem for the axisymmetric case – was obtained by a trial and error approach such as was also used by Robinson [68] to generalise the no-hair theorem to include a source free electromagnetic field. The extension for the electromagnetic case to finite differences remained out of reach until the introduction of more elegant and sophisticated methods by Mazur [69, 70] and Bunting [71–73]. The upshot of this work is that the black hole boundary condition problem for the elliptic system governed by (4.37) has only the Kerr solution, as characterised – with $\lambda = r - m$, $\mu = \cos\theta$ – by the mass and horizon angular velocity values

$$m = \sqrt{c^2 + a^2}, \qquad \Omega = a/2m(m + c)\,.$$

## References


[1] Kerr, R. P. (1963), "Gravitational field of a spinning mass as an example of algebraically special metrics", *Phys. Rev. Lett.* **11**, 237–238.

[2] Kerr, R. P. (1965), "Gravitational collapse and rotation", in *Quasi-stellar Sources and Gravitational Collapse, including the Proceedings of the First Texas Symposium on Relativistic Astrophysics*, ed. I. Robinson, A. Schild & E. L. Schücking (University of Chicago Press, Chicago and London), pp. 99–109.

[3] Kerr, R. P. & Schild, A. (1965), "Some algebraically degenerate solutions of Einstein's gravitational field equations", *Proc. Symp. Appl. Math.* **17**, 199–209.

[4] Debney, G. C., Kerr, R. P. & Schild, A. (1969), "Solutions of the Einstein and Einstein–Maxwell equations", *J. Math. Phys.* **10**, 1842–1854.

[5] Kerr, R. P. & Debney, G. C. (1970), "Einstein spaces with symmetry groups", *J. Math. Phys.* **11**, 2807–2817.

[6] Newman, E. T., Couch, E., Chinnapared, K., Exton, A., Prakash, A. & Torrence, R. (1965), "Metric of a rotating charged mass", *J. Math. Phys.* **6**, 918–919.

[7] Carter, B. (1968), "Global structure of the Kerr family of gravitational fields", *Phys. Rev.* **174**, 1559–1571.

[8] Sen, A. (1992), "Rotating charged black hole solution in heterotic string theory", *Phys. Rev. Lett.* **69**, 1006–1009 [arXiv:hep-th/9204046].

[9] Yazadjiev, S. (2000), "Newman–Janis method and rotating dilaton axion black hole", *Gen. Rel. Grav.* **32**, 2345–2352 [arXiv:gr-qc/9907092].

[10] Burinskii, A. (2002), "Supersymmetric superconducting bag as a core of Kerr spinning particle", *Grav. Cosmol.* **8**, 261–271 [arXiv:hep-th/0008129].

[11] Burinskii, A. (2003), "Orientifold string in the source of the Kerr spinning particle", *Phys. Rev.* **D 68**, 105004 [arXiv:hep-th/0308096].

[12] Pfister, H. & King, M. (2003), "The gyromagnetic factor in electrodynamics, quantum theory and general relativity", *Class. Quantum Grav.* **20**, 205–213.

[13] Perjés, Z. (1973), "Classification of stationary space-times", *Int. J. Theor. Phys.* **10**, 217–227.

[14] Simon, W. (1984), "Characterisations of the Kerr metric", *Gen. Rel. Grav.* **16**, 465–476.

[15] Mars, M. (1999), "A space-time characterization of the Kerr metric", *Class. Quant. Grav.* **16**, 2507–2523 [arXiv:gr-qc/9904070].

[16] Mars, M. (2000), "Uniqueness properties of the Kerr metric", *Class. Quantum Grav.* **17**, 3353–3373 [arXiv:gr-qc/0004018].
[17] Mars, M. (2001), "The Wahlquist–Newman solution", *Phys. Rev.* **D 63**, 064022 [arXiv:gr-qc/0101021].
[18] Senovilla, J. M. M. (1987), "Stationary axisymmetric perfect-fluid metrics with $q + 3p = \text{const}$", *Phys. Lett.* **A 123**, 211–214.
[19] Wahlquist, H. D. (1968), "Interior solution for a finite rotating body of perfect fluid", *Phys. Rev.* **D 172**, 1291–1296.
[20] Carter, B. (1968), "A new family of Einstein spaces", *Phys. Lett.* **A 26**, 399–400.
[21] Carter, B. (1968), "Hamilton–Jacobi and Schrödinger separable solutions of Einstein's equations", *Commun. Math. Phys.* **10**, 280–310.
[22] Gibbons, G. W. & Hawking, S. W. (1977), "Cosmological event horizons, thermodynamics, and particle creation", *Phys. Rev.* **D 15**, 2738–2751.
[23] Charmousis, C., Langlois, D., Steer, D. & Zegers, R. (2007), "Rotating spacetimes with a cosmological constant", *JHEP* **02**, 064 [arXiv:gr-qc/0610091].
[24] Rogatko, M. (2004), "Uniqueness theorem for generalized Maxwell electric and magnetic black holes in higher dimensions", *Phys. Rev.* **D 70**, 044023 [arXiv:hep-th/0406041].
[25] Myers, R. C. & Perry, M. J. (1986), "Black holes in higher dimensional space-times", *Ann. Phys. (N.Y.)* **172**, 304–347.
[26] Carter, B. (1973), "Black hole equilibrium states", in *Black Holes (1972 Les Houches Lectures)*, ed. C. DeWitt & B. S. DeWitt (Gordon and Breach, New York), pp. 57–210.
[27] Hawking, S. W., Hunter, C. J. & Taylor-Robinson, M. M. (1999), "Rotation and the AdS/CFT correspondence", *Phys. Rev.* **D 59**, 064005 [arXiv:hep-th/9811056].
[28] Gibbons, G. W., Lu, H., Page, D. N. & Pope, C. N. (2004), "Rotating black holes in higher dimensions with a cosmological constant", *Phys. Rev. Lett.* **93**, 171102 [arXiv:hep-th/0409155].
[29] Frolov, V. & Kubiznak, D. (2008), "Higher-dimensional black holes: hidden symmetries and separation of variables", *Class. Quantum Grav.* **25**, 154005 [arXiv:0802.0322].
[30] Frolov, V., Krtous, P. & Kubiznak, D. (2007), "Separability of Hamilton–Jacobi and Klein Gordon equations in general Kerr–NUT–AdS spacetimes", *JHEP* **0702**, 005 [arXiv:hep-th/0611245].
[31] Krtous, P., Kubiznak, D., Page, D. N. & Frolov, V. (2007), "Killing–Yano tensors, rank-2 Killing tensors, and conserved quantities in higher dimensions", *JHEP* **0702**, 004 [arXiv:hep-th/0612029].
[32] Papapetrou, A. (1966), "Champs gravitationnels stationnaires à symmetrie axiale", *Ann. Inst. H. Poincaré* **4A**, 83–85.
[33] Boyer, R. H. & Lindquist, R. W. (1967), "Maximal analytic extension of the Kerr metric", *J. Math. Phys.* **8**, 265–281.
[34] Marck, J. A. (1983), "Solution to the equations of parallel transport in Kerr geometry; tidal tensor", *Proc. Roy. Soc. Lond.* **A 385**, 431–438.
[35] Hameury, J. M., Marck, J. A. & Pelat, D. (1994), "$e^+$–$e^-$ annihilation lines from accretion discs around Kerr black holes", *Astron. Astroph.* **287**, 795–802.
[36] Unruh, W. (1973), "Separability of the neutrino equation in a Kerr background", *Phys. Rev. Lett.* **31**, 1265–1267.
[37] Teukolsky, S. A. (1973), "Perturbations of a rotating black hole, I: Fundamental equations for gravitational and electromagnetic perturbations ", *Astroph. J.* **185**, 635–647.

[38] Chandrasekhar, S. (1976), "The solution of Dirac's equation in Kerr geometry", *Proc. Roy. Soc. Lond.* **A 349**, 571–575.
[39] Page, D. N. (1976), "Dirac equation around a charged rotating black hole", *Phys. Rev.* **D 14**, 1509–1510.
[40] Marck, J. A. (1983), "Parallel-tetrad on null geodesics in Kerr–Newman spacetime", *Phys. Lett.* **A 97**, 140–142.
[41] Walker, M. & Penrose, R. (1970), "On quadratic first integrals of the geodesic equations for type $\{22\}$ spacetimes", *Commun. Math. Phys.* **18**, 265–274.
[42] Penrose, R. (1973), "Naked singularities", *Ann. N.Y. Acad. Sci.* **224**, 125–134.
[43] Carter, B. (1987), "Separability of the Killing–Maxwell system underlying the generalised angular momentum constant in the Kerr–Newman black hole metrics", *J. Math. Phys.* **28**, 1535–1538.
[44] Carter, B. (1977), "Killing tensor quantum numbers and conserved quantities in curved space", *Phys. Rev.* **D 16**, 3395–3414.
[45] Carter, B. & McLenaghan, R. G. (1979), "Generalised total angular momentum operator for the Dirac equation in curved space-time", *Phys. Rev.* **D 19**, 1093–1097.
[46] Press, W. H. & Teukolsky, S. A. (1973), "Perturbations of a rotating black hole. I. Dynamical stability of the Kerr metric ", *Astroph. J.* **185**, 649–673.
[47] Whiting, B. (1989), "Mode stability of the Kerr black hole", *J. Math. Phys.* **30**, 1301–1305.
[48] Kamran, N. & Marck, J. A. (1986), "Parallel propagated frame along the geodesics of metrics admitting a Killing–Yano tensor", *J. Math. Phys.* **27**, 1598–1591.
[49] Luminet, J. P. & Marck, J. A. (1985), "Tidal squeezing of stars by Schwarzschild black holes", *Mon. Not. R. Astr. Soc.* **212**, 57–75.
[50] Frolov, V. P., Khokhlov, A., Novokov, I. D. & Pethic, C. J. (1993), "Relativistic tidal interaction of a white dwarf star with a massive black hole", *Astroph. J.* **432**, 680–689.
[51] Marck, J. A., Lioure, A. & Bonazzola, S. (1996), "Numerical study of the tidal interaction of a star and a massive black hole", *Astron. Astroph.* **306**, 666–674.
[52] Connors, P. A., Stark, R. F. & Piran, T. (1980), "Polarization features of X-ray radiation emitted near black holes", *Astroph. J.* **235**, 224–244.
[53] Boyer, R. H. & Price, T. G. (1965), "An interpretation of the Kerr metric in general relativity", *Proc. Camb. Phil. Soc.* **61**, 531–534.
[54] Carter, B. (1966), "Complete analytic extension of the symmetry axis of Kerr's solution of Einstein's equations", *Phys. Rev.* **141**, 1242–1247.
[55] Boyer, R. H. (1969), "Geodesic orbits and bifurcate Killing horizons", *Proc. Roy. Soc. Lond.* **A 311**, 245–252.
[56] Carter, B. (1971), "An axisymmetric black hole has only two degrees of freedom", *Phys. Rev. Lett.* **26**, 331–333.
[57] Israel, W. (1967), "Event horizons in static vacuum spacetimes", *Phys. Rev.* **164**, 1776–1779.
[58] Vishveshwara, C. V. (1970), "Stability of the Schwarschild metric", *Phys. Rev.* **D 1**, 2870–2879.
[59] Hawking, S. W. (1972), "Black holes in general relativity", *Commun. Math. Phys.* **25**, 152–156.
[60] Friedrich, H., Racz, I. & Wald, R. M. (1999), "On the rigidity theorem for spacetimes with a stationary event horizon or a compact Cauchy horizon", *Commun. Math. Phys.* **204**, 691–707 [arXiv:gr-qc/9811021].
[61] Robinson, D. C. (1975), "Uniqueness of the Kerr black hole", *Phys. Rev. Lett.* **34**, 905–906.

[62] Carter, B. (1992), "Mechanics and equilibrium geometry of black holes, membranes, and strings", in *Black Hole Physics, (NATO ASI C364)*, ed. V. de Sabbata & Z. Zhang (Kluwer, Dordrecht), pp. 283–357 [arXiv:hep-th/0411259].
[63] Sudarsky, D. & Wald, R. M. (1993), "Mass formulas for stationary Einstein–Yang–Mills black holes and a simple proof of two staticity theorems", *Phys. Rev.* **D 47**, 5209–5213 [arXiv:gr-qc/9305023].
[64] Chruściel, P. T. & Wald, R. M. (1994), "Maximal hypersurfaces in stationary asymptotically flat spacetimes", *Commun. Math. Phys.* **163**, 561–604 [arXiv:gr-qc/9304009].
[65] Chruściel, P. T. & Wald, R. M. (1994), "On the topology of stationary black holes", *Class. Quantum Grav.* **11**, L147–152 [arXiv:gr-qc/9410004].
[66] Carter, B. (1999), "Has the black hole equilibrium problem been solved?" in *Proc. 5th Marcel Grossmann Meeting on General Relativity, Gravitation, and Relativistic Field Theories*, ed. T. Piran (World Scientific, Singapore), pp. 136–165 [arXiv:gr-qc/9712 028].
[67] Ernst, F. J. (1968), "New formulation of the axially symmetric gravitational field problem", *Phys. Rev.* **167**, 1175–1178.
[68] Robinson, D. C. (1974), "Classification of black holes with electromagnetic fields", *Phys. Rev.* **D 10**, 458–460.
[69] Mazur, P. O. (1982), "Proof of uniqueness of the Kerr–Newman black hole solution", *J. Phys.* **A 15**, 3173–3180.
[70] Mazur, P. O. (1984), "Black hole uniqueness from a hidden symmetry of Einstein's gravity", *Gen. Rel. Grav.* **16**, 211–215.
[71] Bunting, G. (1983), "Proof of the uniqueness conjecture for black holes"(PhD Thesis, University of New England, Armidale N.S.W., Australia).
[72] Carter, B. (1985), "Bunting identity and Mazur identity for non-linear elliptic systems including the black hole equilibrium problem", *Commun. Math. Phys.* **99**, 563–591.
[73] Dain, S. (2006), "Angular momentum–mass inequality for axisymmetric black holes", *Phys. Rev. Lett.* **96**, 101101; "Proof of the angular momentum-mass inequality for axisymmetric black holes" [arXiv:gr-qc/0606105].

# 5

# Four decades of black hole uniqueness theorems

David C. Robinson

## 5.1 Introduction

It is approaching 40 years since Werner Israel announced the first black hole uniqueness theorem at a meeting at King's College London [1]. He had investigated an interesting class of static asymptotically flat solutions of Einstein's vacuum field equations. The solutions had regular event horizons, and obeyed the type of regularity conditions that a broad class of non-rotating equilibrium black hole metrics might plausibly be expected to satisfy. His striking conclusion was that the class was exhausted by the positive mass Schwarzschild family of metrics. This result initiated the research on black hole uniqueness theorems, which continues today. Israel's investigations and all immediately subsequent work on uniqueness theorems were carried out, explicitly or implicitly, within the astrophysical context of gravitational collapse. In the early years attention was centred mainly on four-dimensional static or stationary black holes that were either purely gravitational or minimimally coupled to an electromagnetic field. More recently, developments in string theory and cosmology have encouraged studies of uniqueness theorems for higher dimensional black holes and black holes in the presence of numerous new matter field combinations.

In an elegant article Israel [2] has described the background and influences that led him to formulate his theorem and the immediate reactions, including his own, to his result. Historically flavoured accounts, which include discussions of the evolution of research on black holes and the uniqueness theorems, have also been written by Kip Thorne and Brandon Carter [3, 4]. In the 1960s observational results such as the discovery of quasars and the microwave background radiation stimulated a new interest in relativistic astrophysics. There was increased activity, and more sophistication, in the modelling of equilibrium end states of stellar systems and

The Kerr Spacetime: Rotating Black Holes in General Relativity. Ed. David L. Wiltshire, Matt Visser and Susan M. Scott. Published by Cambridge University Press. 

gravitational collapse [5]. The pioneering work on spherically symmetric gravitational collapse carried out in the 1930s [6] was extended to non-spherical collapse, see e.g. [7, 8]. Significant strides were made in the use of new mathematical tools to study general relativity. Especially notable amongst these, as far as the early theory of black holes was concerned, were the constructions of the analytic extensions of the Schwarzschild and Reissner–Nordström solutions [9–11], the analyses of congruences of null geodesics and the optics of null rays, the precise formulation of the notion of asymptotic flatness in terms of a conformal boundary [12], and the introduction of trapped surfaces. In addition, novel approaches to exploring Einstein's equations, such as the Newman–Penrose formalism [13], were leading to new insights into exact solutions and their structure.

In 1963, by using a null tetrad formalism in a search for algebraically special solutions of Einstein's vacuum field equations, Roy Kerr found an asymptotically flat and stationary family of solutions, metrics of a type that had eluded discovery for many years [14]. He identified each member of the family as the exterior metric of a spinning object with mass $m$ and angular momentum per unit mass $a$. The final sentence of his brief paper announcing these solutions begins: 'It would be desirable to calculate an interior solution to get more insight into this' (the multipole moment structure). Completely satisfying model interiors to the Kerr metric have not yet been constructed, but the importance of these metrics does not reside in the fact that they might model the exterior of some rather particular stellar source. The Kerr family of metrics are the most physically significant solutions of Einstein's vacuum field equations because they contain the Schwarzschild family in the limit of zero angular momentum and because they are believed to constitute, when $a^2 \leq m^2$, the unique family of asymptotically flat and stationary black hole solutions. Within a few years of Kerr's discovery the maximal analytic extension of the Kerr solution was constructed and many of its global properties elucidated [15–17].

In current astrophysics the equilibrium vacuum black hole solutions are regarded as being the stationary exact solutions of primary relevance, with accreting matter or other exterior dynamical processes treated as small perturbations. However, exact black hole solutions with non-zero energy momentum tensors have always been studied, despite the fact that direct experimental or observational support for the gravitational field equations of the systems is often weak or non-existent. In particular, results obtained for the vacuum spacetimes are often paralleled by similar results for electrovac systems describing the coupling of gravity and the source-free Maxwell field. In 1965 Ted Newman and graduate students in his general relativity class at the University of Pittsburgh published a family of electrovac solutions containing three parameters $m$, $a$ and total charge $e$ [18]. It was found by considering a complexification of a null tetrad for the Reissner–Nordström solution, making a complex coordinate transformation, and then imposing a reality condition to recover a real metric. The Kerr–Newman metrics reduce to the Reissner–Nordström solutions when $a = 0$ and to the Kerr family when $e = 0$ and are asymptotically flat and stationary. Their Weyl tensors are algebraically special, of Petrov type D, and the metrics are of Kerr–Schild type so they can be written as the sum of a flat metric and the tensor product of a null vector with itself. When $0 < e^2 + a^2 \leq m^2$ they represent rotating, charged, asymptotically flat and stationary black holes. Later it was realized that this family could easily be extended to a four-parameter family by adding a magnetic charge $p$. The general theory of the equilibrium states of asymptotically flat black holes is based on concepts and structures which were first noted in investigations of the Kerr and Kerr–Newman families. In 1968 Israel extended his vacuum uniqueness theorem to static asymptotically flat electrovac spacetimes [19]. He showed that the unique black hole metrics, in the class he considered, were members of the Reissner–Nordström family of solutions with charge $e$ and $e^2 < m^2$. This result, while not unexpected, physically or mathematically, required ingenious extensions of the calculations in his proof of the first theorem.

Until 1972 the only known asymptotically flat stationary, but not static, solutions of Einstein's vacuum equations with positive mass were members of the Kerr family. In that year further stationary vacuum solutions, which also happened to be axisymmetric, were published [20]. Nevertheless, sufficient was known about the Kerr solution by the time Israel published his uniqueness theorem for him to be able to ask, towards the end of his paper, if in the time independent but rotating case a similar uniqueness result might hold for Kerr–Newman metrics. This question developed into what for a while came to be referred to as the 'Carter–Israel conjecture'. This proposed that the Kerr–Newman solutions with $a^2 + e^2 + p^2 \leq m^2$ were the only stationary and asymptotically flat electrovac solutions of Einstein's equations that were well-behaved from infinity to a regular black hole event horizon. More broadly it was conjectured that, irrespective of a wide range of initial conditions, the vacuum spacetime outside a sufficiently massive collapsed object must settle down so that asymptotically in time its metric is well approximated by a member of the Kerr (or Kerr–Newman) family. The emergence of these conjectures was significantly influenced by John Wheeler with his 'black holes have no hair' conjecture [21] and by Roger Penrose and the wide ranging paper he published in 1969 [22]. Amongst the topics Penrose discussed in this paper was the question of whether or not singularities that form as a result of gravitational collapse are always hidden behind an event horizon. He raised the question of the existence of a 'cosmic censor' that would forbid the appearance of 'naked' singularities unclothed by an event horizon. Subsequently there have been many investigations of what has become termed 'the weak cosmic censorship hypothesis' [23]. Roughly speaking this says that, generically, naked singularities visible to distant observers do not arise in gravitational collapse. Although numerous models providing examples and possible counter-cxamples have been studied and formal theorems proven, the extent of the validity of the hypothesis is still not settled. Nevertheless, in the proofs of the uniqueness theorems it is always assumed that there are no singularities exterior to the event horizon.

In this article a broad introduction to the way in which uniqueness theorems for equilibrium black holes have evolved over the decades will be presented. Obviously the point of view and selection of topics is personal and incomplete. Fortunately more detail and other perspectives are available in various review papers. These will be referred to throughout this article. Comprehensive overviews of the four-dimensional black hole uniqueness theorems, and related research such as hair and no hair investigations, can be found in a monograph and subsequent electronic journal article by Markus Heusler [24, 25]. Attention will be centred here on classical (bosonic) physics and the uniqueness theorems for both static and stationary black hole solutions of Einstein's equations in dimensions $d \geq 4$. Supersymmetric black holes and models in manifolds with dimension less than four, for instance, will not be discussed.

Israel's first theorem will be reviewed in the next section and some of the issues raised by it will be noted. These will be addressed in subsequent sections, decade by decade. The third section deals with the 1970s when the foundations of the general theory and the basic uniqueness results for static and stationary black hole spacetimes were established. This is followed in the next section by a discussion of the progress made in the 1980s. During that period novel approaches to the uniqueness problems for rotating and non-rotating black holes were introduced and new theorems were proven. In addition that decade saw the construction of black hole solutions of Einstein's equations in higher dimensions and the investigation of systems with more complicated matter configurations. The motivation for much of this research came from various approaches to the problem of unifying gravity with the other fundamental forces, rather than from astrophysical considerations. The fifth section considers developments in the 1990s. Once again there were two rather distinct strands of activity. There was a rigorous reconsideration of the mathematical foundations of the theory of four dimensional equilibrium black holes laid down in the 1970s. As a result of this research a number of gaps in the early work have now been filled and mathematically more complete theorems have been established. There was also a vigorous continuation and extension of the research on black holes and new matter field combinations that draws much of its inspiration from the study of gauge theories, thermodynamics and string theory. Uniqueness theorems for higher dimensional black holes and black holes in the presence of a non-zero cosmological constant are discussed in the sixth section.

This work, stimulated by string theory and cosmology, has shown how changing the spacetime dimension, or the structure of the field equations and the boundary conditions, affects uniqueness theorems. The most notable new result has been the recent demonstration that in five dimensions the higher dimensional Kerr black holes are not the only stationary rotating vacuum black hole solutions.

It is a pleasure to acknowledge the contributions made by Roy Kerr to general relativity. The metrics bearing the names of Kerr, Newman, Nordström, Reissner and Schwarzschild have been central to the study of black hole spacetimes. One looks forward to the time when all the theoretical studies will be tested in detail by observations and experiments. There are compelling questions to be answered. What is the relationship between the observed astrophysical black holes and the Kerr and Schwarzschild black hole solutions [26]? More speculatively, what, if anything, will the Large Hadron Collider (LHC) being constructed at CERN reveal about black holes [27]?

The sign conventions of reference [28] are followed, and $c = G = 1$. Unless it is explicitly stated otherwise it will be assumed that the cosmological constant is zero and the spacetimes considered are asymptotically flat and four dimensional.

## 5.2 Israel's 1967 theorem and issues raised by it

In his paper *Event Horizons in Static Vacuum Space-Times,* published in 1967, Israel investigated four-dimensional spacetimes, satisfying Einstein's vacuum field equations [1]. The spacetimes are static; that is, there exists a time-like hypersurface orthogonal Killing vector field, $k^\alpha$,

$$k^\alpha k_\alpha < 0; \; k_{[\alpha} \nabla_\beta k_{\gamma]} = 0, \tag{5.1}$$

and an adapted coordinate system $(t, x^a)$, such that $k^\alpha = (1, 0, 0, 0)$, and the four-dimensional line element is

$$ds^2 = -v^2 \, dt^2 + g_{ab} \, dx^a \, dx^b. \tag{5.2}$$

Here $v$ and the Riemannian 3-metric $g_{ab}$ are independent of $t$. In this coordinate system the vacuum field equations take the form

$$\begin{aligned} {}^{(4)}R_{tt} &\equiv v D^a D_a v = 0 \\ {}^{(4)}R_{ta} &\equiv 0 \\ {}^{(4)}R_{ab} &\equiv R_{ab} - v^{-1} D_a D_b v = 0, \end{aligned} \tag{5.3}$$

where $R_{ab}$ and $D_a$ denote, respectively, the Ricci tensor and covariant derivative corresponding to $g_{ab}$. The class of static spacetimes considered by Israel were required to satisfy the following conditions.

On any $t =$ constant space-like hypersurface, $\Sigma$, maximally consistent with $k^\alpha k_\alpha < 0$:

(i) $\Sigma$ is regular, empty, non-compact and 'asymptotically Euclidean', with the Killing vector $k^\alpha$ normalized so that $k^\alpha k_\alpha \to -1$ asymptotically.
(ii) The invariant ${}^{(4)}R_{\alpha\beta\gamma\delta}\,{}^{(4)}R^{\alpha\beta\gamma\delta}$ formed from the four-dimensional Riemann tensor is bounded on $\Sigma$.
(iii) If $v$ has a vanishing lower bound on $\Sigma$, the intrinsic geometry (characterized by ${}^{(2)}R$) of the 2-spaces $v = c$ is assumed to approach a limit as $c \to 0^+$ corresponding to a closed regular 2-space of finite area.
(iv) The equipotential surfaces in $\Sigma$, $v =$ constant $> 0$, are regular, simply connected closed 2-spaces.

Conditions (i)–(iii) aim to enforce asymptotic flatness and geometrical regularity on and outside the boundary of the black hole given by $v \to 0$. The latter is assumed to be a connected, compact 2-surface with spherical topology, so a single black hole is being considered. Condition (iv), the assumption that the equipotential surfaces of $v$ do not bifurcate, and hence have spherical topology, implies the absence of points where the gravitational force acting on a test particle is zero. The status of this assumption is different from the others and its significance and implications were unclear. It was of central technical importance in the proof of the theorem because it allowed $\Sigma$ to be covered by a single coordinate system with $v$ as one of the coordinates. Using this coordinate system Israel constructed a number of identities from which he was able to deduce that the only static four-dimensional vacuum spacetime satisfying (i)–(iv) is Schwarzschild's spherically symmetric vacuum solution,

$$\begin{aligned} \mathrm{d}s^2 &= -v^2\,\mathrm{d}t^2 + v^{-2}\mathrm{d}r^2 + r^2\,\mathrm{d}\Omega_2^2, \\ v^2 &= (1 - 2mr^{-1}),\ 0 < 2m < r < \infty. \end{aligned} \tag{5.4}$$

In 1968 Israel published the proof of a similar theorem for static electrovac spacetimes [19]. By making similar assumptions and taking a similar approach, but extending the calculations of his vacuum proof in a non-trivial and ingenious way, he obtained an analogous uniqueness theorem for the Reissner–Nordström black hole solutions with $e^2 < m^2$.

Israel's theorem prompted a number of questions. Some arose immediately. Others became of interest later, mainly through the influence of string theory and cosmology. They include the following.

- What is the appropriate global four-dimensional formulation of black hole spacetimes? What are the possible topologies of the two-dimensional surface of the black hole? In the equilibrium case what is the relationship between the Lorentzian four geometry and the 'reduced Riemannian' uniqueness problem of the type studied by Israel?
- Are the Kerr and Kerr–Newman families the unique equilibrium vacuum and electrovac black hole solutions when rotation is permitted?
- What is the significance of the equipotential condition (iv) in Israel's two theorems and how restrictive is it?
- Could an equilibrium, static or stationary, spacetime contain more than one black hole? In other words, could the assumption that the equilibrium black hole horizon has only one connected component be dropped and uniqueness theorems still be proven?
- How mathematically rigorous could the uniqueness theorems be made?
- Would uniqueness theorems still hold when matter fields other than the electromagnetic field were considered?
- What would be the effect of changing the dimension of spacetime or the field equations by, say, introducing a non-zero cosmological constant?

Some of these questions continue to be addressed today.

## 5.3 The 1970s – laying the foundations

During the 1970s the basic framework and theorems which have shaped or influenced all subsequent research on black hole uniqueness theorems were formulated or established.

A paper by Stephen Hawking, published in 1972, initiated the detailed global analysis of four-dimensional, asymptotically flat, stationary black hole systems [29]. In this paper he drew on previous work on the global structure of spacetime, primarily by Penrose, Robert Geroch and himself, to describe the causal structure exterior to black holes. His lectures at the influential 1972 Les Houches summer school also dealt with these investigations [30]. These results were presented in more detail in the 1973 monograph *The Large Scale Structure of Space-Time* [31]. In these works asymptotic flatness is imposed by using Penrose's definition of weakly asymptotically simple spacetimes [32, 33]. In Hawking's paper it is assumed that the spacetime $M$ can be conformally embedded in a manifold $\widetilde{M}$ with boundaries, future and past null infinity $\mathscr{I}^+$ and $\mathscr{I}^-$, providing end points for null geodesics that propagate to asymptotically large distances to the future or from the past. The boundary of the region from which particles or photons can escape to future null infinity, that is the boundary of the set of events in the causal past of future null infinity, defines the future event horizon $H^+$. In the general setting this is not assumed to be connected, allowing for the possibility of systems with more than one black hole. The event horizon, generated by null geodesic segments, forms the boundary between the black holes region and the asymptotically flat region exterior to the black holes.The two-dimensional surface formed by the intersection of a connected component of $H^+$ and a suitable space-like hypersurface, defining a moment in time, corresponds to (the surface of) the black hole at that time. By changing the time orientation a past event horizon, $H^-$, and white hole may be similarly defined. The manifold $M$ is required to satisfy a condition, asymptotic predictability, which ensures that there are no naked singularities.

In the equilibrium case, it is assumed that the spacetime is stationary. This means, in the black hole context, that there exists a one parameter group of isometries generated by a Killing vector field, $k^\alpha$, that in the asymptotically flat region approaches a unit time-like vector field at infinity. When this time-like Killing vector field is hypersurface orthogonal the spacetime is not only stationary but is also static. Henceforth in this article attention will be mainly confined to the equilibrium situation and the domain of outer communications $\langle\langle M\rangle\rangle$. This is the set of events from which there exist both future and past directed curves extending to arbitrary large asymptotic distances. The Killing vector field $k^\alpha$ cannot be assumed to be time-like in all of $\langle\langle M\rangle\rangle$ as this would disallow ergo-regions, as in Kerr–Newman black holes, where $k^\alpha$ is space-like. All the uniqueness theorems apply to $\langle\langle M\rangle\rangle$.

Hawking used this framework to show, justifying one of Israel's assumptions, that the topology of the two-surface of an equilibrium black hole is spherical. More precisely, he used the Gauss–Bonnet theorem to establish that, when the dominant energy condition is satisfied, the two-dimensional spatial cross-section of each connected component of the horizon (in this article often just called the boundary surface or horizon of a black hole) must have spherical or toroidal topology. He then provided an additional argument aimed at eliminating the possibility of toroidal topology.

Hawking also introduced the strong rigidity theorem, for analytic manifolds and metrics, when matter in the spacetime is assumed to satisfy the energy condition and well behaved hyperbolic equations. This theorem relates the teleologically defined event horizon to the more locally defined concept of a Killing horizon [34–36]. A null hypersurface whose null generators coincide with the orbits of a one parameter group of isometries is called a Killing horizon. According to the strong rigidity theorem the event horizon of a stationary black hole is the Killing horizon of a Killing vector $l^\alpha$. The horizon is called rotating if this Killing vector field does not coincide with $k^\alpha$. When the horizon is rotating Hawking concluded that there must exist a second Killing vector field $m^\alpha$. He then argued that the domain of outer communications of a rotating black hole had to be axially symmetric, with the axial symmetry generated by a Killing vector field $m^\alpha$.

The relation between the appropriately normalized Killing vector fields can be written, $l^\alpha = k^\alpha + \Omega m^\alpha$, where the non-zero constant $\Omega$ is the angular velocity of the horizon. When $\Omega$ is zero (and $m^\alpha$ is undefined) so that the event horizon is a Killing horizon for the asymptotically time-like Killing vector $k^a$, it was argued that the domain of outer communications had to be static. In other words, according to this staticity argument an asymptotically flat and stationary black hole which is not rotating must have a static domain of outer communications and therefore $k^\alpha$ must be time-like and hypersurface orthogonal in $\langle\langle M \rangle\rangle$ [37]. Part of the proof of this result was based on unsatisfactory heuristic considerations and it was not until the 1990s that the staticity theorem was firmly established. This theorem will be briefly discussed again later.

Hawking's calculations employed the assumption that the spacetimes, horizons and metrics considered were analytic so that analytic continuation arguments could be used. A theorem proven by Henning Müller zum Hagen, based on the elliptic nature of the relevant equations, provides the justification for the assumption of local analyticity in stationary systems [38]. However, analytic continuations are not necessarily unique. The complete elimination of certain analyticity assumptions, probably more of mathematical than physical significance but desirable nevertheless, remains to be effected, see for example [39].

The strong rigidity and staticity theorems are important because they permit the equilibrium uniqueness problems to be reduced from problems in four-dimensional Lorentzian geometry to two distinct types of lower dimensional Riemannian boundary value problems. In the rotating case the system may be taken to be stationary and axially symmetric and hence the uniqueness problem may be reduced to a two-dimensional Riemannian problem. In the non-rotating case the system may be taken to be static and hence the problem may be reduced to a three-dimensional Riemannian problem. In the remainder of this article most attention will be focused on these dimensionally reduced uniqueness problems.

On each connected component of the horizon of an equilibrium black hole the normal Killing vector field $l^\alpha$ satisfies the equation $\nabla_\alpha(l^\beta l_\beta) = -2\kappa l_\alpha$. According to the first law of black hole mechanics $\kappa$, the surface gravity, is constant there. A connected component of the horizon is called non-degenerate if its surface gravity is non-zero and degenerate otherwise. The connected component of a non-degenerate future horizon can be regarded, in a precisely defined sense, as comprising a branch of a bifurcate Killing horizon. This is a pair of Killing horizons, for the same Killing vector field, which intersect on a compact space-like bifurcation two-surface where the Killing vector vanishes. Old arguments for this technically important result were superseded by better ones in the 1990s [40]. The early uniqueness theorems applied only to black holes with non-degenerate horizons, satisfying the bifurcation property, as in the non-extremal Kerr–Newman black holes. The first attacks on these reduced Riemannian problems also assumed that the horizon was connected so that there was only one black hole. Subsequently, uniqueness theorems for static systems with non-connected horizons have been proven, and comparatively recently theorems that rigorously include the possibility of degenerate horizons have also been constructed. While the physical plausibility of stable equilibrium systems of more than one black hole may be questionable, and the realizability in nature of degenerate horizons is moot, dealing with them mathematically has brought its own rewards. Reconsiderations of the agenda setting global analyses of equilibrium black hole spacetimes will be discussed in a later section.

Carter also presented a series of important lectures at the 1972 Les Houches summer school [37]. In his lecture notes he collected together and extended results he had obtained over a number of years and presented a systematic analysis of asymptotically flat, stationary and axisymmetric black holes. Subsequently he has reconsidered and extended this material in a number of reviews and lecture series [41–43]. A major topic in his lectures was the reduction of the uniqueness problems for stationary, axisymmetric vacuum and Einstein–Maxwell spacetimes to two-dimensional boundary value problems. It was well known that in local coordinates adapted to the symmetries, certain of the Einstein and other field equations for such systems may be reduced to a small number of non-linear elliptic equations with a small number of metric and field components as dependent variables. The remaining field and metric components are then derivable from these variables by quadratures [24, 28]. Carter showed that this could be done globally on the domain of outer communications with the regularity and black hole boundary conditions formulated in a comparatively simple way. He dealt with domains of outer communication for which each connected component of the future boundary $H^+$ of $\langle\langle M\rangle\rangle$ is non-degenerate and, by Hawking's theorem, topologically $R \times S^2$. He also made a natural causality requirement, that off the axisymmetry axis $X = m^\alpha m_\alpha > 0$ in $\langle\langle M\rangle\rangle$. For vacuum and electrovac systems, in particular, he demonstrated that, apart from the axis of symmetry where $X$ is zero, a simply connected domain of outer communications could be covered by a single coordinate system $(t, x, y, \phi)$ in which the metric took a Papapetrou form. In these coordinates, $k^\alpha = (1, 0, 0, 0)$ and $m^\alpha = (0, 0, 0, 1)$. He showed that the axisymmetric stationary black hole metric on $\langle\langle M\rangle\rangle$ may be written in the form

$$\begin{aligned} ds^2 &= -V dt^2 + 2W\, dt\, d\varphi + X\, d\varphi^2 + \Xi\, d\sigma^2, \\ XV + W^2 &= (x^2 - c^2)(1 - y^2), \end{aligned} \tag{5.5}$$

where for a single black hole, $0 < c < x$, $-1 < y < 1$, $c = \kappa A/4\pi$, and $A$ is the black hole area. Carter then reduced the vacuum and Einstein–Maxwell uniqueness problems for single black holes to boundary value problems for systems of elliptic partial differential equations on a two-dimensional manifold $D$ with global prolate spheroidal coordinates $(x, y)$ and metric

$$d\sigma^2 = \frac{dx^2}{x^2 - c^2} + \frac{dy^2}{1 - y^2}. \tag{5.6}$$

Attention will now be confined to the vacuum case where $c = m - 2\Omega J$, $m$ being the mass and $J$ the angular momentum about the axisymmetry axis. The relevant Ernst-like vacuum field equations on $D$ can be conveniently written in terms of a single complex field $E = X + iY$, where $Y$ is a potential for $W$, and derived from a Lagrangian density

$$L = \frac{\nabla E \cdot \nabla \overline{E}}{(E + \overline{E})^2}, \tag{5.7}$$

with $\nabla$ denoting the covariant derivative with respect to the two-metric. The complex field equation is

$$\nabla \left( \frac{\rho \nabla E}{(E + \overline{E})^2} \right) + \frac{2\rho \nabla E . \nabla \overline{E}}{(E + \overline{E})^3} = 0. \tag{5.8}$$

All the metric components are uniquely determined by $E$ and the boundary conditions. (When the metric is not only axisymmetric but also static, $Y = W = 0$, and the field equation reduces to the linear equation $\nabla(\rho \nabla \ln X) = 0$.) For a black hole solution $E$ is required to be regular when $x > c > 0$, and $-1 < y < 1$. Boundary conditions on $E$ and its derivatives ensure regularity on the axis of symmetry as $y \to \pm 1$ and regularity of the horizon as $x \to c > 0$. The conditions, $x^{-2} X = (1 - y^2) + O(x^{-1})$, $Y = 2Jy(3 - y^2) + O(x^{-1})$ as $x \to \infty$, ensure asymptotic flatness.

In 1971 Carter was able to prove, within this framework, that stationary axisymmetric vacuum black hole solutions must fall into discrete sets of continuous families, each depending on at least one and at most two parameters [44]. The unique family admitting the possibility of zero angular momentum is the Kerr family with $a^2 < m^2$. This was a highly suggestive but not conclusive result. Since the theorem was deduced by considering equations and solutions linearized about solutions of Eq. (5.8) it was not at all clear if, or how, the full non-linear theory could be handled. However, in 1975 I constructed a proof of the uniqueness of the Kerr black hole by using Carter's general framework [45]. Two possible black hole solutions $E_1$ and $E_2$ were used to construct a non-trivial generalized Green's identity of the form *divergence = positive terms mod field equations*. This was integrated over the two-dimensional manifold $D$. Stokes' theorem and the boundary conditions were then used to show that the integral of the left hand side was zero.

Consequently each of the positive terms on the right-hand side had to be zero and this implied that $E_1 = E_2$. Hence, Kerr black holes, with metrics on the domain of outer communication given, in Boyer–Lindquist coordinates, by

$$ds^2 = -V\,dt^2 + 2W\,dt\,d\varphi + X\,d\varphi^2 + \frac{\Sigma}{\Delta}\,dr^2 + \Sigma\,d\theta^2, \tag{5.9}$$

where $0 \le a^2 < m^2$; $r_+ = m + (m^2 - a^2)^{1/2} < r < \infty$, and

$$\begin{aligned}
V &= \frac{\Delta - a^2 \sin^2\theta}{\Sigma}, \qquad W = -\frac{a \sin^2\theta (r^2 + a^2 - \Delta)}{\Sigma}, \\
X &= \frac{(r^2 + a^2)^2 - \Delta a^2 \sin^2\theta}{\Sigma} \sin^2\theta, \\
\Sigma &= r^2 + a^2 \cos^2\theta, \qquad \Delta = r^2 + a^2 - 2mr,
\end{aligned} \tag{5.10}$$

are the only stationary, axially symmetric, vacuum black hole solutions with non-degenerate connected horizons. According to Hawking's rigidity theorem, 'axially symmetric' can be removed from the previous sentence.

In a separate development in the early 1970s Israel's theorems for static black holes were reconsidered by Müller zum Hagen, Hans Jurgen Seifert and myself. First we looked at static, single black hole vacuum spacetimes [46]. For these the event horizon is connected and, by the generalized Smarr formula [37], necessarily non-degenerate. In a somewhat technical paper we were able to avoid using both Israel's assumption (iv) about the equipotential surfaces of $v$, and his assumption about the spherical topology of the horizon. Our extension of Israel's theorem made use of the fact that the spacial part of the Schwarzschild metric, $g_{ab}$, is conformally flat. Indeed all asymptotically Euclidean, spherically symmetric three-metrics are locally conformally flat. Now a three-metric is conformally flat if and only if the Cotton tensor

$$R_{abc} \equiv D_b \left( R_{ac} - \frac{1}{4} R g_{ac} \right) - D_c \left( R_{ab} - \frac{1}{4} R g_{ab} \right) \tag{5.11}$$

is zero. By using a three dimensionally covariant approach we were able to show that the Cotton tensor had to vanish and thence to conclude that the only static vacuum black holes in four dimensions, with connected horizons, were Schwarzschild black holes. I was soon able to simplify and improve this proof [47]. To illustrate this approach an outline, based mainly on the latter paper but also containing results from [46], is presented in the appendix.

Using similar techniques we also extended Israel's static electrovac theorem to prove uniqueness of the Reissner–Nordström black hole when the horizon was again assumed to be connected [48]. The Smarr formula does not imply that the horizon is non-degenerate in this case, and satisfactory rigorous progress with degenerate electrovac horizons was not made until the late 1990s. In this paper we also noted that solutions of the Einstein–Maxwell equations might exist for which the metric was static but the Maxwell field was time dependent. We identified the form of these Maxwell fields, and the reduced equations they had to satisfy. However we were only able to construct a plausibility argument against such black hole solutions. Subsequently it has been shown that Einstein–Maxwell solutions of this type, albeit not asymptotically flat solutions since they are cylindrically symmetric, do exist [49]. Further investigation of this type of non-inherited symmetry for other fields might be of interest. I also managed to generalize Carter's no-bifurcation result from the vacuum case considered by him to stationary Einstein–Maxwell spacetimes [50]. I showed that black hole solutions, with connected non-degenerate horizons, formed discrete continuous families, each depending on at most four parameters (effectively the mass, angular momentum, electric and magnetic charge). Furthermore, of these only the four parameter Kerr–Newman family contained members with zero angular momentum.

Investigations of Weyl metrics corresponding to static, axially-symmetric, multi-black hole configurations, with non-degenerate horizons, were undertaken by Müller zum Hagen and Seifert [51], and independently by Gary Gibbons [52]. The type of method that Hermann Bondi [53] had used to tackle the static, axially symmetric two body problem was employed. It was shown that the condition of elementary flatness failed to hold everywhere on the axis of axial symmetry. Hence it was concluded that static, axially symmetric configurations of more than one black hole in vacuum, or of black holes and massive bodies which do not surround or partially surround a black hole, do not exist. Jim Hartle and Hawking appreciated that things were different when the black holes were charged [54]. They showed that completed Majumdar–Papapetrou electrovac solutions [55, 56], derivable from a potential with discrete point sources, could be interpreted as static, charged multi-black hole solutions. Each of the black holes has a degenerate horizon and a charge with magnitude equal to its mass. The electrostatic repulsion balances the gravitational attraction and the system is in neutral equilibrium. The single black hole solution is the $e^2 = m^2$ Reissner–Nordström solution. While these multi-black hole solutions are physically artificial, their existence showed that mathematically complete uniqueness theorems for electrovac systems had to take into account both the Kerr–Newman and the Majumdar–Papapetrou solutions and systems with horizons that need not be connected and could be degenerate. When static axisymmetric electrovac spacetimes were considered, and each black hole was assumed to have the same mass to charge ratio, Gibbons concluded that the solutions had to be Majumdar–Papapetrou black holes [57].

Studies of black holes with other fields, such as scalar fields, were also initiated. Working within the same framework as Israel, J. E. Chase showed that the only black hole solution of the static Einstein-scalar field equations, when the massless scalar field was minimally coupled, was the Schwarzschild solution [58]. In other words the scalar field had to be constant. A similar conclusion was reached by Hawking when he considered stationary Brans–Dicke black holes [59]. His calculation was a very simple one using, in a mathematically standard way, just the linear scalar field equation in the Einstein gauge. Interestingly this calculation, and a similar one by Jacob Bekenstein, did not depend heavily on all the detailed properties of the horizon.

Wheeler's 'black holes have no hair' conjecture inspired a number of investigations of matter in equilibrium black hole systems. According to the original no hair conjecture collapse leads to equilibrium black holes determined uniquely by their mass, angular momentum and charge (electric and/or magnetic), asymptotically measurable conserved quantities subject to a Gauss law, and with no other independent characteristics (hair) [21, 60]. The linear stability analyses, see e.g. [61], and Richard Price's observation of a late time power law decay in perturbations of the Schwarzschild black hole [62], provided support for both the weak cosmic censorship hypothesis and the no hair conjecture.

Other early investigations also supported the no hair conjecture. For instance, Bekenstein showed that the domains of outer communication of both static and stationary black holes could not support minimally coupled scalar fields whether massive or massless, massive spin-1 or Proca fields, nor massive spin-2 fields [63, 64]. He was able to draw his conclusions without using the Einstein equations; only the linear matter field equations and boundary conditions were needed, not the details of the gravitational coupling. Bekenstein also studied a black hole solution, with a conformally coupled scalar field, that had scalar hair [65, 66]. It turned out that this solution has unsatisfactory features, the scalar field diverges on the horizon and the solution is unstable. Nevertheless such work was the forerunner of many later hair and no-hair investigations.

By the mid 1970s the uniqueness theorems for static and stationary black hole systems discussed above had been constructed and the main thrust of theoretical interest in black holes had turned to the investigation of quantum effects. While not all of the results obtained in this decade, and discussed above, were totally satisfactory or complete [4] they provided the foundations and reference points for all subsequent investigations. At the end of the decade the main gap in the uniqueness theorems appeared to be the lack of a proof of the uniqueness of a single charged stationary black hole. It seemed clear that the uniqueness proof for the Kerr solution was extendable to a proof of Kerr–Newman uniqueness. However the technical detail of my electrovac no-bifurcation result was sufficiently complicated to make the prospect of trying to construct a proof rather unpalatable, unless a more systematic way of attacking the problem could be found.

## 5.4 The 1980s – systematization and new beginnings

The 1980s saw both the introduction of new techniques for dealing with the original stationary and static black hole uniqueness problems and the investigation of new systems of black holes. The interest in the latter was grounded not so much in astrophysical considerations as in renewed attempts to develop quantum theories that incorporated gravity. It included the construction of higher dimensional black hole solutions and the investigation of systems such as Einstein–dilaton–Yang–Mills black holes.

The uniqueness problem for stationary, axially symmetric electrovac black hole spacetimes was independently reconsidered, within the general framework set up by Carter, by Gary Bunting and Pawel Mazur. The reduced two-dimensional electrovac uniqueness problem is formally similar to the vacuum problem outlined above, but there are four equations and dependent variables instead of two, so the system of equations is more complicated. It had long been realized that the Lagrangian formulation of these equations might play an important role in the proof of the uniqueness theorems. In fact, I had used the Lagrangian for the vacuum equations given by Eq. (5.7), which is positive and quadratic in the derivatives, in a reformulation of Carter's no-bifurcation result [67]. However there are more productive interpretations of the Lagrangian formalism. It had been known since the mid 1970s that the Euler–Lagrange equations corresponding, as in Eq. (5.8), to the basic Einstein equations for stationary axisymmetric metrics, could be interpreted as harmonic map equations [68]. In addition, in the 1970s there was a growth of interest in generalized sigma models; that is, in the study of harmonic maps from a Riemannian space $M$ to a Riemannian coset space $N = G/H$, where $G$ is a connected Lie group and $H$ is a closed sub-group of $G$. Influenced by these developments Bunting and Mazur used these interpretations of the Lagrangian structure of the equations. Bunting's approach was more geometrically based, and in fact applied to a general class of harmonic mappings between Riemannian manifolds. He constructed an identity which implied that the harmonic map was unique when the sectional curvature of the target manifold was non-positive [69, 70]. Mazur on the other hand focused on a non-linear sigma model interpretation of the equations, with the target space $N$ a Riemannian symmetric space. Exploiting the symmetries of the field equations, he constructed generalized Green's identities when $N = SU(p, q)/S(U(p) \times U(q))$. When $p = 1$, $q = 2$ he obtained the identity needed to prove the uniqueness of the Kerr–Newman black holes. This is a generalization of the identity used in the proof of the uniqueness of the Kerr black hole which corresponds to the choice $N = SU(1,1)/U(1)$ [71–73]. Bunting and Mazur's systematic approaches provided computational rationales lacking in the earlier calculations, and enabled further generalizations to be explored within well-understood frameworks. In summary, Bunting and Mazur succeeded in proving that stationary axisymmetric black hole solutions of the Einstein–Maxwell electrovac equations, with non-degenerate connected event horizons, are necessarily members of the Kerr–Newman family with $a^2 + e^2 < m^2$.

In another interesting development Bunting and Masood-ul-Alam constructed a new approach to the static vacuum black hole uniqueness problem [74]. They used results from the positive mass theorem, published in 1979, to show, without the simplifying assumption of axial symmetry used in earlier multi-black hole calculations, that a non-degenerate event horizon of a static black hole had to be connected. In other words, there could not be more than one such vacuum black hole in static equilibrium. The thrust of their proof was to show, again, that the three metric $g_{ab}$ was conformally flat. However their novel method of proving conformal flatness did not make use of the Cotton tensor and so was not so tied to use in only three dimensions. In addition their approach relied much less on the details of the field equations. Consequently it subsequently proved much easier to apply it to other systems. Their proof that the constant time three-manifold $\Sigma$ with 3-metric $g_{ab}$ must be conformally flat proceeded along the following lines. Starting with $(\Sigma, g_{ab})$, as in Eq. (5.2), they constructed an asymptotically Euclidean, complete Riemannian three-manifold $(N, \gamma_{ab})$ with zero scalar curvature and zero mass. This was done by first conformally transforming the metric,

$$g_{ab} \rightarrow \mp\gamma_{ab} = \frac{1}{16}(1 \mp v)^4 g_{ab} \tag{5.12}$$

on two copies of $(\Sigma, g_{ab})$ so that $(\Sigma, +\gamma_{ab})$ was asymptotically Euclidean with mass $m = 0$, and $(\Sigma, -\gamma_{ab})$ "compactified the infinity". Then the 2 copies of $\Sigma$ were pasted along their boundaries to form the complete three-manifold $(N, \gamma_{ab})$. They then utilized the following corollary to the positive mass theorem proven in 1979 [75, 76].

Consider a complete oriented Riemannian three-manifold which is asymptotically Euclidean. If the scalar curvature of the three-metric is non-negative and the mass is zero, then the Riemannian manifold is isometric to Euclidean 3-space with the standard Euclidean metric.

From this result it follows that $(N, \gamma_{ab})$ has to be flat and therefore $(\Sigma, g)$ is conformally flat. Thus, as in the earlier uniqueness proofs, the metric must be spherically symmetric. Therefore the exterior Schwarzschild spacetime exhausts the class of maximally extended, asymptotically flat and static vacuum spacetimes with non-degenerate, but not necessarily connected, horizons.

Further uniqueness theorems for static electrovac black holes were also proven [77]. In particular, Bunting and Masood-ul-Alam's type of approach was used to construct a theorem showing that a non-degenerate horizon of a static electrovac black hole had to be connected, and hence the horizon of a Reissner–Nordström black hole with $e^2 < m^2$ [78].

In this decade new exact stationary black hole solutions, some with more complicated matter field configurations than had been considered in the past, were increasingly studied. These studies were often undertaken as contributions to ambitious programmes for unifying gravity and other fundamental forces. They shed new light on the no hair conjecture and the extent to which black hole uniqueness theorems might apply. For example, generalizations of the four-dimensional Einstein–Maxwell equations, which typically arise from Kaluza–Klein theories, and stationary black hole solutions were studied and uniqueness theorems constructed [79]. Investigation of a model naturally arising in the low-energy limit of $N = 4$ supergravity led Gibbons to find a family of static, spherically symmetric Einstein–Maxwell-dilaton black hole solutions in four dimensions [80, 81]. These have scalar hair but carry no independent dilatonic charge. In 1989 static spherically symmetric non-abelian SU(2) Einstein–Yang–Mills black holes, with vanishing Yang–Mills charges and therefore asymptotically indistinguishable from the Schwarzschild black hole, were found. The solutions form an infinite discrete family and are labelled by the number of radial nodes of the Yang–Mills potential exterior to a horizon of given size. Hence there is not a unique static black hole solution within this Einstein–Yang–Mills class [82–84]. Although these latter solutions proved unstable, such failures of the no hair conjecture and uniqueness encouraged the subsequent investigation of numerous black hole solutions with new matter configurations.

Interest in higher dimensional black holes also started to increase. Higher dimensional versions of the Schwarzschild and Reissner–Nordström solutions had been found in the 1960s [85] and in 1987 Robert Myers found the higher dimensional analogue of the static Majumdar–Papapetrou family [86]. The metric for the Schwarzschild black hole with mass $m$ in $d > 3$ dimensions is

$$\mathrm{d}s^2 = -v^2\,\mathrm{d}t^2 + v^{-2}\,\mathrm{d}r^2 + r^2\,\mathrm{d}\Omega^2_{d-2}, \tag{5.13}$$

where $v^2 = (1 - \frac{C}{r^{d-3}}) > 0$, $\mathrm{d}\Omega^2_{d-2}$ is the metric on the $(d-2)$-dimensional unit radius sphere which has area $A_{d-2}$ and $C = \frac{16\pi m}{A_{d-2}(d-2)}$. As the form of this metric suggests, these static higher dimensional black holes have similar properties to the four-dimensional solutions. Myers and Malcolm Perry found and studied the the $d$-dimensional generalizations of the Kerr metrics [87]. In general, the Myers–

Perry metrics are characterized by $[(d-1)/2]$ angular momentum invariants and the mass.

The family of metrics with a single spin parameter $J$ is given by

$$\begin{aligned} \mathrm{d}s^2 &= -\mathrm{d}t^2 + \Delta(\mathrm{d}t + a\sin^2\theta\,\mathrm{d}\phi)^2 + \Psi\mathrm{d}r^2 \\ &\quad + \rho^2\,\mathrm{d}\theta^2 + (r^2+a^2)\sin^2\theta\,\mathrm{d}\phi^2 + r^2\cos^2\theta\,\mathrm{d}\Omega^2_{(d-4)}, \\ \rho^2 &= r^2 + a^2\cos^2\theta, \qquad \Delta = \frac{\mu}{r^{d-5}\rho^2}, \; \Psi = \frac{r^{d-5}\rho^2}{r^{d-5}(r^2+a^2)-\mu}, \\ m &= \frac{(d-2)A_{(d-2)}}{16\pi}\mu, \qquad J = \frac{2ma}{(d-2)}. \end{aligned} \tag{5.14}$$

When $a = 0$ the metric reduces to the metric given in Eq. (5.13) and when $d = 4$ the metric reduces to the Kerr metric. When $d > 4$, there are three Killing vector fields. If $d = 5$ there is a horizon if $\mu > a^2$ and no horizon if $\mu \leq a^2$. If $d > 5$ a horizon exists for arbitrarily large spin. Further interesting properties are discussed in their paper.

## 5.5 The 1990s – rigour and exotic fields

During the last decade of the twentieth century two rather different lines of research on uniqueness theorems were actively pursued. On the one hand there was a renewed effort to improve and extend the scope and rigour of the uniqueness theorems for four-dimensional black holes. Here the approach was more mathematical in nature and emphasized rigorous geometrical analysis. On the other hand activity in theoretical physics related to string theory, quantum gravity and thermodynamics encouraged the continued, less formal, investigations of black holes with new exterior matter fields.

First it should be noted that further progress was made on eliminating the possibility of static multi-black hole spacetimes in a number of four-dimensional systems. Bunting and Masood-ul-Alam's approach to proving conformal flatness, which does not require the assumption that the horizon is connected, was used in a new proof that the exterior Reissner–Nordström solutions with $e^2 < m^2$ are the only static, asymptotically flat electrovac spacetimes with non-degenerate horizons [88]. Proofs of the uniqueness of the family of static Einstein–Maxwell-dilaton metrics, originally found by Gibbons [80, 81], were also constructed by using the same general approach [89, 90]. Stationary axially symmetric black holes with non-degenerate horizons that are not connected were also studied, but no definitive conclusions have yet been reached [91–93]. Whether regular stationary black hole spacetimes exist in which repulsive spin–spin forces between black holes are strong enough to balance the attractive gravitational forces remains unknown. To date, no uniqueness theorems dealing with stationary, but not static, black holes that may possess degenerate horizons, such as the extreme Kerr and the extreme Kerr–Newman horizons, have been proven.

Since the early 1990s significant progress has been made, particularly by Bob Wald, Piotr Chruściel and their collaborators, in tidying up and improving the global framework, erected in the 1970s, on which the uniqueness theorems rest. Mathematical shortcomings in the earlier work, of varying degrees of importance, were highlighted by Chruściel in a 1994 review, challengingly entitled *"No Hair" theorems: Folklore, Conjectures, Results* [94]. Statements, definitions and theorems from the foundational work have, where necessary, been corrected, sharpened and extended, and this line of rigorous mathematics has now been incorporated into a programme of classification of static and stationary solutions of Einstein's equations [95–97]. Mention can be made of the more important results. Chruściel and Wald obtained improved topological results by employing the topological censorship theorem [98]. For a globally hyperbolic and asymptotically flat spacetime satisfying the null energy condition the topological censorship theorem states that every causal curve from $\mathscr{I}^-$ to $\mathscr{I}^+$ is homotopic to a topologically trivial curve from $\mathscr{I}^-$ to $\mathscr{I}^+$ [99]. Chruściel and Wald showed that when it applied, the domain of outer communications had to be simply connected. They also gave a more complete proof of the spherical topology of the surface of stationary black holes. Basically, if the horizon topology is not spherical there could be causal curves, outside the horizon but linking it, that were not deformable to infinity, thus violating the topological censorship theorem [100].

An improved version of the rigidity theorem for analytic spacetimes, with horizons that are analytic submanifolds but not necessarily connected or non-degenerate, was constructed by Chruściel. A more powerful and satisfactory proof of the staticity theorem, that non-rotating stationary black holes with a bifurcate Killing horizon must be static, was constructed by Daniel Sudarsky and Wald [101, 102]. The new proof made the justifiable use of a slicing by a maximal hypersurface, and supersedes earlier proofs which had unsatisfactory features. Mention should also be made of the establishment, by István Rácz and Wald, of the technically important result, referred to earlier, concerning bifurcate horizons [40]. These authors considered non-degenerate event (and Killing) horizons with compact cross-sections, in globally hyperbolic spacetimes containing black holes but not white holes. This is the appropriate setting within which to consider the equilibrium end state of gravitational collapse. They showed that such a spacetime could be globally extended so that the image of the horizon in the enlarged spacetime is a proper subset of a regular bifurcate Killing horizon. They also found the conditions under which matter fields could be extended to the enlarged spacetime, thus providing justification for hypotheses made, explicitly or implicitly, in the earlier uniqueness theorems. In the late 1990s Chruściel extended the method of Bunting and Masood-al-Alam and the proof of the uniqueness theorem for static vacuum spacetimes [103]. He considered horizons that may not be connected and may have degenerate components on which the surface gravity vanishes, and constructed the most complete black hole theorem to date. The statement of his main theorem, which applies to black holes solutions with asymptotically flat regions (ends) in four dimensions, is the following:

**Theorem 5.1** *Let $(M, g)$ be a static solution of the vacuum Einstein equations with defining Killing vector $k^\alpha$. Suppose that $M$ contains a connected space-like hypersurface $\Sigma$, whose closure $\overline{\Sigma}$ is the union of a compact interior and a finite number of asymptotically flat ends, such that: (a) $g_{\mu\nu}k^\mu k^\nu < 0$ on $\Sigma$, and (b) the topological boundary $\partial\Sigma = \overline{\Sigma}\backslash\Sigma$ of $\Sigma$ is a non-empty topological manifold with $g_{\mu\nu}k^\mu k^\nu = 0$ on $\partial\Sigma$. Then $\Sigma$ is diffeomorphic to $R^3$ minus a ball, so that it is simply connected, it has only one asymptotically flat end, and its boundary $\partial\Sigma$ is connected. Further there exists a neighbourhood of $\Sigma$ in $M$ which is isometrically diffeomorphic to an open subset of the Schwarzschild spacetime.*

An analogous, although less complete, theorem for static electrovac spacetimes that included the possibility of non-connected, degenerate horizons was also constructed [104, 105]. It was shown that if the horizon is connected then the spacetime is a Reissner–Nordström solution with $e^2 \leq m^2$. If the horizon is not connected and all the degenerate connected components of the horizon with non-zero charge have charges of the same sign then the spacetime is a Majumdar–Papapetrou black hole solution.

In the work more oriented towards the study of black holes and high energy physics, there was a proliferation of research into 'exotic' matter field configurations such as dilatons, Skyrmions and sphalerons, into various types of non-minimal scalar field couplings and into fields arising in low energy limits of string theory. This type of research continues today. The immediate physical relevance of the Lagrangian systems considered is often of less importance than the contribution their study makes to deciding the extent to which black hole solution spaces can be parameterized by small numbers of global charges, or to deciding whether or not a class of systems admits stable solutions with non-trivial hair. Gravitating non-abelian gauge theories and gravity coupled scalar fields have featured prominently in this research. It has been shown, for example, that black holes in non-abelian gauge theories, and in theories with appropriately coupled scalar fields, can have very different hair properties from black holes in the originally studied Einstein–Maxwell or minimally coupled scalar field theories. Such research has also provided models that demonstrate the effect of varying the assumptions made in the early uniqueness theorems. It effectively includes many constructive proofs of existence and/or non-uniqueness. For instance, the existence of Einstein–Yang–Mills black holes that have zero angular momentum but need not be static has been established [106], and static black holes that need not be axially symmetric, let alone spherically symmetric, have been shown to exist [107]. Uniqueness theorems for self-gravitating harmonic mappings and discussions of Einstein–Skyrme systems can be found in reference [24], and further information about black holes in the presence of matter fields can be found, for example, in references [25, 108–110].

## 5.6 The 2000s – higher dimensions and the cosmological constant

The important role of black holes in string theory, and recent conjectures that black hole production may occur and be observable in high-energy experiments (TeV gravity) at the LHC [27], have stimulated investigations of black holes in higher dimensional spacetimes. In addition observational results in cosmology, and theoretical speculations in string theory, have encouraged the continued development of earlier work on black hole solutions of Einstein's equations with a non-zero cosmological constant $\Lambda$.

Uniqueness theorems for asymptotically flat black holes with static exteriors have, not unexpectedly, been extended to higher dimensions. In fact Seungsu Hwang showed in 1998 that the Schwarzschild–Tangherlini family, Eq. (5.13), is the unique family of static vacuum black hole metrics with non-degenerate horizons [111]. Subsequently other four-dimensional uniqueness theorems for static black holes with non-degenerate horizons have been extended to dimension $d > 4$ [112–117]. All these calculations deal with the relevant reduced $(d-1)$-dimensional Riemannian problem. They all follow the approach introduced in four dimensions by Bunting and Masood-ul-Alam, and need higher dimensional positive energy theorems (a topic still being explored) to show that the exterior $(d-1)$-dimensional Riemannian metric must be conformally flat. Appropriate arguments are then employed to show that the conformally flat Riemannian metrics, and the spacetime metrics and fields, must be spherically symmetric and members of the relevant known family of solutions. The stability of certain static higher dimensional black holes, such as the Schwarzschild family, has also been investigated and confirmed [118].

It obviously follows from the uniqueness theorems above that those black hole spacetimes have horizons that are topologically $S^{d-2}$, as do the Myers–Perry black holes. However the general methods used to restrict horizon topologies in four dimensions cannot be used in the same way in higher dimensions. Although, unlike the Gauss–Bonnet theorem, a version of topological censorship holds in any dimension it does not restrict the horizon topology as much when $d > 4$ [119].

Furthermore it is clear that a rigidity theorem in higher dimensions would not by itself imply the existence of sufficient isometries to allow the construction of generalizations of harmonic map or sigma model formulations of the equations governing stationary black hole exterior geometries. These differences were highlighted in 2002 when it was shown that in five dimensions, in addition to the Myers–Perry black hole family with rotation in a single plane, there is another asymptotically flat, stationary, vacuum black hole family characterized by its mass $m$ and spin $J$. This black ring family, as it was termed by its discoverers Roberto Emparan and Harvey Reall, also has three Killing vector fields [120, 121]. However its horizon is topologically $S^1 \times S^2$ whereas the Myers–Perry black holes have $S^3$ horizon topology. Moreover there is a range of values for its mass and spin for which there exist two black ring solutions as well as a Myers–Perry black hole. Hence there is not a unique family of stationary black hole vacuum solutions in five dimensions, and the global parameters $m$ and $J$ do not identify a unique rotating black hole. The Emparan–Reall family has many interesting properties and there are charged and supersymmetric analogues. It suffices to note here that it contains no static and spherically symmetric limit black hole. Furthermore, analysis of perturbations off the spherically symmetric vacuum solution suggests that the Myers–Perry solutions are the only regular black holes near the static limit. The full discussion of this remark and more details about stability, including the cases where $\Lambda$ is non-zero, can be found in references [118, 122].

It is natural to ask if uniqueness theorems can be constructed when the class of solutions considered is restricted by further conditions. A couple of results have shown that this is possible in five dimensions at least [123, 124]. When it is assumed that there are two commuting rotational Killing vectors, in addition to the stationary Killing vector field, and that the horizon is topologically $S^3$, it has been shown that vacuum black holes with non-degenerate horizons, must be members of the Myers–Perry family. The additional assumptions enable the appropriate extensions of the four dimensional uniqueness proofs for stationary black holes to be constructed. In the vacuum case, for example, the uniqueness problem is formulated as a $N = SL(3, R)/SO(3)$ non-linear sigma model boundary value problem and the corresponding Mazur identity is constructed. However, as is pointed out in reference [123] this approach does not appear to be extendable to higher dimensions. In six dimensions, for instance, the Myers–Perry black hole has only two commuting space-like Killing vector fields. However the direct generalization of the sigma model formulation used in four and five dimensions requires the six-dimensional spacetime to admit three such Killing vector fields.

When $d > 4$, the full global context has not, by 2004, been explored in the same depth as it has been in four dimensions. Differences from the four-dimensional case, another example being the failure of conformal null infinity to exist for radiating systems in odd dimensions [125], suggest further failures of uniqueness. Indeed Reall has conjectured that when $d > 4$, in addition to the known solutions, there exist stationary asymptotically flat vacuum solutions with only two Killing vector fields [126].

In conclusion, a brief comment should be made about black hole solutions of Einstein's equations with a non-zero cosmological constant $\Lambda$. The Kerr–Newman family of metrics admits generalizations which include a cosmological constant, and these provide useful black hole reference models [37, 127]. Both (locally) asymptotically de Sitter ($\Lambda > 0$) and asymptotically anti-de Sitter ($\Lambda < 0$) black hole models have been studied quite extensively, mainly since the 1990s. Topological and hair results may change when $\Lambda$ is non-zero; examples of papers which include general overviews of investigations of these topics are cited in references [122, 128–130]. Not so much is known about uniqueness theorems when $\Lambda$ is non-zero. There are non-existence [131] and uniqueness results for static black holes solutions with $\Lambda < 0$. Broadly stated, it has been shown that a static asymptotically AdS single black hole solution with a non-degenerate horizon must be a Schwarzschild–AdS black hole solution if it has a certain $C^2$ conformal spatial completion [132].

## Acknowledgments

I would like to thank Robert Bartnik, Malcolm MacCallum, B. Robinson and David Wiltshire for their kind assistance.

## Appendix: a simple proof of the uniqueness of the Schwarzschild black hole

Consider the static metric and vacuum field equations given by Eqs. (5.1–5.3). The conditions which isolated single black hole solutions must satisfy are formulated on a regular hypersurface $\Sigma$, $t = const$, where $0 < v < 1$. They are:

(a) asymptotic flatness which here is formulated on $\Sigma$ as the requirement of asymptotically Euclidean topology with the usual boundary conditions, given in asymptotically Euclidean coordinates, $x^a$, by

$$g_{ab} = (1 + 2mr^{-1})\delta_{ab} + h_{ab}; \quad v = 1 - mr^{-1} + \mu; \quad m \text{ constant}; \tag{5.15}$$

where $h_{ab}$ and $\mu$ are all $O(r^{-2})$, with first derivatives $O(r^{-3})$, as $r = (\delta_{ab}x^a x^b)^{1/2} \to \infty$.

(b) Regularity of the horizon of the single black hole which is formulated here as the requirement that the intersection of the future and past horizons constitute a regular compact, connected, two-dimensional boundary $B$ to $\Sigma$ as $v \to 0$. It can be shown that the extrinsic curvature of $B$ in $\overline{\Sigma}$, with respect to $g_{ab}$, must vanish, and also that the function

$$w \equiv -\frac{1}{2}\nabla_{[\alpha}k_{\beta]}\nabla^{\alpha}k^{\beta} = g^{ab}v,_a v,_b \tag{5.16}$$

is constant on $B$. The latter constant, denoted, $w_0$, is the square of the surface gravity. It is necessarily non-zero (that is the horizon is necessarily non-degenerate) since the horizon is assumed connected.

Using the vacuum field equations, Eqs. (5.3), the following identities can be constructed,

$$D_a(vD^a v) = D_a v D^a v, \tag{5.17}$$

$$D_a(v^{-1}D^a w) = 2vR_{ab}R^{ab}. \tag{5.18}$$

By integrating Eq. (5.17) over $\Sigma$ and using the boundary conditions, it can be seen that the mass $m$ is non-negative, and zero if and only if $v$ is constant and $g_{ab}$ and the four-metric are flat. In a similar way integration of the first of Eqs. (5.3) leads to the recovery, in this framework, of the generalized Smarr formula

$$4\pi m = w_0^{1/2} S_0, \tag{5.19}$$

where $S_0$ is the area of $B$. Integration of Eq. (5.18) and the use of the Gauss–Bonnet theorem on $B$ gives

$$w_0^{1/2}\int_B {}^{(2)}R \, \mathrm{d}S = 8\pi w_0^{1/2}(1-p) \geq 0, \tag{5.20}$$

with equality if and only if the three metric $g_{ab}$ has zero Ricci tensor and is therefore flat. It follows that the genus $p$ must be zero and the topology of $B$ must be spherical. Now by using the field equations to evaluate the Cotton tensor $R_{abc}$, given in Eq. (5.11), it can be shown that

$$R_{abc}R^{abc} = 4v^{-4}wD_aD^a w - 4v^{-5}wD^a wD_a v - 3v^{-4}D_a wD^a w. \tag{5.21}$$

Therefore at critical points of the harmonic function $v$ on $\Sigma$, where $w = 0$, the Cotton tensor and the gradient of $w$ must vanish. This expression can be used to construct the identities

$$\begin{aligned} D_a(pv^{-1}D^a w + qwD^a v) = {} & +\frac{3}{4}v^{-1}w^{-1}p[D_a w + 8wv(D_a v)(1-v^2)^{-1}]^2 \\ & +\frac{p}{4}v^3 w^{-1}R_{abc}R^{abc}, \end{aligned} \tag{5.22}$$

where $p(v) = (cv^2 + d)(1 - v^2)^{-3}$ and

$$q(v) = -2c(1 - v^2)^{-3} + 6(cv^2 + d)(1 - v^2)^{-4}$$

and both $c$ and $d$ are real numbers.

It follows from Eqs. (5.21) and (5.22) that the divergence on the left-hand side of Eq. (5.22), which is overtly regular everywhere on $\Sigma$, is non-negative on $\Sigma$ when $c$ and $d$ are chosen so that $p$ is non-negative. Making two such sets of choices $c = -1, d = +1$ and $c = +1, d = 0$, integrating over $\Sigma$, and using the boundary conditions and Gauss's theorem gives two inequalities,

$$\begin{aligned} w_0 S_0 &\leq \pi, \\ w_0^{3/2} S_0 &\geq \frac{\pi}{4m}. \end{aligned} \tag{5.23}$$

It is straightforward to see that these inequalities and Eq. (5.19) are compatible if and only if equality holds in Eq. (5.23). For this to be the case the right hand side of Eq. (5.22) must vanish. Hence $R_{abc}$ must be zero, so the three-geometry must be conformally flat and $w = w_0(1 - v^2)^4$. Since $w$ has no zeroes on $\Sigma$ coordinates $(v, x^A)$ like those used by Israel can now be introduced on $\Sigma$. The three-metric on $\Sigma$ then takes the form

$$\mathrm{d}s^2 = w_0^{-1}(1 - v^2)^{-4}\mathrm{d}v^2 + g_{AB}\mathrm{d}x^A\mathrm{d}x^B. \tag{5.24}$$

The conformal flatness of this metric can be shown to imply that the level surfaces of $v$ are umbilically embedded in $\Sigma$ [46]. It now follows quickly from the field equations that the four-metric in Eq. (5.2), is the Schwarzschild metric.

## References


[1] Israel, W. (1967), *Phys. Rev.* **164**, 1776.
[2] Israel, W. (1987), in *300 Years of Gravitation*, ed. S. W. Hawking & W. Israel (Cambridge University Press, Cambridge), pp. 199–276.
[3] Thorne, K. S. (1994), *Black Holes and Time Warps* (Norton, New York).
[4] Carter, B. (1999), in *Proceedings of the 8th Marcel Grossmann Meeting*, ed. T. Piran & R. Ruffini (World Scientific, Singapore), pp. 136–165 [arXiv:gr-qc/9712038].
[5] Harrison, B. K., Thorne, K. S., Wakano, M. & Wheeler, J. A. (1965), *Gravitation Theory and Gravitational Collapse* (University of Chicago Press, Chicago).
[6] Oppenheimer, J. R. & Snyder, H. (1939), *Phys. Rev.* **56**, 455.
[7] Doroshkevich, A. G., Zel'dovich, Ya. B. & Novikov, I. D. (1966), *Sov. Phys. JETP* **22**, 122.
[8] Penrose, R. (1965), *Phys. Rev. Lett.* **14**, 57.
[9] Kruskal, M. D. (1960), *Phys. Rev.* **119**, 1743.

[10] Szekeres, G. (1960), *Publ. Math. Debrecen.* **7**, 285. Reprinted: *Gen. Rel. Grav.* **34**, 2001 (2002).
[11] Graves, J. C. & Brill, D. R. (1960), *Phys. Rev.* **120**, 1507.
[12] Penrose, R. (1963), *Phys. Rev. Lett.* **10**, 66.
[13] Newman, E. T. & Penrose, R. (1962), *J. Math. Phys.* **3**, 566.
[14] Kerr, R. (1963), *Phys. Rev. Lett.* **11**, 237.
[15] Carter, B. (1966), *Phys. Rev.* **141**, 1242.
[16] Boyer, R. H. & Lindquist, R. N. (1967), *J. Math. Phys.* **8**, 265.
[17] Carter, B. (1968), *Phys. Rev.* **174**, 1559.
[18] Newman, E. T., Couch, E., Chinnapared, R., Exton, A., Prakash, A. & Torrence, R. (1965), *J. Math. Phys.* **6**, 918.
[19] Israel, W. (1968), *Commun. Math. Phys.* **8**, 245.
[20] Tomimatsu, A. & Sato, H. (1972), *Phys. Rev. Lett.* **29**, 1344.
[21] Ruffini, R. & Wheeler, J. A. (1971), *Phys. Today* **24**, 30.
[22] Penrose, R. (1969), *Riv. Nuovo Cimento, Numero Speziale* **I**, 257. Reprinted: *Gen. Rel. Grav.* **34**, 1141 (2002).
[23] Wald, R. M. (1997), "Gravitational collapse and cosmic censorship" [arXiv:gr-qc/9710068].
[24] Heusler, M. (1996), *Black Hole Uniqueness Theorems* (Cambridge University Press, Cambridge).
[25] Heusler, M. (1998), "Stationary black holes: uniqueness and beyond", *Living Rev. Rel.* **1**, 6 [http://www.livingreviews.org/lrr-1998-6].
[26] Rees, M. (2003), *The Future of Theoretical Physics and Cosmology,* ed. G. W. Gibbons, E. P. S. Shellard & S. J. Rankin (Cambridge University Press, Cambridge), pp. 217–235.
[27] Giddings, S. B. (2002), *Gen. Rel. Grav.* **34**, 1775.
[28] Wald, R. M. (1984), *General Relativity* (The University of Chicago Press, Chicago).
[29] Hawking, S. W. (1972), *Commun. Math. Phys.* **25**, 152.
[30] Hawking, S. W. (1973), *Black Holes (Proceedings of the 1972 Les Houches Summer School)*, ed. C. DeWitt & B. S. DeWitt (Gordon and Breach, New York), pp. 1–55.
[31] Hawking, S. W. & Ellis, G. F. R. (1973), *The Large Scale Structure of Space-Time* (Cambridge University Press, Cambridge).
[32] Penrose, R. (1965), *Proc. Roy. Soc. Lond.* **A 284**, 159.
[33] Penrose, R. (1968), *Battelle Rencontres*, ed. C. M. DeWitt & J. A. Wheeler (Benjamin, New York), pp. 121–235.
[34] Vishveshwara, C. V. (1968), *J. Math. Phys.* **9**, 1319.
[35] Boyer, R. H. (1969), *Proc. Roy. Soc. Lond.* **A 311**, 245.
[36] Carter, B. (1969), *J. Math. Phys.* **10**, 70.
[37] Carter, B. (1973), *Black Holes (Proceedings of the 1972 Les Houches Summer School*), ed. C. DeWitt & B. S. DeWitt (Gordon and Breach, New York), pp. 57–214.
[38] Müller zum Hagen, H. (1970), *Proc. Camb. Phil. Soc.* **68**, 199.
[39] Chruściel, P. T. (2005), *Acta Phys. Polon.* **B 36**, 17.
[40] Rácz, I. & Wald, R. M. (1996), *Class. Quantum Grav.* **13**, 539.
[41] Carter, B. (1979), *General Relativity, an Einstein Centenary Survey,* ed. S. W. Hawking & W. Israel (Cambridge University Press, Cambridge), pp. 294–369.
[42] Carter, B. (1987), *Gravitation in Astrophysics*, ed. B. Carter & J. Hartle (Plenum, New York), pp. 63–122.
[43] Carter, B. (1992), *Black Hole Physics (NATO ASI C364)*, ed. V. de Sabbata & Z. Zhang (Kluwer, Dordrecht), pp. 283–357.

[44] Carter, B. (1971), *Phys. Rev. Lett.* **26**, 331.
[45] Robinson, D. C. (1975), *Phys. Rev. Lett.* **34**, 905.
[46] Müller zum Hagen, H., Robinson, D. C. & Seifert, H. J. (1973), *Gen. Rel. Grav.* **4**, 53.
[47] Robinson, D. C. (1977), *Gen. Rel. Grav.* **8**, 695.
[48] Müller zum Hagen, H., Robinson, D. C. & Seifert, H. J. (1974), *Gen. Rel. Grav.* **5**, 61.
[49] MacCallum, M. A. H. & Van den Bergh, N. (1985), *Galaxies, Axisymmetric Systems and Relativity: Essays Presented to W.B. Bonnor on his 65th birthday,* ed. M. A. H. MacCallum (Cambridge University Press, Cambridge).
[50] Robinson, D. C. (1974), *Phys. Rev.* **10**, 458.
[51] Müller zum Hagen, H. & Seifert, H. J. (1973), *Int. J. Theor. Phys.* **8**, 443.
[52] Gibbons, G. W. (1974), *Commun. Math. Phys.* **35**, 13.
[53] Bondi, H. (1957), *Rev. Mod. Phys.* **29**, 423.
[54] Hartle, J. & Hawking, S. W. (1972), *Commun. Math. Phys.* **26**, 87.
[55] Papapetrou, A. (1945), *Proc. Roy. Irish Acad.* **51**, 191.
[56] Majumdar, S. D. (1947), *Phys. Rev.* **72**, 390.
[57] Gibbons, G. W. (1980), *Proc. Roy. Soc. Lond.* **A 372**, 535.
[58] Chase, J. E. (1970), *Commun. Math. Phys.* **19**, 276.
[59] Hawking, S. W. (1972), *Commun. Math. Phys.* **25**, 167.
[60] Misner, C. W., Thorne, K. S. & Wheeler, J. A. (1973), *Gravitation* (W. H. Freeman, San Francisco).
[61] Chandrasekhar, S. (1983), *The Mathematical Theory of Black Holes* (Oxford University Press, Oxford).
[62] Price, R. H. (1972), *Phys. Rev.* **D 5**, 2419.
[63] Bekenstein, J. D. (1972), *Phys. Rev.* **D 5**, 1239.
[64] Bekenstein, J. D. (1972), *Phys. Rev.* **D 5,** 2403.
[65] Bocharova, N., Bronnikov, K. & Melnikov, V. (1970), *Vestnik. Moskov. Univ. Fizika. Astron.* **25**, 706.
[66] Bekenstein, J. D. (1975), *Ann. Phys. (N.Y.)* **91**, 72.
[67] Robinson, D. C. (1975), *Proc. Camb. Phil. Soc.* **78**, 351.
[68] Nutku, Y. (1975), *J. Math. Phys.* **16**, 1431.
[69] Bunting, G. L. (1983), *Proof of the Uniqueness Conjecture for Black Holes* (Ph.D. thesis, University of New England, Armadale, N.S.W.).
[70] Carter, B. (1985), *Commun. Math. Phys.* **99**, 563.
[71] Mazur, P. O. (1982), *J. Phys.* **A 15**, 3173.
[72] Mazur, P. O. (1984), *Gen. Rel. Grav.* **16**, 211.
[73] Mazur, P. O. (1987), *Proc. 11th International Conference on General Relativity and Gravitation*, ed. M. A. H. MacCallum (Cambridge University Press, Cambridge), pp. 130–157 [arXiv:hep-th/0101012].
[74] Bunting, G. L. & Masood-ul-Alam, A. K. M. (1987), *Gen. Rel. Grav.* **19**, 147.
[75] Schoen, R. & Yau, S.-T. (1979), *Commun. Math. Phys.* **65**, 45.
[76] Witten, E. (1981), *Commun. Math. Phys.* **80**, 381.
[77] Simon, W. (1985), *Gen. Rel. Grav.* **17**, 761.
[78] Rubak, P. (1988), *Class. Quantum Grav.* **5**, L155.
[79] Breitenlohner, P., Maison, D. & Gibbons, G. W. (1988), *Commun. Math. Phys.* **120**, 295.
[80] Gibbons, G. W. (1982), *Nucl. Phys.* **B207**, 337.
[81] Gibbons, G. W. & Maeda, K. (1988), *Nucl. Phys.* **B 298**, 741.
[82] Volkov, M. S. & Gal'tsov, D. V. (1989), *JETP Lett.* **50**, 346.

[83] Bizon, P. (1990), *Phys. Rev. Lett.* **64**, 2844.
[84] Künzle, H. P. & Masood-ul-Alam, A. K. M. (1990), *J. Math. Phys.* **31**, 928.
[85] Tangherlini, F. R. (1963), *Nuovo Cimento* **27**, 636.
[86] Myers, R. C. (1987), *Phys. Rev.* **D 35**, 455.
[87] Myers, R. C. & Perry M. J. (1986), *Ann. Phys. (N.Y.)* **172**, 304.
[88] Masood-ul-Alam, A. K. M. (1992), *Class. Quantum Grav.* **9**, L53.
[89] Masood-ul-Alam, A. K. M. (1993), *Class. Quantum Grav.* **10**, 2649.
[90] Mars, M. & Simon, W. (2003), *Adv. Theor. Math. Phys.* **6**, 279.
[91] Weinstein, G. (1990), *Comm. Pure Appl. Math.* **43**, 903.
[92] Weinstein, G. (1994), *Trans. Amer. Math. Soc.* **343**, 899.
[93] Weinstein, G. (1996), *Commun. Part. Diff. Eq.* **21**, 1389.
[94] Chruściel, P. T. (1994), *Contemp. Math.* **170**, 23.
[95] Chruściel, P. T. (1996), *Helv. Phys. Acta* **69**, 529.
[96] Chruściel, P. T. (2002), *Proceedings of the Tübingen Workshop on the Conformal Structure of Space-times, Springer Lecture Notes in Physics* **604**, ed. H. Friedrich & J. Frauendiener (Springer, Berlin), pp. 61–102.
[97] Beig, R. & Chruściel, P. T. (2005), "Stationary black holes" [arXiv:gr-qc/0502041].
[98] Chruściel, P. T. & Wald, R. M. (1994), *Class. Quantum Grav.* **11**, L147.
[99] Friedman, J. L., Schleich, K. L. & Witt, D. M. (1995), *Phys. Rev. Lett.* **71**, 1486 (1993). Erratum ibid **75**, 1872.
[100] Jacobson, T. & Venkataramani, S. (1995), *Class. Quantum Grav.* **12**, 1055.
[101] Sudarsky, D. & Wald, R. M. (1992), *Phys. Rev.* **D 46**, 1453.
[102] Sudarsky, D. & Wald, R. M. (1993), *Phys. Rev.* **D 47**, 5209.
[103] Chruściel, P. T. (1999), *Class. Quantum Grav.* **16**, 661.
[104] Heusler, M. (1997), *Class. Quantum Grav.* **14**, L129.
[105] Chruściel, P. T. (1999), *Class. Quantum Grav.* **16**, 689.
[106] Brodbeck, O., Heusler, M., Straumann, N. & Volkov, M. (1997), *Phys. Rev. Lett.* **79**, 4310.
[107] Ridgway, S. A. & Weinberg, E. J. (1995), *Phys. Rev.* **D 52**, 3440.
[108] Moss, I. (1994), "Exotic black holes" [arXiv:gr-qc/9404014].
[109] Bekenstein, J. D. (1996), *Proceedings of the Second International Sakharov Conference on Physics*, ed. I. M. Dremin & A. M. Semikhatov (World Scientific, Singapore), pp. 216–219.
[110] Volkov, M. S. & Gal'tsov, D. V. (1999), *Phys. Rep.* **319**, 1.
[111] Hwang, S. (1998), *Geometriae Dedicata* **71**, 5.
[112] Gibbons, G. W., Ida, D. & Shiromizu, T. (2002), *Phys. Rev. Lett.* **89**, 041101.
[113] Gibbons, G. W., Ida, D. & Shiromizu, T. (2002), *Phys. Rev.* **D 66**, 044010.
[114] Gibbons, G. W., Ida, D. & Shiromizu, T. (2003), *Prog. Theor. Phys. Suppl.* **148**, 284.
[115] Rogatko, M. (2002), *Class. Quantum Grav.* **19**, L151.
[116] Rogatko, M. (2003), *Phys. Rev.* **D 67**, 084025.
[117] Rogatko, M. (2004), *Phys. Rev.* **D 70**, 044023.
[118] Kodama, H. (2004), *J. Korean Phys. Soc.* **45**, 568.
[119] Cai, M. & Galloway, G. J. (2001), *Class. Quantum Grav.* **18**, 2707.
[120] Emparan, R. & Reall, H. S. (2002), *Phys. Rev. Lett.* **88**, 101101.
[121] Emparan, R. & Reall, H. S. (2002), *Gen. Rel. Grav.* **34**, 2057.
[122] Kodama, H. (2004), *Prog. Theor. Phys.* **112**, 249.
[123] Morisawa, Y. & Ida, D. (2004), *Phys. Rev.* **D 69**, 124005.
[124] Rogatko, M. (2004), *Phys. Rev.* **D 70**, 084025.
[125] Hollands, S. & Wald, R. M. (2004), *Class. Quantum Grav.* **21**, 5139.
[126] Reall, H. S. (2003), *Phys. Rev.* **D 68**, 024024.

[127] Gibbons, G. W., Lu, H., Page, D. N. & Pope, C. N. (2005), *J. Geom. Phys.* **53**, 49.
[128] Galloway, G. J., Schleich, K., Witt, D. M. & Woolgar, E. (1999), *Phys. Rev.* **D 60**, 104039.
[129] Radu, E. & Winstanley, E. (2004), *Phys. Rev.* **D 70**, 084023.
[130] Winstanley, E. (2005), *Class. Quantum Grav.* **22**, 2233.
[131] Galloway, G. J., Surya, S. & Woolgar, E. (2003), *Class. Quantum Grav.* **20**, 1635.
[132] Anderson, M. T., Chruściel, P. T. & Delay, E. (2002), *JHEP* **10**, 063.

# 6

# Ray-traced visualisations in asymptotically flat spacetimes: the Kerr–Newman black hole

Benjamin R. Lewis and Susan M. Scott

## 6.1 General relativistic ray-tracing

The aim of general relativistic ray-tracing is to produce images that show how distant objects or portions of the sky would be perceived by an observer in a given spacetime.

Ray-tracing is a technique to accurately simulate photographs taken from particular observers' perspectives. The history of each observed light ray is traced back to its respective source, to deduce the colour and brightness that would be perceived at the corresponding position on the photograph.

### *6.1.1 Metrics*

There are many known exact solutions of general relativity theory, that is, many metrics which satisfy the constraints of the Einstein field equation. We are interested in exploring what would be perceived by observers in various different asymptotically flat spacetimes; this is a study of how local gravitational fields can distort the appearance of the background of distant stars.

Previous investigations have taken advantage of analytical results specific to particular spacetimes. For example, spherical symmetry greatly assists the calculation of photon deflection in the Schwarzschild spacetime (Nemiroff, 1993). In order that this project be applicable to new spacetimes, without relying on analytical results, all necessary information should be numerically obtained directly from the metrics.

### *6.1.2 Geodesics on arbitrary spacetimes*

Using any coordinate chart on a spacetime, the geodesic equation can be expressed as a coupled system of four second-order ordinary differential equations (6.1) and

The Kerr Spacetime: Rotating Black Holes in General Relativity. Ed. David L. Wiltshire, Matt Visser and Susan M. Scott. Published by Cambridge University Press. 

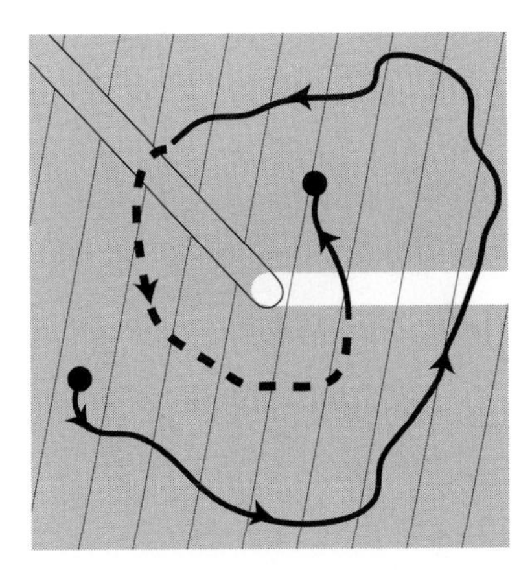

Fig. 6.1. The Schwarzschild spherical polar coordinate chart is a convenient chart with which to work, but it is invalid where the longitudinal coordinate $\phi$ would be zero (or $2\pi$). A second chart (related by some rotation to the original chart) is required to trace orbiting geodesics which do not pass through the main axis of the coordinate systems. The diagram shows a curve initially being traced on the second chart, swapping to the original chart (dashed line), then returning to the second chart.

can be treated as an initial value problem. These equations contain the Christoffel symbol $\Gamma^a_{bc}$ to represent a simple function of the chart metric and of the gradients of that metric. Given the formulae for the metric tensor components in some coordinate chart, the Christoffel symbols can be computed quickly (by using automatic differentiation, see Moylan, Scott, and Searle, 2005) for any point on that chart. Given a tangent vector at some point on the chart (i.e. $\{dx^a/d\lambda\}$ and $\{x^a\}$ at $\lambda_0$) the system of equations can be solved using standard numerical integration algorithms (e.g. the Runge–Kutta or Bulirsch–Stoer methods, see e.g. Press *et al.*, 1992) to determine the path of the corresponding geodesic through that chart,

$$\frac{d^2x^a}{d\lambda^2} + \Gamma^a_{bc}\frac{dx^b}{d\lambda}\frac{dx^c}{d\lambda} = 0. \tag{6.1}$$

For the spacetimes of general relativity theory, a single chart is not normally adequate to cover the entire manifold (see Fig. 6.1). Standard integration algorithms obviously fail where geodesics approach the coordinate chart boundary. By using multiple charts (to cover the manifold) and inter-chart maps (coordinate transformations), software has been developed to automatically switch charts as necessary (Evans, Scott, and Searle, 2002) and thus to integrate any geodesics through the entire spacetime.

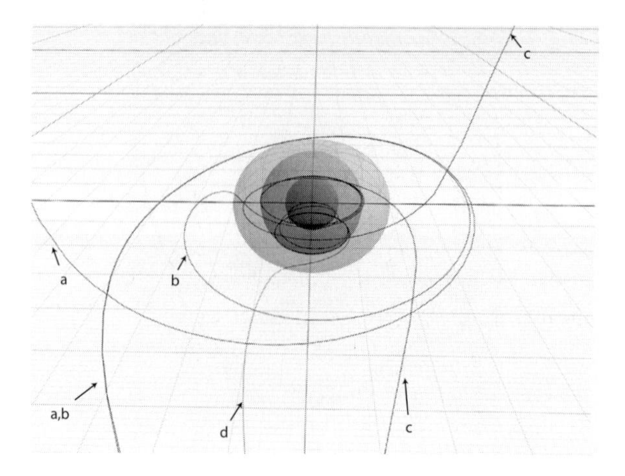


Fig. 6.2. Null geodesics (a,b,c,d) in the Kerr spacetime visualized by GRworkbench, using Boyer–Lindquist coordinates (suppressing the time coordinate), for an angular momentum to mass ratio of 95%. Note surfaces corresponding to the inner event horizon, outer event horizon and the ergoregion static limit.

### 6.1.3 GRworkbench

This project is largely facilitated by GRworkbench, a sophisticated numerical research software package that is currently under development at The Australian National University. GRworkbench mirrors the mathematical concepts of differential geometry within its internal source code organisation (Scott, Evans, and Searle, 2001), simplifying many processes. For example, GRworkbench represents a point on a manifold (or an event in spacetime) by a type of object that (by automatically accessing the available inter-chart maps) can return its own coordinates on any chart that covers the point, so that elsewhere it is unnecessary to maintain track of which charts have been used to define different objects.

In addition to its numerical algorithms, GRworkbench is already a powerful visualisation tool (see Fig. 6.2 for example). Four-dimensional spacetime charts can be mapped, using arbitrary "hypershadows" (Searle, 1999), into three-dimensional space. Graphics hardware is then used to project this onto a two-dimensional display while still allowing the user to rotate and explore the three-dimensional model.

## 6.2 Sources

The gravitational warping of spacetime only redistributes light. Assuming that no local matter distribution is present, which otherwise may scatter or produce light, the flux and spectrum of a particular light ray is determined by tracing the corresponding geodesic far enough back so as to ascertain from where among the distant background objects that light originated.

### 6.2.1 Celestial dome

Early astronomers imagined that the stars occupied fixed positions on an immense sphere (or dome) around the Earth (beyond the Moon and planets). Every star may then be specified completely by its declination ($\Theta$) and right ascension ($\Phi$) coordinates on the two-dimensional sky or celestial surface (analogous to latitude and longitude coordinates on Earth).

This model is now known to be incorrect because of experiments detecting parallax; the apparent ($\Theta$, $\Phi$) positions of stars at varying radii do shift slightly when observed from different points in the Solar System. The closest stars beyond the Solar System are *parsecs* away; observations an astronomical unit (i.e. the average orbital radius of the Earth) apart give various stellar parallax angles of up to an arcsecond. By comparison, however, a typical photograph (e.g. 30 degree field of view and megapixel resolution) only has arcminute precision ($30^\circ \div \sqrt{10^6}$).

In any asymptotically flat spacetime, we can choose a celestial dome surface sufficiently far from the origin of coordinates so as to be located in a region which is arbitrarily close to flat spacetime. Thus, we can (consistently to some controlled precision) define a single continuous inertial frame over this enclosing surface and the outside region, and everything is well defined.

For the asymptotically flat spacetimes that we are investigating, assuming the distance to the visible background stars is much larger than the length scales on which spacetime is strongly warped or traversed by observers, the celestial sphere model is an excellent approximation. Null geodesics can simply be traced from the observer with increasing affine parameter until they reach the radius of the celestial sphere, at which point the position ($\Theta$, $\Phi$) is used to determine from which star or constellation the photons originated.

Ideally the celestial sphere would have infinite radius but, since GRworkbench can only integrate the geodesic equation to finite affine parameter values, pragmatically it is necessary to select a finite radius. The chosen radius must be large enough to accurately predict the asymptotic directions of the geodesics (that is, large enough so that all observed geodesics have negligible non-radial components at that distance), but an excessive radius will decrease performance and may also encourage numerical errors to accumulate (e.g. by successive rounding).

### 6.2.2 Minkowski exterior

The basic celestial sphere method only considers the *position* of the geodesic after integration. The next order approximation (see Fig. 6.3) is to also consider the final *direction* tangent vector of the geodesic (rather than assuming it is purely radial), thereby overcoming potential parallax errors. The calculation of sky

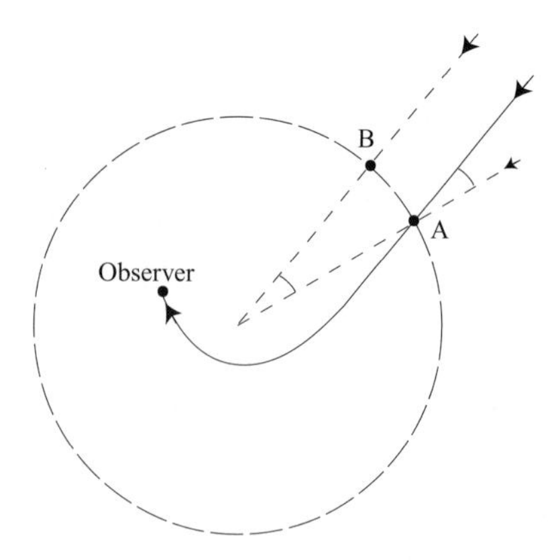


Fig. 6.3. The origin of a photon can be estimated by (A) the position where its geodesic intersects a distant surface, however, a better estimate (B) also accounts for the angle of intersection of the geodesic with the distant surface.

coordinates, based on the direction of the tangent vector of the geodesic beyond a selected "celestial" radius, is equivalent to assuming that the spacetime is completely flat (namely Minkowski space) beyond the selected radius. This is a reasonable assumption for a "sufficiently large" radius in an asymptotically flat spacetime.

In a flat spacetime it is straightforward to extrapolate the direction of a normalised and purely space-like tangent vector $\hat{\mathbf{v}}$ to produce sky coordinates $(\Theta, \Phi)$. Declination is the arcsine of the component of $\hat{\mathbf{v}}$ in the normalised direction $(\hat{\mathbf{z}})$ to the polar star, while the normalised *equatorial* (orthogonal to $\hat{\mathbf{z}}$) component $\hat{\mathbf{v}}_{\perp\hat{\mathbf{z}}}$ determines the right ascension.

Using orthonormal spherical polar coordinates $(t, r, \theta, \phi)$ and letting $\pi_\theta$ and $\pi_\phi$ represent the $(\theta, \phi)$ coordinates of the *position* on the celestial sphere associated with a tangent vector, the sky coordinates $(\Theta, \Phi)$ to which the vector points are given as follows:

$$\sin\,\Theta = \hat{\mathbf{v}} \cdot \hat{\mathbf{z}} \qquad \tan\,(\Phi - \pi_\phi) = \frac{\hat{\mathbf{v}}_{\perp\hat{\mathbf{z}}} \cdot \hat{\phi}}{\hat{\mathbf{v}}_{\perp\hat{\mathbf{z}}} \cdot \hat{\mathbf{r}}_{\perp\hat{\mathbf{z}}}},$$

where $\hat{\mathbf{r}}_{\perp\hat{\mathbf{z}}}$ and $\hat{\phi}$ are both normalised.

Integration of the geodesic equation to determine position coordinates will also automatically provide tangent vectors, and the extra computation of sky coordinates is negligible compared with the processing cost of the integration, so this method does not decrease performance. Indeed the selected radius (to which geodesics are integrated) need only be large enough that spacetime is approximately flat outside, not so large as to approximate observed geodesics as radial, so this method can produce the necessary accuracy from a shorter integration.

### *6.2.3 Colour*

The energy of a particle, measured in the frame of a particular observer, is the time-like component of the momentum of the particle. Tangent vectors $\{dx^a/d\lambda\}$ along a geodesic path remain proportional to the corresponding momentum (Rainich, 1928). Consequently the fractional frequency shifts of *all* photons traversing a given geodesic path is determined by calculating the projection of the geodesic tangent vector onto the "stellar" time axis (which is defined in the region sufficiently far from the origin), having chosen the geodesic parameterisation (see Section 6.3.2) such as to have unit time-like component in the observer's frame.

The important colour shift in the Schwarzschild spacetime is the blue-shift corresponding to dilation of time for observers near a gravitational mass. This visual effect is uniform in all directions (assuming the mass and observer are stationary with respect to the distant stars). If another observer at the same point is moving at velocity with respect to the background stars, this second perspective will differ simply according to the special relativistic transformation of the *plenoptic function* (e.g. giving rise to colour shifts in the direction of motion).

The interaction between the mass-energy of each photon and the curvature of spacetime can, in principle (however negligibly), alter the frequency of a photon (e.g. by global momentum conservation arguments). This is not calculated using these techniques, which consider only geodesics of the background metric.

### *6.2.4 Brightness*

Surface brightness is a measure of the radiation flux, from an extended source, observed per solid angle. In Newtonian optics all observers agree on the surface brightness of a given object. This follows from Liouville's theorem (conservation of volume in phase space), that is, at points where the total flux is comparatively increased (e.g. closer to the source or behind a converging lens) the source always subtends a proportionally larger solid angle. Ray-tracing is simplified by the invariance of surface brightness because it implies that ray-traced images can correctly be composed of trivially rescaled (i.e. resampled) portions of the "texture" images.

In general relativity the surface brightness is not completely invariant; sources that are rapidly receding, or located lower in a gravitational well, are observed to be darker (as well as redder). Surface brightness varies only in proportion to $(1+z)^{-4}$, where $1+z$ is the observed fractional frequency shift (Perlick, 2004). Therefore, where brightness readjustment is required, it is still (following colour shift) uniform for all colours. In this sense "stationary" observers in the static Schwarzschild spacetime do not see directional brightness variations.

## 6.3 Initial values

The geodesics being studied must be null and have well-understood space-like projections where they reach the observer.

### 6.3.1 Camera vantage

The appearance of a photograph depends both on the camera itself and on the vantage from which the image is taken. Important camera variables are the field of view (e.g. fisheye versus zoom lens) and the shape of the film (i.e. height to width ratio).

Classically (in the limit of small aperture and fast shutter) the camera vantage is completely described by the position of the photographer (and the time) when the image was taken, the direction the camera was pointing and which way up the camera was held. Relativistically, observers with different velocities would still obtain different images.

The observer's position, in space and time, is implicitly contained within all tangent vectors from the *event* at which the image is taken. The camera's vantage can be specified by an ordered set of four orthogonal tangent vectors. These are the time-like tangent vector to the camera's world line, and space-like directions indicating where the camera is facing, which way is up, and which way is left.

Classically, and in many possible spacetimes, the tangent vector corresponding to left (versus right) could be calculated consistently from the first three vectors (by considering the chirality of the coordinate charts). In general, however, the manifold may not be *orientable*. For example, there is no way to define left and right continuously over the surface of a Möbius strip or Klein bottle.

### 6.3.2 Tangent vectors

The appearance of each spot in the photograph depends on the light ray that reaches the corresponding spot on the simulated camera film.

Each spot can be referenced using Cartesian coordinates (some $x$ and $y$). The light travelling from the (pinhole) aperture, to a spot on the film, traverses a straight line in the local Lorentz frame of the camera (being small compared to the scale over which spacetime curvature varies). Using an orthonormal basis of tangent vectors ($\mathbf{dt}$, $\mathbf{dz}$, $\mathbf{dy}$, $\mathbf{dx}$ corresponding to those in Section 6.3.1) the direction from aperture to film is $\mathbf{u} \equiv x\,\mathbf{dx} + y\,\mathbf{dy} - \mathbf{dz}$.

The light ray observed (by the camera) to be moving in space-like direction $\mathbf{u}$ with unit energy (i.e. unit time-like component) must have tangent vector $\mathbf{v} \equiv \mathbf{u}\sqrt{||\mathbf{u}^2||}^{-1} + \mathbf{dt}$. Note $||\mathbf{v}^2|| = 0$.

By integrating back the geodesic with tangent vector **v**, the origin of any photons striking the film at the given spot $(x, y)$ is determined, and appropriate colouring of the corresponding part of the image can be deduced.

### 6.3.3 Discretisation

For each picture it is only practical to trace a small number of rays compared with the number of photons a camera would capture, so the optimal choice of rays must be considered. Since the digital image is expressed as a grid of pixel units (with constant brightness and colour over the area of each pixel), the most obvious (naïve) method is to trace one ray for each pixel. Each ray is incident on the centre of a corresponding pixel and the entire pixel is coloured according to whichever exact point in the sky from which that ray is determined to originate.

This method (having limited scope for error) produces images that are physically acceptable but appear grainy (i.e. *aliased*). Whereas a real camera colours each pixel by averaging over many photons, this method sometimes produces a lighter or darker pixel according to whether the pixel's central ray happens to fall exactly centred on a small star; this defect is most pronounced in directions where large constellations are compressed into a small solid angle. There can also be Moiré pattern artefacts; the regular sampling frequency may interfere with the background sky image.

The accuracy can be improved by tracing a larger number of rays for every pixel (i.e. *over-sampling*). Each increase incurs a drastic (eventually prohibitive) computational cost, however, and produces relatively small (and diminishing) improvement to the final image.

Different methods can produce tangible improvements without any significant processing cost. Instead of colouring according to the exact centres of the pixels, a slightly smoother (artistically *anti-aliased*) image can be produced by colouring each pixel according to the average colour of the four corners. This offset does not affect the number of rays that need tracing but reduces the graininess problem. Moiré patterning can be disrupted by individually adding a random small variation to the direction in which the ray is traced for each pixel.

There is a high degree of correlation between geodesics with similar initial directions. Often all corners of a pixel will correspond to nearby points on the background image; each square pixel will map to some area of the background image. This area of the background can be approximated by the quadrilateral with corners at these four known points, so the average colour of this quadrilateral provides a good estimate of what colour the pixel should take. Although not increasing the ray-tracing computations, such increased image accuracy requires much greater post-processing complexity than the previous methods, particularly to

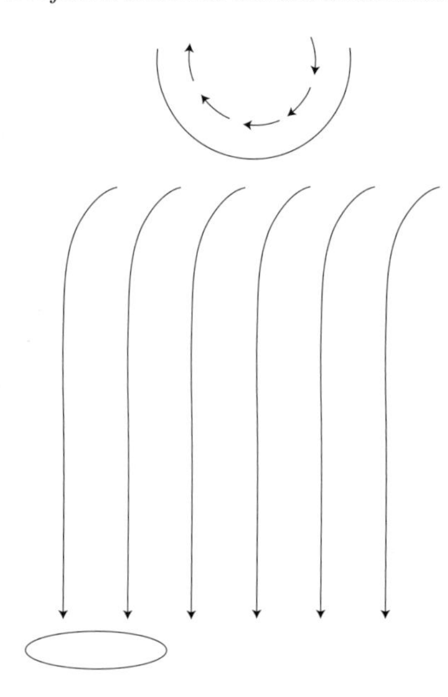

Fig. 6.4. The first-order effect of frame dragging is to displace the shadow subtended by the event horizon.

maintain improvement (and avoid visible artefacts) in directions corresponding to edges of the internal representation (e.g. a Mercator projection image) of the entire background sky. The method fails in directions of strong deflection gradients; for example, amid the series of Einstein rings against the edge of a black hole, pixels may map to large areas of sky which the quadrilateral represents poorly. This mode of failure, however, nonetheless produces better images than the naïve method since such pixels are best approximated by the average colour of any large area of sky.

The directional correlation can also be employed to reduce the number of geodesic integrations. For example, images of Mandelbrot fractals are frequently drawn using *solid guessing* algorithms. Like tracing null geodesics around black holes, these fractals are drawn by tracing orbits (solving an initial value problem, $Z_{n+1} = Z_n^2 + C$ rather than Eq. (6.1)) to determine the position at which (if ever) the orbit escapes beyond some radius. There is a similar correlation between orbits corresponding to nearby points on the image. The solid guessing algorithm recursively divides the image into smaller squares. If the points corresponding to all four corners of some (potentially large) square return the same result (e.g. if none of the four orbits escape to the celestial radius) then it is *assumed* that all points within the square would also give the same result, so every pixel within that entire solid area is coloured in the same manner. Alternatively, if not all four points return the same result, then the square is subdivided into smaller squares again (until squares the size of pixels are produced). Solid guessing algorithms greatly improve performance (e.g. of those geodesics which lead to the event horizon, the majority would not need to be traced). The process can be further sped up in combination with a sophisticated version of the quadrilateral method above. For example, if large sections of the image correspond to light rays that are insignificantly deflected by the spacetime curvature then it should not be necessary to explicitly integrate every individual geodesic.

The obvious disadvantage of solid guessing is that the algorithms do sometimes make incorrect guesses, producing completely unacceptable (and possibly misleading) images. This potential is highly undesirable if attempting to study perspectives in some previously uninvestigated spacetime. This project has employed only the *least* sophisticated method, making no assumptions about the spacetime.

### 6.4 Kerr–Newman geometry

Kerr–Newman geometry is the general solution of the Einstein field equation for a rotating, charged mass in vacuum; in the limit this geometry corresponds to a black hole (with mass $M$, angular momentum $Ma$ and charge $Q$). To an outside observer it takes infinite time for light to reach the event horizon, and hence no radiation is observed from a black hole.

The spacetime metric is

$$\begin{aligned} \mathrm{d}s^2 = &-\left(\frac{\Delta}{\rho^2}\right)\left(\mathrm{d}t - a\ \sin^2\theta\ \mathrm{d}\phi\right)^2 + \left(\frac{\rho^2}{\Delta}\right)\mathrm{d}r^2 \\ &+\rho^2\,\mathrm{d}\theta^2 + \frac{\sin^2\theta}{\rho^2}\left((r^2+a^2)\,\mathrm{d}\phi - a\ \mathrm{d}t\right)^2, \end{aligned} \tag{6.2}$$

where

$$\begin{aligned} \Delta &\equiv (r-M)^2 - \xi^2, \\ \xi^2 &\equiv M^2 - a^2 - Q^2, \\ \rho^2 &\equiv r^2 + a^2\ \cos^2\theta, \end{aligned} \tag{6.3}$$

and

$$\begin{aligned} &t \in \mathbb{R}; \qquad \theta \in (0,\pi); \qquad \phi \in (0, 2\pi); \\ &0 < r < M-\xi, \quad M-\xi < r < M+\xi, \quad \text{or } r > M+\xi. \end{aligned} \tag{6.4}$$

The Kerr–Newman geometry is represented here using Boyer–Lindquist charts. This spherical polar form simplifies the expression of the metric and of the mapping between chart coordinates and the celestial sphere.

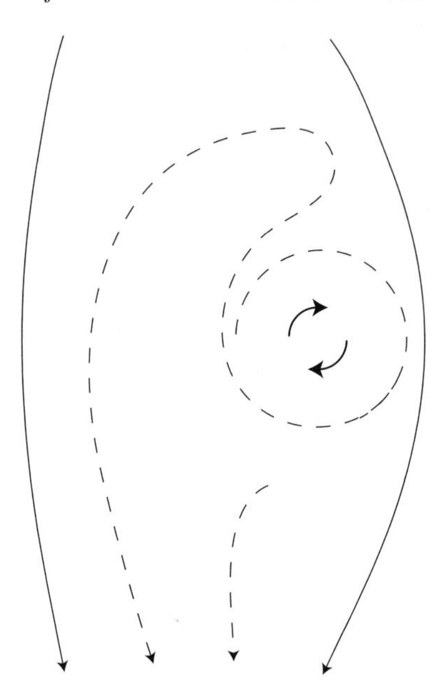

Fig. 6.5. Geodesics passing a Kerr black hole, in the direction of rotation, are effectively boosted by inertial frame dragging and so can often escape despite passing comparatively close to the black hole. Geodesics passing on the other side are opposed by the rotation and tend to be captured even for larger impact parameters; photons observed from this 'downstream side' have not passed comparatively close to the black hole.

The single Boyer–Lindquist chart has a significant border, or gap, where the azimuthal coordinate approaches $2\pi$ or zero. To permit orbiting geodesics to be traced (see Fig. 6.1), a second Boyer–Lindquist chart, related to the first by $180^\circ$ axial rotation, is added to the atlas. Note that this still neglects the polar axis itself (where the azimuthal coordinate is not well defined); these irregularities would be avoided by adding a chart based on Cartesian (rather than spherical polar) coordinates, such as the Kerr–Schild chart.

Boyer–Lindquist charts cannot extend across the event horizons. Adding the ingoing Kerr charts would allow determination of the view from the black hole interior but, for the purposes of simulating the view from outside a black hole, it is more convenient to omit interior charts from the atlas. Geodesic integration fails upon reaching the edge of the atlas, providing a simple mechanism to detect the event horizon's shadow.

The ray-tracer may simply attempt to integrate every geodesic to a prespecified large affine parameter. If photons are observed arriving along a geodesic (i.e. the geodesic does not originate from the event horizon) then the integration will succeed and the integration endpoint will usually be far enough from the black hole so as to provide an accurate estimate of from whence the photons originated.

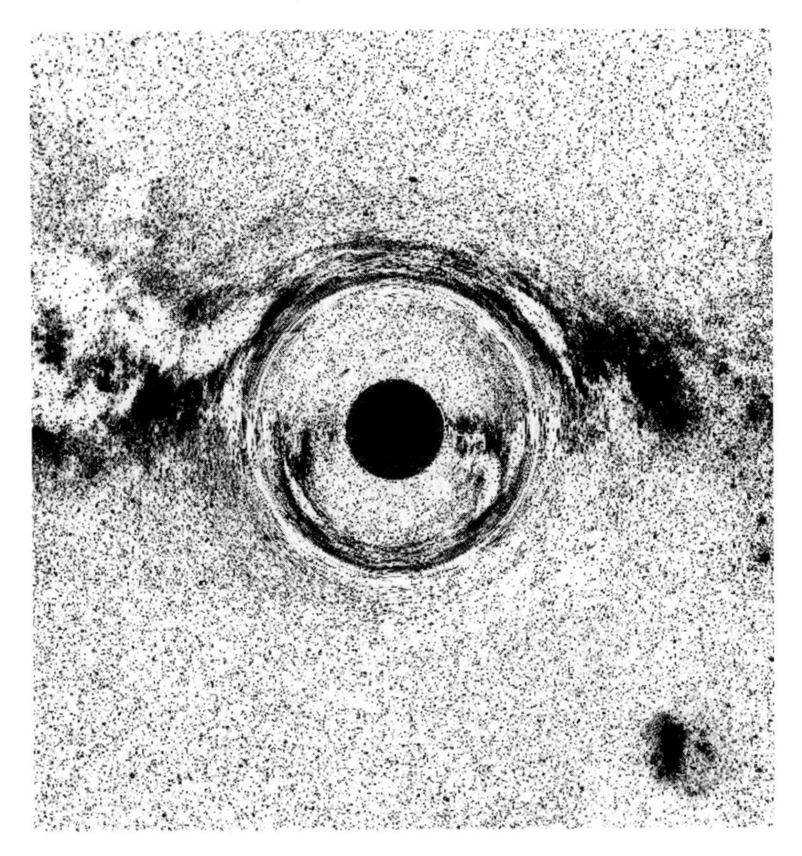

Fig. 6.6. Ray-traced image of the Schwarzschild black hole (with Milky Way background). Note the first Einstein ring and the shadow of the event horizon. This image is from the perspective of an observer at rest with respect to the black hole and background stars. Refer to (Lewis, 2005) for the original colour images.

A small proportion of geodesics neither fall into the black hole nor escape within any computationally reasonable (or even finite) affine parameter; light may orbit a black hole indefinitely. The origin of any such borderline geodesics cannot be accurately estimated but, by pre-specifying a large enough integration distance, these geodesics are few and are confined to an extremely defocussed area of the observer's sky such that accuracy is unnecessary.

Under ray-traced visualisation the Kerr–Newman geometry (see Fig. 6.7) retains the same basic visual features as the simpler Schwarzschild geometry (see Fig. 6.6), i.e. non-zero angular momentum or electric charge does not drastically alter the optical appearance of a black hole and, indeed, the magnitude of angular momentum and charge is bounded relative to the mass of the black hole due to cosmic censorship. Far from the black hole most of the sky appears the same as for flat space. In the direction of the singularity a perfectly black disc shaped shadow is observed. Rather than obstructing the view of part of the sky, constellations

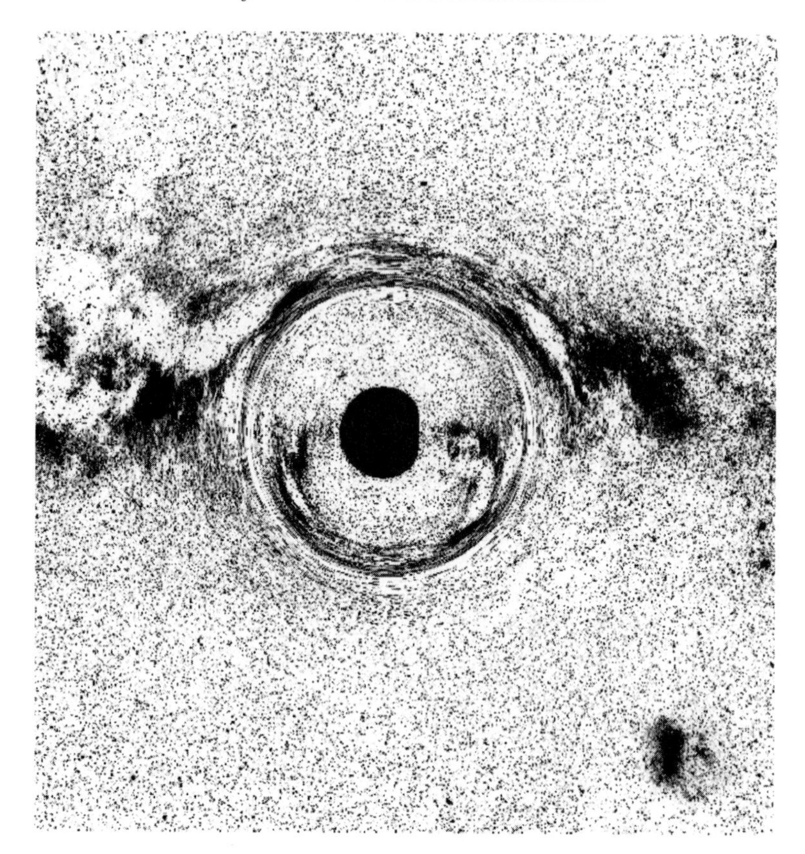

Fig. 6.7. Ray-traced image of the Kerr–Newman black hole (with Milky Way background) which is near critical with electric charge and angular momentum to mass ratios of 0.707. Note that the shadow of the event horizon is shifted, smaller and is asymmetric, whereas the Einstein ring is not significantly perturbed. The direction of angular momentum is such that an infalling test mass would be swept leftwards. Compare with Fig. 6.6.

expected in the shadow's general direction will appear to have been pushed outwards so as to remain visible. If any bright star was expected in precisely the same direction as the singularity, it will instead appear stretched into the shape of a bright halo around the shadow (the first, or outermost, "Einstein ring"). With sufficient resolution one would observe an infinite sequence of concentric Einstein rings, each one exponentially closer to the shadow, separated by compressed images of the entire sky.

A spinning black hole causes inertial frame dragging (in a similar manner to how a car speedometer needle is dragged around by a rapidly spinning magnet in close proximity). For observers away from the axis of rotational symmetry, to a first approximation, increasing angular momentum displaces the image of the horizon's shadow to one side (see Fig. 6.4). Angular momentum also breaks the observed symmetry (see Fig. 6.5); photons from one side are boosted by frame dragging and so can pass closer to the singularity, whereas photons from the other side are opposed by the rotation. Electrically charging the black hole has the visual effect of shrinking (and accentuating any asymmetry of) the shadow of the event horizon.

## 6.5 Conclusions

This ray-tracer was developed to produce visualisations from vantage points in any asymptotically flat spacetime. Various general relativistic capabilities have been incorporated in other ray-tracing software (see e.g. Weiskopf *et al.*, 2006), however, our approach is based on directly employing convenient differential geometry abstractions (using GRworkbench).

The ray-tracer has been demonstrated on specific Kerr–Newman geometries, and is also intended to assist the investigation of other solutions of the Einstein field equation. Current work is progressing to make GRworkbench (including the ray-tracer) a user-friendly and practical tool for future applications in research and education.

## References


Evans, B. J. K., Scott, S. M. & Searle, A. C. (2002), "Smart geodesic tracing in GRworkbench", *Gen. Rel. Grav.* **34**, 1675.

Lewis, B. R. (2005), "Visualising general relativity" (Honours Thesis, Australian National University, Canberra) [http://hdl.handle.net/1885/43269].

Moylan, A. J., Scott, S. M. & Searle, A. C. (2005), "Functional programming framework for GRworkbench", *Gen. Rel. Grav.* **37**, 1517.

Nemiroff, R. (1993), "Visual distortions near a neutron star and black hole", *Am. J. Phys.* **61**, 619.

Perlick, V. (2004), "Gravitational lensing from a spacetime perspective", *Liv. Rev. Relat.* **7**, 9 [http://www.livingreviews.org/lrr-2004-9].

Press, W. H., Teukolsky, S. A., Vetterling, W. T. & Flannery, B. P. (1992), *Numerical recipes in C* (Cambridge University Press, Cambridge).

Rainich, G. Y. (1928), "Corpuscular theory of light and gravitational shift", *Phys. Rev.* **31**, 448.

Scott, S. M., Evans, B. J. K. & Searle, A. C. (2001), "GRworkbench: a computational system based on differential geometry", in *Ninth Marcel Grossmann Conference*, ed. V. G. Gurzadyan, R. T. Jantzen & R. Ruffini (World Scientific, Singapore), p. 458.

Searle, A. C. (1999), "GRworkbench" (Honours Thesis, Australian National University, Canberra).

Weiskopf, D., Borchers, M., Ertl, T. *et al.* (2006), "Explanatory and illustrative visualization of special and general relativity", *IEEE Trans. Visual. Comp. Graph.* **12**, 522.

# Part II

## Astrophysics: The ongoing observational revolution in our understanding of rotating black holes

# 7

# The ergosphere and dyadosphere of the Kerr black hole

Remo Ruffini

## 7.1 Introduction

On most days I seem to use Roy Kerr's name ten times or more, primarily because the empty Einstein space constructed by him in Kerr (1963) is so important to relativistic field theories. This should not come as a surprise. If one looks on Google for "Kerr AND metric OR space" one will find at least $1.33 \times 10^6$ citations!

Sometimes this arises during the experimental verification of Einstein's theory using advanced space technologies. For instance, I recently had great pleasure seeing the launch of NASA's *Gravity Probe B* Mission. This is directly related to the Kerr solution (see e.g. Fairbank *et al.*, 1988).[1] This work also appears in explanations of the electrodynamical aspects of accretion disks for binary X-ray sources, extragalactic jets from active galactic nuclei and microquasars in our galaxy, (see e.g. Giacconi, 2002; Giacconi & Ruffini, 1978; Punsly, 2001).

For us Kerr spacetime was visualized by numerically integrating the trajectories of five test particles leading to a splendid new image. This became the logo of our Centers for Relativistic Astrophysics, ICRA and ICRANet (see Fig. 7.1 and Johnston & Ruffini, 1974), and a beautiful sculpted version is given triennially to the recipients of the Marcel Grossmann Awards. (See http://www.icra.it/MG/awards/.)

I will try to give a few examples of each of these applications in my talk. They can all be taken as examples of the following principle: *Humans have always been too conservative in their imagination. They have never been able to reach by imagination alone the realities discovered by logic and scientific endeavor based*

[1] For the relation of gyroscope motion to the Kerr solution, see e.g. Ohanian & Ruffini 1994; Ruffini & Sigismondi 2003, and references therein.

The Kerr Spacetime: Rotating Black Holes in General Relativity. Ed. David L. Wiltshire, Matt Visser and Susan M. Scott. Published by Cambridge University Press. 

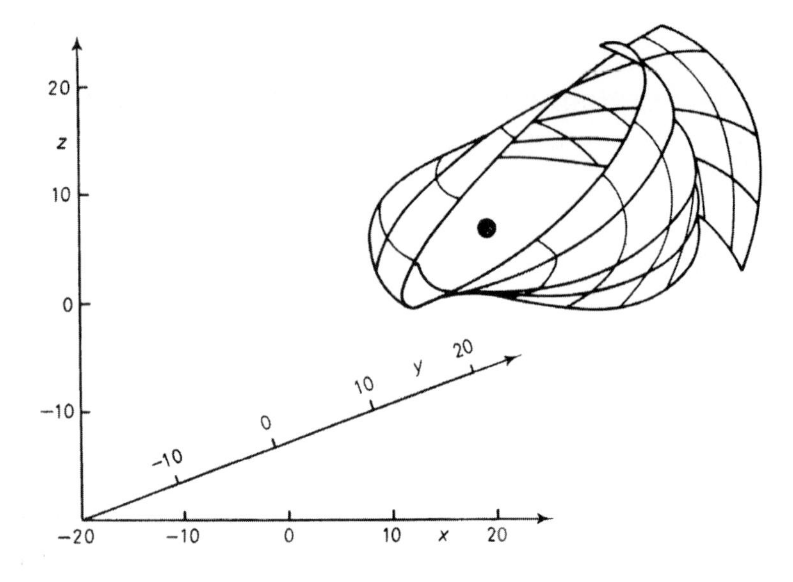


Fig. 7.1. Motion of uncharged cloud of particles corotating about an extreme spinning charged black hole. The orbits are stable. The vertical lines indicate isochronous points (as seen from infinity). Details in Johnston & Ruffini (1974).

*on the necessary mathematical formalism.*[2] This may be the reason why new ideas are not always easily accepted by the scientific establishment. There may be much strife before the previously inconceivable becomes the universally accepted.

S. Chandrasekhar and I had many discussions where we compared the difficulties each had getting his ideas on white dwarfs and mine on black holes accepted. The observational properties of white dwarfs were well known before Chandrasekhar (1935) explained them by applying degenerate Fermi statistics to stars (see Chandrasekhar, 1939). This theory was forcefully rejected in Eddington (1935), a paper that was published in the same journal and even immediately preceded that of Chandrasekhar!

The situation has been even more difficult for black holes. We had to struggle against the preconceived notions of two quite independent groups, theoretical physicists and observational astronomers. Firstly, there was the resistance of some physicists against accepting certain properties of these new objects, e.g. their horizons, their mass–energy formulae and the amazing power being generated by

[2] Examples of two theoretical predictions that could not have been imagined are those by Maxwell of electromagnetic waves, (see e.g. Maxwell (1986)) and Dirac of antimatter simply from their celebrated equations. Dirac himself remarked in one of his talks that his equation was more intelligent than its author (see, e.g. http://physics.indiana.edu/~sg/p622/lecture1quotes.html). Similarly, the concept of a critical mass for gravitational collapse, constructed by Oppenheimer from Einstein equations, was a complete surprise to John Archibald Wheeler (see "Discussion of Wheeler's report" in Institut International de Physique Solvay, 1958, pp. 147–148) and even Einstein himself (Einstein, 1939). Both were dubious at first but changed their minds later.

them. Secondly, astronomers were unable to explain the huge amounts of energy being released by the newly observed binary X-ray sources, but they still did not recognize the need for radically new ideas. Nuclear forces had been used for decades to explain the energy from stars but were clearly unable to explain the X-ray emission from these binaries since the energy released was both too sudden and too great (see below). The radically new idea of the conversion of gravitational energy by accretion processes around a gravitationally collapsed star was needed for this (see e.g. Giacconi & Ruffini, 1978, and references therein).

The Kerr metric has been crucial in this long debate. Although Roy's discovery was initially ignored by astrophysicists, the almost simultaneous discovery of quasars triggered the interest of a small minority in gravitational collapse (see e.g. Robinson, Schild & Schücking, 1965). This solution has become an essential mathematical tool, creating the theoretical framework needed to interpret the flood of observations from the newly built X-ray and $\gamma$-ray detectors in space, as well as from the corresponding optical and radio ones on the Earth's surface. The first major triumph for the new theories was the identification of Cygnus–X1 as a black hole inside our galaxy. (See Giacconi & Ruffini (1978) and Giacconi (2002).)

Even bigger challenges have confronted relativistic astrophysics in the last thirty years. These have included how to test the black hole mass–energy formula which predicts that up to 29% (50%) of its energy can be rotational (electromagnetic) and how to show that this energy, in principle extractable, could fuel the most ultra-relativistic and energetic phenomena ever observed in nature, Gamma-Ray Bursts (GRBs). The difficulties in bridging the communication gap between these new concepts and the traditional ones of the physical, astronomical and astrophysical communities have been simply enormous, much bigger than the corresponding ones for white dwarfs and binary X-ray sources. We will mention three of these major challenges.

The first has been for astronomers. They had successfully understood astronomical systems by observing their evolution over periods of many years. It was difficult for them to accept that significant observations in GRBs can occur over periods as short as a fraction of a millisecond because of the rate at which the source evolved. An almost instantaneous spectrum changing on such a short time scale is no surprise to a physicist but it was to astronomers. They tend to associate a fixed and constant spectral characteristic to any given source, not a rapidly changing one.

The second challenge has been for physicists. They needed to understand the properties of the extreme gravitational field around black holes and also some of the latest developments of relativistic quantum field theory. The concept of the dyadosphere (see below) has not been accepted easily. It needs simultaneously detailed knowledge of Kerr–Newman geometry as well as of quantum field theory

Fig. 7.2. Albert Einstein, Hideki Yukawa and John Archibald Wheeler, with the dedication of John Wheeler on April, 5th, 1968.

as developed in recent decades through the study of heavy ion collisions and high powered laser sources.

The last challenge has been for both groups. These systems move ultra-relativistically with a Lorentz $\gamma$-factor starting as high as 500 and then dropping all the way down to 1. The observed arrival times of the emitted photons are not what matters, just the corresponding rates of emission at the source, and to calculate these the entire past world-line of the source and its gravitational potential must be known (Ruffini *et al.*, 2001a)!

## 7.2 The maximum binding energy in Kerr geometry

After a period spent with Pasqual Jordan in Hamburg, I was invited to Princeton as a postdoctoral fellow by John Archibald Wheeler,[3] starting September 1st, 1967. Those were very active days for the astrophysical community. Pulsars had just been discovered by Jocelyn Bell and Tony Hewish (1967), and many theorists were actively trying to explain them as rotating neutron stars (see Gold, 1968, 1969; Pacini, 1968; Finzi & Wolf, 1968). These had already been predicted by George Gamow using Newtonian physics (Gamow, 1938) and by Robert Julius Oppenheimer and

[3] Or Johnny, as the members of our group called him (see Fig. 7.2).

students using general relativity (Oppenheimer & Serber, 1938; Oppenheimer & Volkoff, 1939; Oppenheimer & Snyder, 1939). The crucial evidence confirming that pulsars were neutron stars came when their energetics was understood (Finzi & Wolf, 1968). The following relation was established from the observed pulsar period $P$ and its positive first derivative $dP/dt$:

$$\left(\frac{dE}{dt}\right)_{obs} \simeq 4\pi^2 \frac{I_{NS}}{P^3}\frac{dP}{dt}, \tag{7.1}$$

where $(dE/dt)_{obs}$ is the observed pulsar bolometric luminosity and $I_{NS}$ is its moment of inertia derived from the neutron star theory. This has to be related to the observed pulsar period. This equation not only identifies the role of neutron stars in explaining the nature of pulsars, but clearly indicates that the neutron star's rotational energy is the pulsar energy source. This success exemplifies how to understand any astrophysical system we must first explain its energetics. We will return to this point later.

Wheeler decided to change the focus of his group from the physics of neutron stars, which had already been the subject of a celebrated book by him and coworkers (Harrison *et al.*, 1965), to the total gravitational collapse of a star with a mass larger than the critical neutron star mass and at the endpoint of its thermonuclear evolution. This far more extreme general relativistic system had already been conceived by Robert Oppenheimer and his students. It was frozen in both time and temperature due to its gravitational redshift becoming infinite at the corresponding Schwarzschild horizon. We did not like the name "frozen star" given to it by our Soviet colleagues, and instead decided to follow Wheeler's suggestion and call it a "black hole" (Ruffini & Wheeler, 1971a). This emphasized its two most extreme characteristics, the infinite gravitational redshift at its event horizon and its inevitable gravitational collapse.

At that time Princeton had a very pleasant scientific ambiance because of the fortunate interaction between its bright students and the outstanding scientists at both the Institute for Advanced Study and the University. These included Kurt Gödel, Eugene Wigner, Freeman Dyson, Martin Schwarzschild, David Wilkinson and Tullio Regge. In the following years, as a Member of the Institute for Advanced Study and an instructor and later assistant professor at Princeton University, I enjoyed many discussions with them and learned from their experiences and their understanding of physics.

Johnny Wheeler and Tullio Regge had developed a powerful mathematical physics technique (Regge & Wheeler, 1957) and we used this formalism, as later completed by the seminal work of Frank Zerilli (Zerilli, 1970, 1974), to describe the physical processes in a Schwarzschild solution. Our group studied a variety

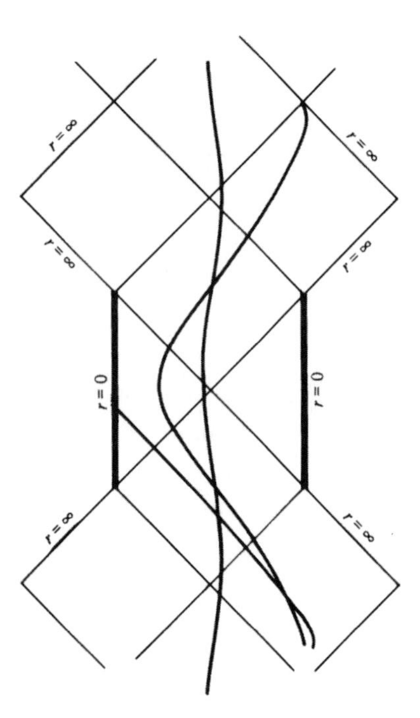


...Although the symmetries provide only three constants of the motion, a fourth one turns out to be obtainable from the unexpected separability of the Hamilton–Jacobi equations... The equations of charged particle orbits can be integrated completely in terms of explicit quadrature.

Carter, Phys. Rev., 174, 1559 (1968)

Fig. 7.3. Brandon Carter and the introduction of his famous fourth constant of the motion.

of problems on the emission of gravitational radiation in these highly relativistic regimes, using the Regge–Wheeler–Zerilli tensorial harmonics techniques.

My first encounter with the Kerr solution occurred in Princeton when Brandon Carter visited our group. He had just published a most remarkable paper (Carter, 1968, see Fig. 7.3) on the separability of the Hamilton–Jacobi equations for the trajectories of charged test particles in Kerr–Newman spaces. This classic work is a mandatory text for all students in my course on theoretical physics at "La Sapienza", the University of Rome. In this paper Brandon sets the mathematical foundations for the new physics in the spacetime described by the Kerr solution and its electromagnetic generalization. It was Johnny's idea that instead of attempting to integrate the first order equations derived by Brandon we should apply a well established effective potential technique to them. The simplest version of this technique has been well known since the classic works of Jacobi in classical mechanics (see Jacobi, 1995) and its use in discussions of the radially separated Schrödinger equation in quantum mechanics (see e.g. Landau & Lifshitz, 1981). It had also been extended to the more complex classical motion of charged particles in the Earth's magnetosphere by Carl Størmer (see Størmer, 1934, and references therein). It proved to be just as useful for the Kerr spacetime, even though the physical conditions there were very different from those in the original applications.

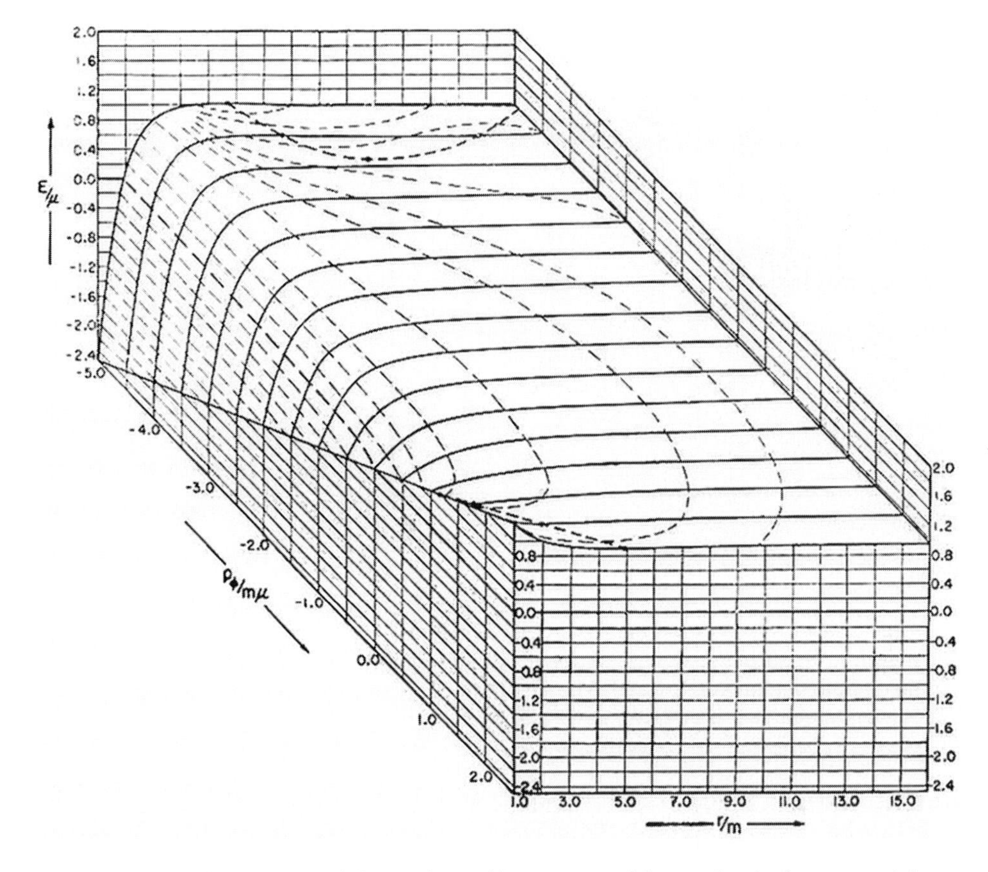


Fig. 7.4. Effective potential experienced by a test particle moving in the equatorial plane of an extreme Kerr black hole. For corotating orbits, with positive values of the angular momentum, the maximum binding of 42.35% of the rest mass of the test particle is reached at the horizon.

I still remember Johnny's suggestion that we draw the effective potential on the largest possible diagram, thereby minimizing the imperfections in the final printed version. Unlike today when diagrams can be almost instantly constructed and plotted by a computer, back then each value had to be calculated using stacks of punched cards on a computer, and then plotted on the final diagram. This was particularly impressive for our case – the diagram measured three by two meters (see Fig. 7.4)! A byproduct of preparing such a meticulous diagram was the time it gave us to think about the underlying physical process. It was during this numerical work that we realized that co-rotating orbits in the Kerr solution were much more tightly bound than counter-rotating ones. It was very gratifying when my good friend Evgeny Lifshitz found these results so important that he mentioned the Kerr solution extensively in the text of the last edition of the Landau and Lifshitz treatise,

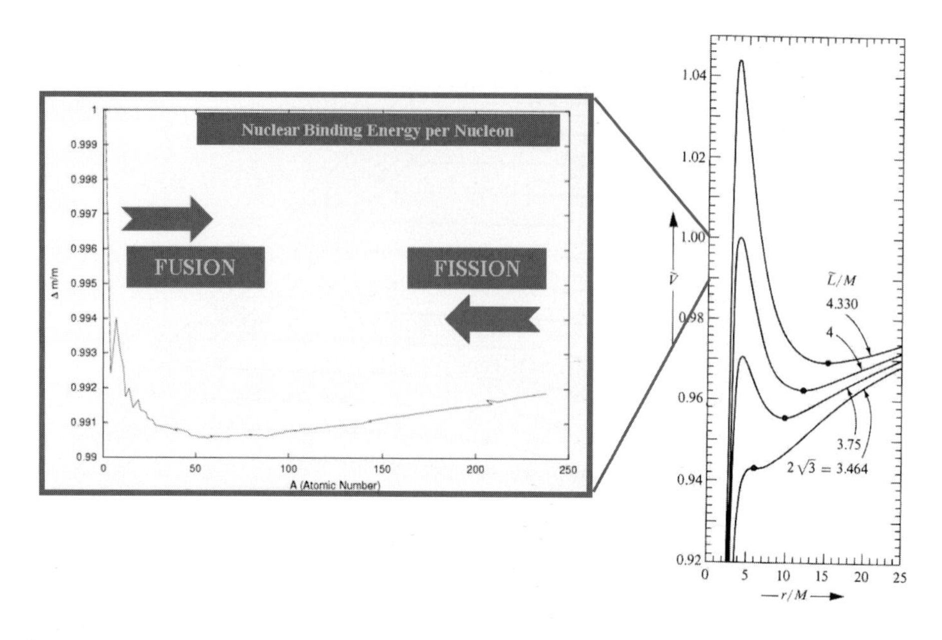


Fig. 7.5. Nuclear versus gravitational binding energy in a Schwarzschild black hole compared and contrasted. The gravitational binding energy in a Kerr metric is even bigger (see Fig. 7.4).

together with both Brandon's work and the results of Wheeler and myself as named problems for bright students!

## 7.3 The significance and magnitude of binding energy

One of the greatest feats of twentieth century science was understanding the nuclear binding energy of the elements. The precise value of the mass defect was calculated as a function of the atomic number for all known elements (see Fig. 7.5). This has generated some of the most striking conceptual, technological, and cultural changes in the history of mankind. The life cycle of our solar system and even life on our planet depend essentially on the nuclear binding energy released when elements are transformed. Fusion processes occur when elements lighter than iron are converted to heavier ones with a larger nuclear binding energy per nucleon. Conversely, fission processes occur when atoms heavier than iron split into smaller constituents, again with more binding energy per nucleon.

Jean Perrin (1920) and Arthur Eddington (1920) were the first to point out, independently, that the fusion of four hydrogen nuclei into one helium nucleus could explain the energy production in stars. This idea was put on a solid theoretical base by Robert Atkinson and Fritz Houtermans (1929a,b) using George Gamow's quantum theory of barrier penetration (Gamow & Houtermans, 1928) and was further developed by C. F. von Weizsäcker (1937, 1938). The monumental theoretical work by Hans Bethe (1939), and later by Burbidge *et al.* (1957), completed the understanding of the basic role of fusion processes in the stars. Together with Fermi (1949), they realized that the relative abundances of the elements in our entire solar system, including the planets, depend universally on nuclear burning. The presence of heavy elements also proved that these processes had already occurred in a previous generation of stars. All forms of energy which make life on our planet possible are derived from the sun, an enormous but relatively simple nuclear fusion reactor dominated by less than one hundred different nuclear reactions.

Fermi's work also led to our understanding of the fundamental role of fission and to the first chain reaction in Chicago (Fermi *et al.*, 1942). A significant fraction of electric power on our planet is now generated by fission reactors, and progress has been made to design a viable controlled fusion process to generate an alternative and secure energy source. The latter may prove to be essential if we are to maintain our desired quality of life. It is well known that both fission and fusion processes have been used in military research, and that some of the people who have contributed to the development of relativistic astrophysics, e.g. John Archibald Wheeler, Ya. B. Zel'dovich and A. Sakharov, had also previously made significant contributions to this field.

The huge difference between the sizes of mass defects due to nuclear binding energy and the mass defects of particles around a black hole due to its gravitational binding energy (see Fig. 7.5) shows that the energy generation processes are far larger in the general relativistic scenario.

A major step in proving the importance of such deep gravitational fields in astrophysical systems came with the identification of the first known black hole in our galaxy, Cygnus–X1 (Giacconi & Ruffini, 1978). A further step was made by the introduction of the blackholic energy and the possibility of thereby explaining some of the most energetic processes in our Universe.

## 7.4 The ergosphere of a Kerr spacetime

The central tools used in our research into the new physics of the Kerr spacetime were the separability of the associated Hamilton–Jacobi equation, discovered by Brandon Carter, and the effective potential technique. Achille Papapetrou had written to Johnny Wheeler, telling him of a sixteen year old high school student, Demetrios Christodoulou from Athens, who appeared to be specially gifted in physics and general relativity. When Wheeler examined him in Paris he was so impressed that he used Princeton University's freedom as a private institution to enroll him immediately as an undergraduate. Demetrios was not allowed to return

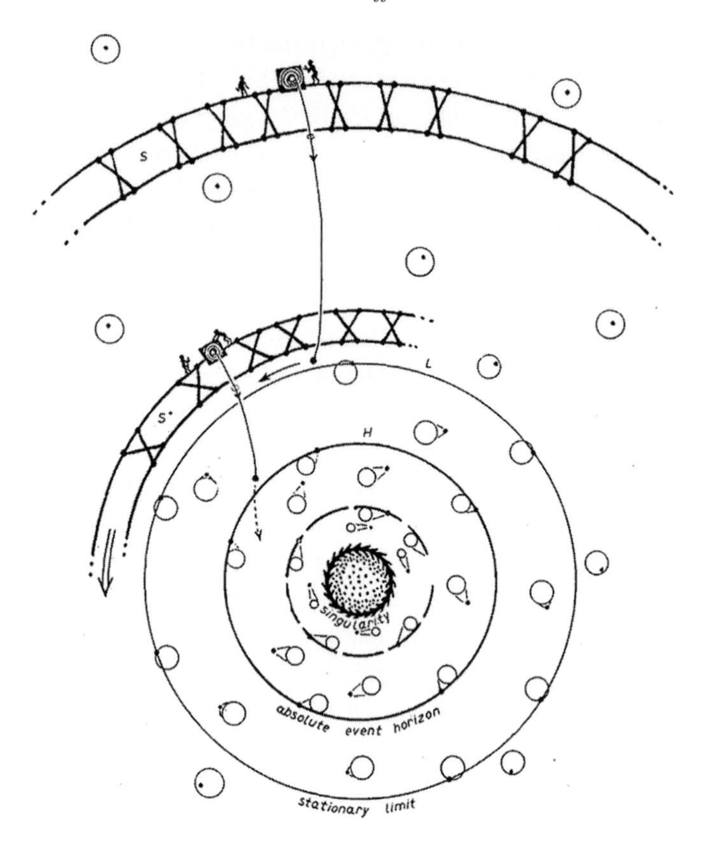


Fig. 7.6. Extraterrestrial civilization as idealized by Penrose (1969).

to Athens where he had been quarreling with his high school teachers. Instead he compressed a four-year undergraduate program into a single year. He was then enrolled in graduate school at the age of 17. Wheeler was officially his thesis advisor and assigned him the study of the collapse of a spherically symmetric massless scalar field, which later became chapter 1 of his thesis. I started by investigating with him the capture of test particles, both charged and uncharged, in a Kerr–Newman spacetime using the effective potential technique I had developed with Johnny (see Fig. 7.4). This became the remaining chapter of his thesis.

In 1969 I attended the first meeting of the European Physical Society in Florence. In those days the universities in Europe and elsewhere were in a very agitated state and so I was not surprised when this meeting was disrupted. I found myself sitting on the steps of the Palazzo della Signoria with Roger Penrose, discussing some aspects of a provocative talk he had just presented. In this he considered the possibility of an advanced civilization extracting energy from a Kerr spacetime by lowering tethered particles toward the singularity (see Fig. 7.6). He also considered a ballistic method: an object splitting into two pieces, one crossing the horizon and the other escaping with more mass–energy than the original body. However, there were many aspects of his lecture which were not clear to me, and which were not cleared up by our discussion.

Returning to Princeton, Johnny and I began examining the details of particle decay around a black hole. We showed that, as claimed by Penrose, energy could indeed be extracted. A particle with positive energy and positive angular momentum in a Kerr spacetime could come from infinity with a finite impact parameter and then split into two separate particles. One, counter-rotating and in a negative energy state as seen from infinity, would be captured by the horizon. The other, co-rotating and more energetic than the initial particle, would escape to infinity. We also identified the region in which such a process could occur as that between the horizon and the infinite redshift surface. I decided to call this the "energosphere" since energy extraction processes could exist there. Johnny said that this name was too clumsy, and that it should have a shorter more concise name. He suggested "ergosphere". After thinking a moment and recalling that the word *ergon* exists in Greek and means work, I agreed with him. This name has since become very popular. Originally I had mixed feelings toward this energy extraction process, was intrigued by its construction, and thought that proving its viability conceptually would be very interesting. However, the rest masses of the particles had to be reduced very significantly in the process making it hardly achievable from a physical point of view (see Fig. 7.7).

## 7.5 The mass formula of a black hole

While this exercise of looking at the decay of a particle in the ergosphere was continuing, Demetrios and I started a systematic analysis of all possible trajectories for test particles near a Kerr black hole. The solution for circular orbits had previously proved to be very elegant, depending on certain old theorems on the algebraic solvability of some special polynomials of sixth degree. We were fortunate to find these in the treatise of Paolo Ruffini (1803), a copy of which was contained in the main library at Princeton.

These polynomials had some rather fortunate non-physical factors and they could therefore be reduced to fourth-order ones with classic algebraic solutions. In doing this analysis we became aware of a very particular subset of trajectories corresponding to a limiting capture process. These occurred on the horizon, had zero radial kinetic energies and very specific angular momenta. A peculiarity of these paths is that the energy for counter-rotating particles is negative when seen from an observer at infinity. The corresponding capture process, with zero radial kinetic energy on the horizon, leads to a decrease of the total energy and angular

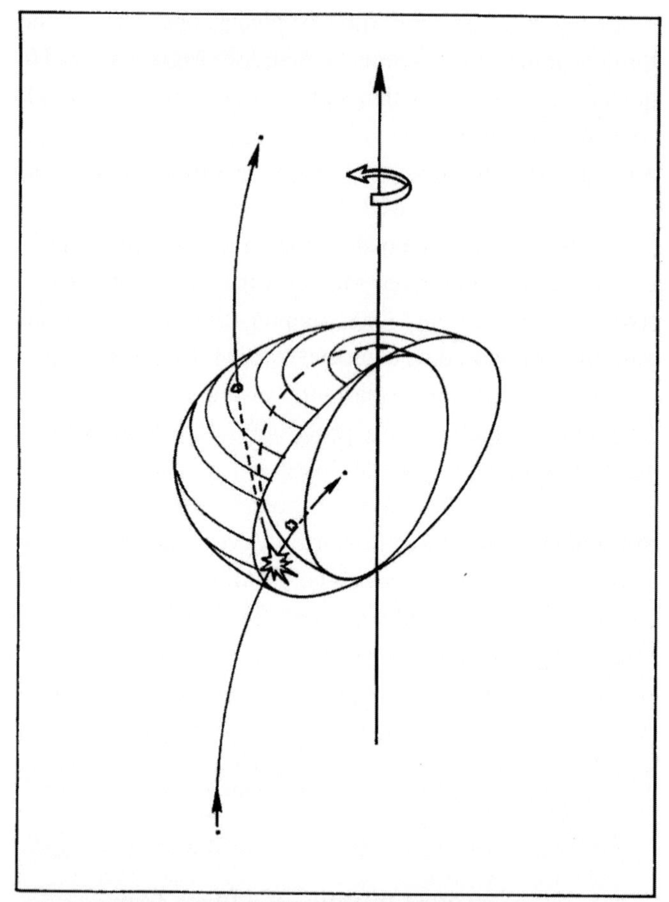

**Ergosphere of a rotating black hole.** The region between the surface of infinite redshift (outer) and the event horizon (inner), here shown in a cutaway view, is called the "ergosphere." When a particle disintegrates in this region and one of the fragments falls into the black hole, the other fragment can escape to infinity with more rest plus kinetic energy than the original particle. Figure 5

Fig. 7.7. The ergosphere. Reproduced from Ruffini & Wheeler (1971a).

momentum of the black hole. The same capture process for a particle with the same rest mass but with the opposite value for the angular momentum (i.e. co-rotating) leads to a positive contribution to both the total energy and angular velocity of the black hole. Most remarkably, the capture of the two particles with equal and opposite angular momenta leads to a black hole with its original total angular momentum. What was truly unexpected was that the total energy of the black hole was unchanged by the succession of two of these capture processes. We called these very special pairs of limiting transformations the reversible transformations of a black hole. All the other capture processes, with non-zero kinetic energy or occurring off the horizon, lead to irreversible transformations where the total mass energy of the black hole must increase.

In those days Johnny was very involved with other members of the physics department in what he considered a far more fundamental problem, the Teichmüller space of what he called "superspace". Such Teichmüller spaces, with their Riemannian non-Finslerian metrics, were meaningless to me from a physical standpoint. Nevertheless, we were able to talk to him about those thermodynamical analogies which were surfacing from the physics of black holes. It was clear to me that we were dealing with a very new situation in which the rotational energy of the Kerr black hole could be increased or decreased at will. There had to be some new and underlying quantity characterizing the black hole. I was convinced that it should be possible to split its total energy into rotational and Coulomb energy. I suggested this problem to Demetrios. He was indeed able to express the infinitesimal limit on the horizon of the capture process I had examined with Wheeler and to integrate the corresponding differential equations. Next morning he came in smiling and visibly satisfied, saying "It is true. As you expected the rotational energy contribution to the Kerr solution can be split from its total energy by integrating our reversible transformations. There is a formula which relates these quantities to the non-rotating rest mass of the black hole". I gave the name "irreducible mass" ($m_{ir}$)[4] to this black hole rest mass, since it can never decrease: it is left unchanged by reversible transformations and increases monotonically for all irreversible ones,

$$\Delta m_{ir} \geq 0. \tag{7.2}$$

The same evening Johnny and I were walking back to the Institute through the woods bordering the golf course and swimming pool of the Institute. He had been very busy all day on certain fundamental issues of superspace. When I told him the result that Demetrios and I had just obtained, Johnny said that it was very important. It was then sent for publication to Physical Review Letters. I insisted that the sole author of the letter should be Demetrios since it was he who had solved the mathematical equations. At the same time Johnny was delaying the publication of our results on both the ergosphere and the decay process and I therefore decided to insert in the letter a figure which included both the definition of the ergosphere and the details of the computations carried out with Johnny (see Fig. 7.8). These had been clearly propaedeutic to the Demetrios result. The Editor objected to having our unpublished material in such a short letter when we were not coauthors. I promptly asked him if this objection would stand if the two authors of the figure would volunteer to ask him in writing to accept the Demetrios letter in that format. Finally, he accepted it and the results were published on 30 November, 1970 (Christodoulou, 1970). This letter recorded our propaedeutical analysis on

[4] From the Italian word "irriducibile".

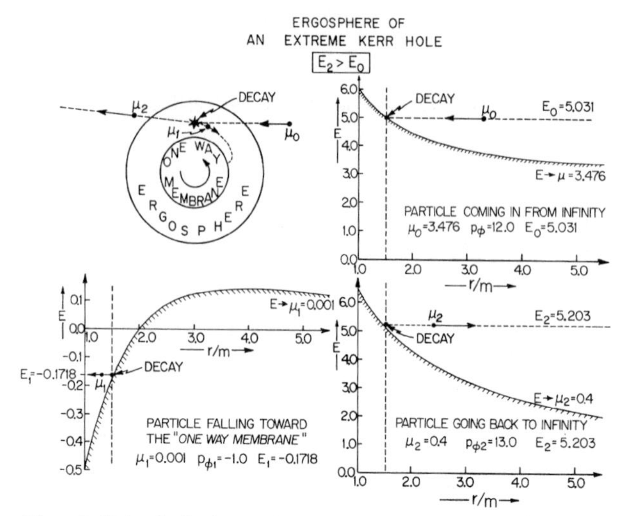


FIG. 2. (Reproduced from Ruffini and Wheeler, Ref. 4, with their kind permission.) Decay of a particle of rest-plus-kinetic energy $E_0$ into a particle which is captured into the black hole with positive energy as judged locally, but negative energy $E_1$ as judged from infinity, together with a particle of rest-plus-kinetic energy $E_2 > E_0$ which escapes to infinity. The cross-hatched curves give the effective potential (gravitational plus centrifugal) defined by the solution $E$ of Eq. (2) for constant values of $p_\varphi$ and $\mu$.

Fig. 7.8. The particle decay process in the field of a black hole as reproduced by Christodoulou (1970).

the ergosphere, the decay process and the mass energy formula for a Kerr black hole.

A few months later Johnny sent me a copy of a preprint by Penrose and Floyd giving an example of energy extraction from a spinning black hole,[5] but we were not interested in assessing priorities for such a contrived gedanken experiment. I was particularly concerned by the necessary reduction of the rest mass of the particles in such a decay process and we were working with Demetrios toward a more general formula for a Kerr–Newman black hole.[6] The mass formula was finally reached in July 1971 by Demetrios and myself (Christodoulou & Ruffini, 1971):

$$m^2 = \left(m_{ir} + \frac{e^2}{4m_{ir}}\right)^2 + \frac{L^2}{4m_{ir}^2}, \tag{7.3}$$

$$S = 16\pi m_{ir}^2, \tag{7.4}$$

$$\frac{L^2}{4m_{ir}^4} + \frac{e^4}{16m_{ir}^4} \leq 1, \tag{7.5}$$

where $m$ is the total mass–energy of the black hole, $m_{ir}$ the irreducible mass, $S$ the surface area and $e$ and $L$ are the black hole charge and angular momentum.

[5] See Fig. 7.9 with the handwriting of Wheeler and myself.

[6] Brandon Carter and Werner Israel had conjectured this to be the most general black hole, and were attempting to prove the uniqueness of its geometry.

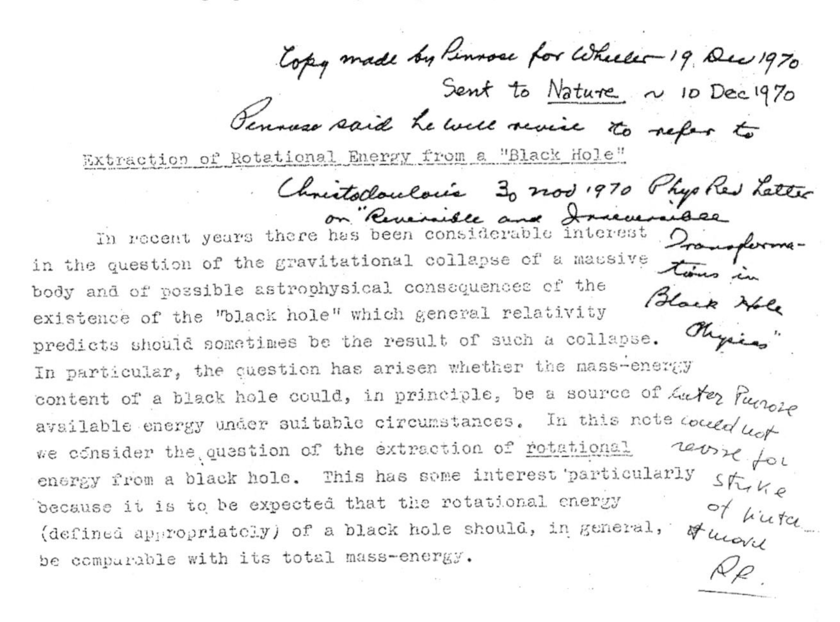

Copy made by Penrose for Wheeler 19 Dec 1970
Sent to Nature ~ 10 Dec 1970
Penrose said he will revise to refer to

Extraction of Rotational Energy from a "Black Hole"

Christodoulou 30 Nov 1970 Phys Rev Letter on "Reversible and Irreversible Transformations in Black Hole Physics"

In recent years there has been considerable interest in the question of the gravitational collapse of a massive body and of possible astrophysical consequences of the existence of the "black hole" which general relativity predicts should sometimes be the result of such a collapse. In particular, the question has arisen whether the mass-energy content of a black hole could, in principle, be a source of available energy under suitable circumstances. In this note we consider the question of the extraction of rotational energy from a black hole. This has some interest particularly because it is to be expected that the rotational energy (defined appropriately) of a black hole should, in general, be comparable with its total mass-energy.

Later Penrose could not revise for strike of [illegible]
RR.

Fig. 7.9. First page of a preprint by Floyd and Penrose annotated with the handwriting of Johnny and myself (see Floyd & Penrose, 1971).

The inequality in Eq. (7.5) gives the maximum possible values consistent with the existence of an horizon. This implies that up to 29% (50%) of the total black hole mass–energy could be in principle extracted using reversible transformations to reduce its rotational (Coulomb) energy. From Eqs. (7.4) and (7.2) it then follows that the surface area of the Kerr–Newman black hole must necessarily increase in any capture process:

$$\delta S = 32\pi m_{ir} \delta m_{ir} \geq 0. \tag{7.6}$$

In the meantime, S. Hawking (1971) had also derived this inequality from a different and possibly more general viewpoint, limited to the Kerr case. Demetrios finally defended his Ph.D. thesis at the age of 19, answering a splendid set of questions by the external examiners: Eugene Wigner for the theory and David Wilkinson for the experiments (see Fig. 7.10).

In front of us there was now a vast horizon to be explored dealing with the energetics of the black holes. This will be outlined in the following sections.

## 7.6 Introducing the black hole

Hermann Bondi, who was then Director General of the European Space Research Organization, had in the meantime invited Francis Everitt, Martin Rees, Leonard Schiff, Johnny, myself and a few others to Interlaken Switzerland to discuss the possibilities for fundamental research from space platforms.

Fig. 7.10. Demetrios Christodoulou being addressed by Eugene Wigner during his Ph.D. thesis defense. Sitting, from left to right, David Wilkinson, Eugene Wigner (background), Johnny Wheeler (foreground) and myself.

Some of the seminal ideas discussed in Interlaken were expanded into an extended report by Wheeler and myself (Ruffini & Wheeler, 1971b) and used as the first ten chapters of our book with Martin Rees (Rees, Ruffini & Wheeler, 1976) and also in Misner, Thorne & Wheeler (1973). Our book with Martin and Johnny was considerably delayed in publication, and I therefore decided to add certain later results as appendices (see e.g. the important contribution by Shvartsman, 1971), while leaving the spirit of the earlier work unchanged. When writing this report, Johnny and I became convinced that the field had finally come of age. The study of black holes had moved from being a topic of research in formal general relativity with hypotheses assumed for purely mathematical convenience to being a field of profound physical significance. We therefore decided to write an article addressing the physics community at large. The editor of Physics Today accepted this and was especially helpful. He not only gave us the cover of Physics Today but also commissioned Helmuth Wimmer, an artist working at the Heiden Planetarium in New York, to find an appropriate representation of a black hole.

I recall explaining our research to Helmuth in a two hour session at the Physics Today office in New York. He told me later that he left the discussion totally confused and with a terrible headache. He woke up at 4am the following morning with a mental image that he immediately recorded in a painting. In his words, "This must be what they are talking about". He called me back at 9am. I rushed

Introducing the black hole

According to present cosmology, certain stars end their careers in a total gravitational collapse that transcends the ordinary laws of physics.

Remo Ruffini and John A. Wheeler

Fig. 7.11. Introducing the black hole. From Ruffini & Wheeler (1971a).

to New York that same morning and found the drawing to be quite beautiful. I asked Helmuth to change the sequence of colors in the spectrum but to leave the rest alone. The final picture was perfect and the article created a very favorable reaction in the scientific community (see Fig. 7.11). Helmut was very happy and kindly offered me the painting in recognition of my explanation of our work to him (and also to increase his annual tax deductions due to the donation!). In the end we decided to donate the original to Princeton University, where it still hangs in the mathematical physics library in Jadwin Hall, and I accepted for myself the first proofs of the cover of Physics Today, signed by Helmuth Wimmer.

In the article, Johnny and I decided to emphasize one of the most profound aspects of the physics of black holes, namely that they can be characterized completely by their mass, charge and angular momentum.[7] This was becoming highly important to physics in view of the existence of the mass–energy formula for black holes. New domains of physics were being opened up, showing how nature might extract enormous power from black holes.

We also emphasized the vast number of papers written in the Soviet Union (see e.g. Zel'dovich & Guseynov, 1965, 1965; Shklovsky, 1967). They proposed certain methods for detecting black holes in binary star systems. In those days Saturn was

[7] Many have been working on the mathematical proofs of the "uniqueness theorem" of Carter and Israel (see Robinson's contribution in this volume).

passing in front of the sun so we added a significant sentence: "Of all objects that one can conceive to be traveling through empty space, few offer poorer prospects of detection than a solitary black hole of solar mass. No light comes directly from it. It can not be seen by its lens action or other effect on a more distant star. It is difficult enough to see Venus, 12 000 km in diameter, swimming across the disk of the sun; looking for a 15-km object moving across a far-off stellar light source would be unimaginably difficult". The message was clear: in order to succeed we had to capitalize on binary star systems with a black hole as one of their members.

This brings us to the fundamental work of Riccardo Giacconi and his group. Before doing this I recall a result obtained in Princeton with my second graduate student Clifford Rhoades. Our determination of the absolute upper limit to a neutron star's mass was essential for formulating the paradigm for the identification of the first black hole in our galaxy.

## 7.7 On the maximum mass of a neutron star

With the help of some exceptional students in Princeton we were able to pursue our research much further. Although Johnny's main research interest was still superspace, he was also very interested in the thermodynamics of black holes started by Demetrios and myself with the introduction of reversible and irreversible transformations. He guided Jacob Bekenstein during his Ph.D. thesis on the topic of a seemingly absurd analogy: to identify the surface area of the black hole with a generalized entropy. He also suggested that the problem of dimensionality could be overcome by expressing the black hole surface area in units of the Planck mass squared. This is a pure number like any entropy should be! As Jacob recalls in his recent book (Bekenstein, 2006), I found this proposed identification non-contradictory but also possibly unnecessary.

I decided to concentrate myself on the astrophysical applications of our results with Demetrios, setting the goal to find a place in the Universe where the extraction of energy from a black hole could be observed.

Clifford Rhoades and I achieved a relevant intermediate step. Chandrasekhar (1930) and Landau (1932) had shown clearly and independently that a critical mass of $\sim 1.5 M_{\odot}$ exists for white dwarfs. Any potential white dwarf with a larger mass must collapse gravitationally. The existence of this was traceable back to the extreme special relativistic regimes encountered in the degenerate electron gas responsible for the equilibrium configuration of white dwarfs.

Newtonian gravity had been used in their analysis, the electron gas had been assumed neutral on average and the detailed electromagnetic interactions with each nucleus within the white dwarf had been neglected. The corrections to this basic treatment were found to be negligible to the first approximation (Ruffini,

2001a). There was still a critical mass, although it was slightly smaller. Similarly, if one compares and contrasts the results of the computations performed in Newtonian theory with those in General Relativity the lowest order differences are also negligible.

Neutron stars were introduced by George Gamow (1938) and by Robert Oppenheimer and his students (Oppenheimer & Serber, 1938; Oppenheimer & Volkoff, 1939). It quickly became clear that the treatment developed for white dwarfs could be applied to a system of self-gravitating neutrons. The concept of critical mass can be applied to neutron stars for the same physical reasons as for white dwarfs. The reason for this is traceable back to the extreme special relativistic regimes encountered in the degenerate neutron gas responsible for the equilibrium configuration of neutron stars. The neutron gas was still described as a free gas of fermions in that paper. The general relativistic corrections proved much larger for neutron stars than for white dwarfs. Finally, Oppenheimer estimated the critical mass to be $0.7M_{\odot}$.

It was evident that neutron stars are far more complex than white dwarfs. For the latter the electrons composing the fermion gas supporting the star are subject only to electromagnetic and gravitational interactions. The electromagnetic interactions can be precisely computed within the framework of Maxwell theory, possibly the best tested theory in physics. However, the critical mass for neutron stars occurs at supranuclear densities and so the strong interactions among nuclei cannot be neglected. Unlike for white dwarfs where Maxwell theory is sufficient, neither field equations nor a theoretical description of bulk matter exist for such large densities. Moreover, there is no hope in the near future for laboratory experiments at these pressures. Various phenomenological attempts to estimate the neutron star mass have shown that this could be quite sensitive to strong interactions. If a factor 2 could easily exist, why not a factor 10 or even larger? The formation of a black hole would be avoided by masses less than some critical mass, but the value for this was unknown.

For this reason Clifford Rhoades and I (Rhoades & Ruffini, 1971) used an alternative approach to determine from first principles an absolute upper limit to the neutron star mass. We adopted three criteria: the correctness of general relativity; the existence of a fiducial density up to which a reliable equation of state maximizing the critical mass could exist; and non-violation of causality. For supranuclear densities we assumed an equation of state consistent with causality and the le Chatelier principle. The proof involved the introduction of a variety of extremization techniques. Particularly helpful was the concept of the *domain of dependence* for the values of the critical mass as a function of the chosen fiducial density as introduced in my Les Houches lectures (see page R29 in Ruffini, 1973). The absolute upper limit to the neutron star critical mass was found to be $3.2M_{\odot}$. All observed neutron star masses until today are well within this limit.

Also in my Les Houches lectures, I introduced the concepts of alive black holes and dead black holes, differentiating the ones with charge and angular momentum from the ones uniquely characterized by their irreducible mass.[8]

### 7.8 The *Uhuru* satellite

The launch of the *Uhuru* satellite in 1971 by the group directed by Riccardo Giacconi meant that the universe could be examined for the first time in the X-ray band of the electromagnetic spectrum. It was a fundamental leap forward, creating a tremendous surge in relativistic astrophysics. Simultaneous observations of astrophysical objects could now be made in X-ray wavelengths by UHURU in space and in optical and radio wavelengths by ground based observatories. This unprecedentedly large collaboration generated high quality data on those binary systems where a normal star is stripped of matter by a compact massive companion star, either a neutron star or a black hole.

We had just published our article in Physics Today when Gloria Lubkin, its editor, told me at a Washington meeting of the American Physical Society; "You should listen to Riccardo Giacconi's talk. He claims to have observed some of the phenomena forecast by Wheeler and yourself in your article".[9]

I still recall how our almost daily discussions on the Uhuru observations strongly motivated me to find a method to discriminate between neutron stars and black holes as binary X-ray sources. Following the work with Clifford Rhoades, we were ready to establish the basic paradigm for distinguishing between them.

The observations by Riccardo and his team clearly gave the first unambiguous evidence for the discovery of neutron stars accreting matter from stars evolved out of the main sequence (see Fig. 7.12). Soon after, important contributions on the accretion were presented by Shakura (1972a,b). The two X-ray pulsating sources Hercules–X1 and Centaurus–X3 were typical examples of this phenomenon. All the characteristic parameters of these binary sources could be derived from the data (see R. Giacconi, pages 17–42, in Giacconi & Ruffini, 1978). From the binary period and the Doppler velocities of the main sequence star and from the pulsed X-ray emission of the neutron star it was possible to calculate the neutron star masses for the first time. These proved to be systematically lower than our

[8] The Les Houches school represented a moment of great scientific tension and in fact signaled a division in black hole research. My work on alive black holes encountered strong resistance by Kip Thorne and his group. Also Jacob Bekenstein, presenting a report on his recently discussed Ph.D. thesis in an informal seminar, received strong criticisms from Stephen Hawking, although he was later to change his mind (Bekenstein, 2006).

[9] That was the first time I met Riccardo. I did not know until years later that he was born in Italy and that he got his doctorate at the University of Milan, and for some years I did not realize that he knew Italian. It was only through a rather unorthodox New Year greeting that I found out that he did indeed speak the language. By that time, our collaboration with him and Herbert Gursky, a member of his group, had become very intense.

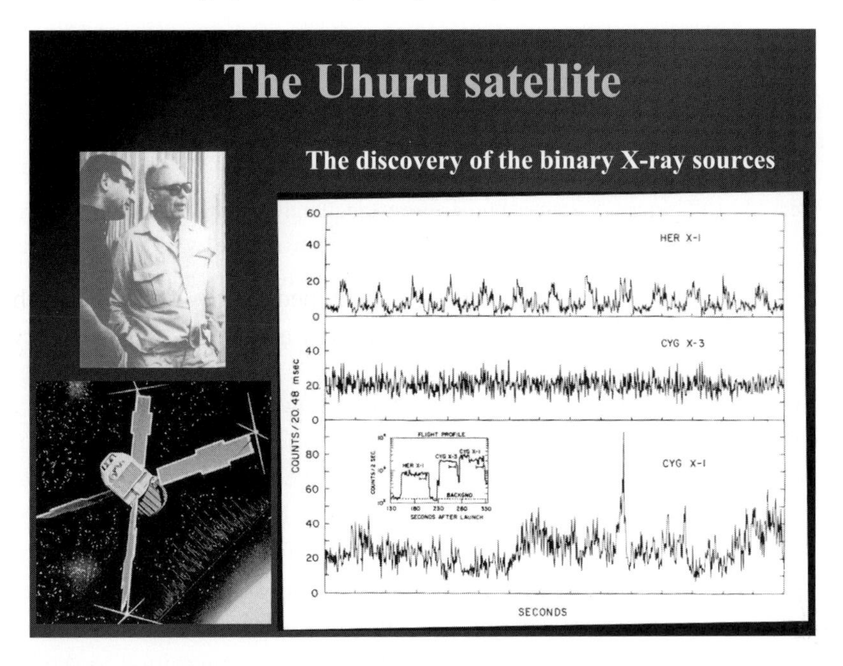


Fig. 7.12. On the left side, Riccardo Giacconi and Luigi Broglio (above) at the launch of the *Uhuru* satellite (below) from the S. Marco platform. On the right side, the time variability of Hercules–X1 compared and contrasted with Cygnus–X1.

absolute upper limit of $3.2M_\odot$. Hercules–X1 and Centaurus–X3 were crucial in differentiating these binary systems from pulsars. Unlike pulsars with their monotonically increasing pulsation periods, these sources had fluctuating ones (see e.g. Fig. 10, p. 194 in Gursky & Ruffini, 1975). It was clear that the rotational energy, used previously as an explanation of the energetics of pulsars, could not be significant for these systems. The source of the very large X-ray luminosity, up to $10^{37}$ erg s$^{-1}$ (i.e. $10^4$ solar luminosity), had to be accretion in the deep gravitational well of a neutron star. For the first time we were witnessing direct evidence for the role of gravitation as the energy source of an extremely energetic astrophysical system!

We were then ready to establish the paradigm for the identification of Cygnus–X1 as the first "black hole". This was observed to be (see Fig. 7.12) a non-pulsating source with significant time structure as short as a few milliseconds. I identified three essential steps to strengthen this identification:

(i) The "black hole uniqueness theorem" implies axial symmetry and the absence of regular pulsations from black holes. However, although this is true for a black hole, it is also true for a neutron star provided that its magnetic field, if any, is aligned with its rotation axis.

(ii) The "effective potential technique", see Fig. 7.4, shows that the percentage of the body's rest energy released by accretion is at most 1% for neutron stars but as much as 42% for an extreme Kerr black hole. This clearly proves the importance of gravitational energy as a generator for binary X-ray sources. Again, this is equally true for either neutron stars or black holes; only the efficiency factor is different. The accretion observed in all binary systems can therefore adequately explain their X-ray emissions.

(iii) The "upper limit on the maximum mass of a neutron star" was indeed the crucial discriminating factor between non-magnetized neutron stars and black holes. If the gravitationally collapsed star in a binary X-ray source is non-pulsating, emits X-rays by accretion, and has a mass larger than 3.2 solar masses, then it must be a black hole.

These results were announced in a widely attended session chaired by John Wheeler at the 1972 Texas Symposium in New York. The evidence for Cygnus–X1 being a black hole was presented and was extensively reported in the New York Times.[10] The New York Academy of Sciences, which hosted the Texas Symposium, had just awarded me their Cressy Morrison Award for my work on neutron stars and black holes. Much to their dismay I did not submit a paper for the proceedings of this conference, the reason being that the substance of my talk at the Texas Symposium was recorded in a Letter that had just been submitted to Ap.J. with Robert Leach, a Princeton undergraduate (see Fig. 7.13 and Leach & Ruffini, 1973).

The formulation of this paradigm did not come easily but slowly matured after innumerable discussions with R. Giacconi and H. Gursky, both face to face and over the phone. I still remember an irate professor of the Physics Department at Princeton pointing out at a faculty meeting my outrageous phone bill of $274 for one month, a scandalous amount for those times. Its size was largely due to my frequent calls to the Smithsonian. Fortunately, the department chairman, Murph Goldberg, had a much more relaxed and sympathetic attitude about this situation.

The results were summarized in my talk at the Sixteenth Solvay Conference on Astrophysics and Gravitation, held at the University of Bruxelles in 1974, and were expanded in the 1975 Enrico Fermi Varenna Summer school directed by Riccardo Giacconi and myself. The title of the school was "On the Physics and Astrophysics of Neutron Stars and Black Holes". The proceedings were published in both hardcover and paperback, and are to be reprinted soon. The conclusion of the story of this great scientific adventure was well told by Riccardo in his Nobel lecture in Stockholm (Giacconi, 2002 and also Giacconi, 2005).

Black holes have played an essential though passive role in these accretion processes, providing the deep potential well used to release the observed X-ray flux. We had to find a different astrophysical system where black holes could play

[10] The acceptance of our paradigm was far from unanimous at this time. Some astrophysicists who were initially amongst my strongest and most irate public objectors later became fervent supporters of my ideas. However, not all of them remembered to quote my results later!

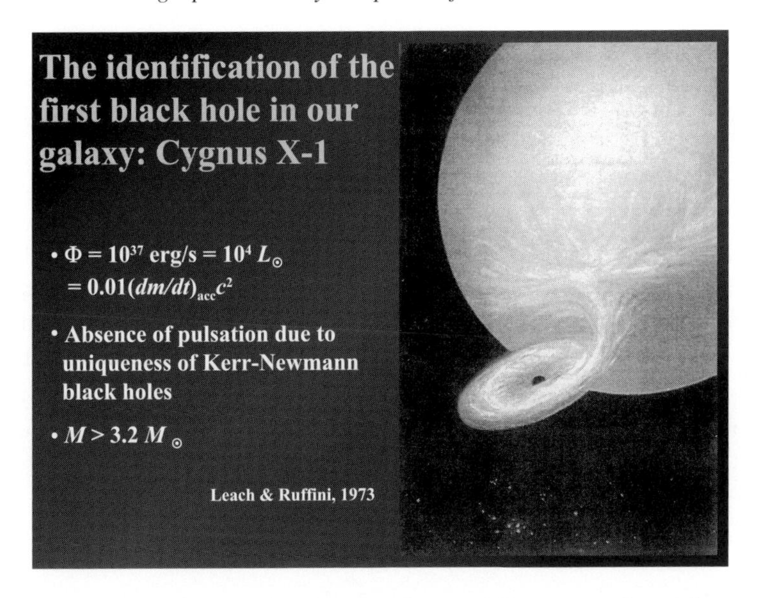


Fig. 7.13. The identification paradigm for Cygnus–X1 versus a pictorial representation of Cygnus–X1 and the companion star HDE-226868, whose binary nature was precisely pointed out by Webster & Murdin (1972); Bolton (1972a,b).

an active role if we were to extract energy from them. This has been my main goal in recent years as I will explain shortly.

## 7.9 Astrophysical "Tokomak" machines

Much attention was still being given in those days to the analogy I have already mentioned between black hole physics and the usual laws of thermodynamics. Following Bekenstein (1973, 1974), Stephen Hawking went a step further (Hawking, 1974, 1975, 1976). He proposed that a black hole radiates with a black body spectrum whose temperature $T$ is defined as the surface gravity at the horizon multiplied by the Planck length squared and divided by $2\pi$. The Hawking idea was of great conceptual interest. As Jacob recalls in his recent book (Bekenstein, 2006), I was concerned about the initial formulation of this program of research.[11]

Anyway, I was not personally very much interested in these processes since they were of marginal interest in relativistic astrophysics. I summarized this point of view in my Varenna lectures (see R. Ruffini, pp. 324–325 in Giacconi & Ruffini, 1978) and, more recently, in Table 1 at p. 788 of Ruffini (2001b). I then turned to the physical processes which might use black holes as realistic sources of energy

[11] Recently, I have been reconsidering some aspects by integrating the equations of a specific example (Ruffini & Vitagliano, 2003).

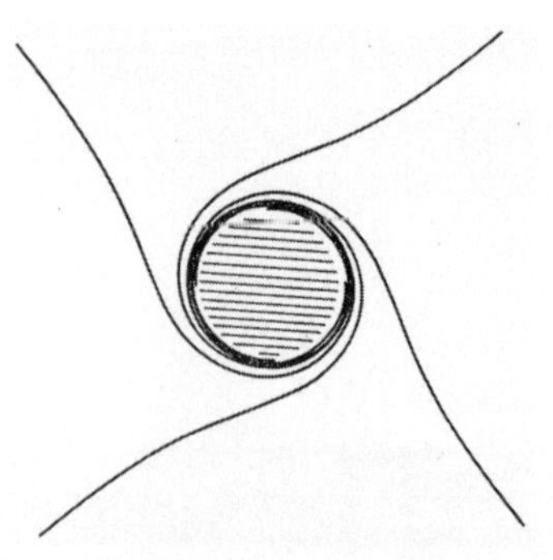

Fig. 7.14. Magnetic lines of force in the equatorial plane of a maximally rotating Kerr black hole accreting magnetized plasma. The winding of the lines of force is due to the dragging of inertial frames. Details in Ruffini & Wilson (1975).

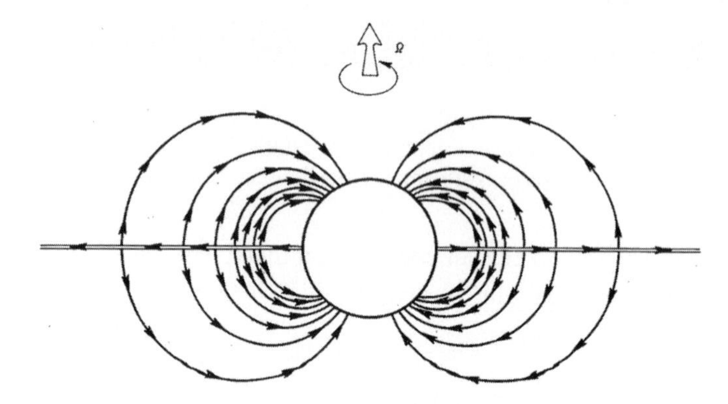

Fig. 7.15. Lines of currents for the magnetosphere showed in Fig. 7.14. Details in Giacconi & Ruffini (1978), pages 338 and following.

for astrophysical processes by using their extractable rotational or electromagnetic energy, what we call today the "blackholic" energy.[12] We first focused on processes which use predominantly the rotational energy. This process can provide as much as 29% of the mass energy of an extreme Kerr black hole (see Eqs. (7.3) and (7.5)). Together with Jim Wilson, we gave a simple analytic example of a Kerr black hole accreting magnetized plasma (see Fig. 7.14 and Ruffini & Wilson, 1975).

We studied the entire electrodynamical system of currents for such an astrophysical circuit (see Fig. 7.15) as well as the general conditions for stability of an accreting plasma in the field of a black hole (Damour *et al.*, 1978). Particularly important were the contributions, motivated by our work, on the torque and momentum transfer in accreting black holes (Damour, 1975), which introduced the

[12] This word is the English translation for the Italian word *energia buconerale*, suggested by Iacopo Ruffini, unhappy to hear continuously the wording "extractable energy from black holes".

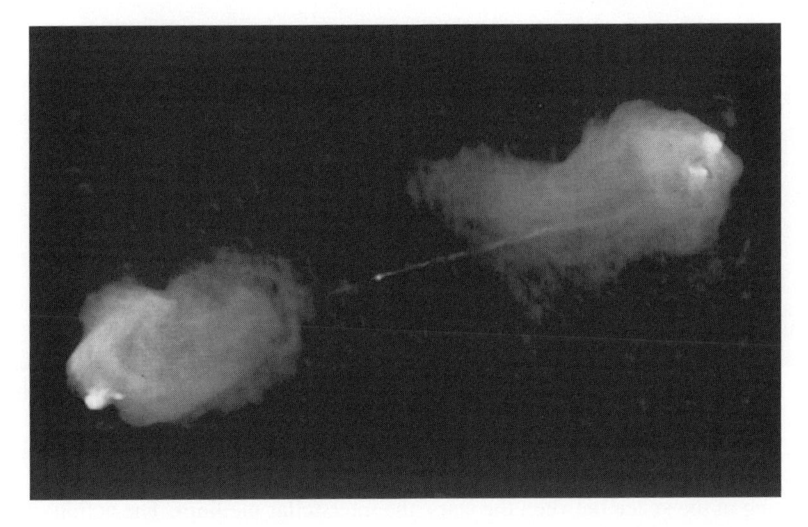

Fig. 7.16. Grayscale representation of the image of Cygnus A at 5 GHz with 0.4 in. resolution made with the Very Large Array in Socorro, NM. The full source extent is 120 arc sec = 120 kpc. North is at the top and west is to the right. Reproduced from Carilli *et al.* (1998).

lines of currents reproduced in Fig. 7.15. Further results, fundamental for the black hole thermodynamics, were presented by Thibault on the black hole eddy currents (Damour, 1978) and on the surface effects in black hole physics (Damour, 1982).

Our paper with Jim was soon followed by a similar paper by Blandford & Znajek (1977) and later by another series of articles by Punsly, and by Punsly and collaborators (see e.g. Punsly, 2001, and references therein). These works promised to help explain extragalactic radio jets. The characteristic time scale for this energy extraction process is typically millions of years (see Fig. 7.16) as can be deduced from the size of the radio lobes and jets in extragalactic radio sources. It is most interesting that recent radio observations made by the Westerbork Synthesis Radio Telescope on Cygnus–X1 have evidenced the existence of a jet in such a system (see Figs. 7.17, 7.18, details in Gallo *et al.*, 2005). The lifetime of such a jet has been estimated to be $\sim$ 0.02–0.32 Myr, which is comparable with the estimated age of the progenitor of the black hole in Cygnus–X1 (see Mirabel & Rodrigues, 2003). The total power dissipated by the jets of Cygnus–X1 in the form of kinetic energy has been estimated to be as high as the bolometric X-ray luminosity of the system (Gallo *et al.*, 2005). It is then fair to say that Cygnus–X1 is even more interesting than what we understood in 1974. It is, beyond any doubt, a Kerr black hole, since we see an active process of rotational energy extraction by the jets, as given by Eq. (7.3).

We looked then to the extraction of electromagnetic energy. This process can provide as much as 50% of the mass energy of an extreme Kerr–Newman black hole

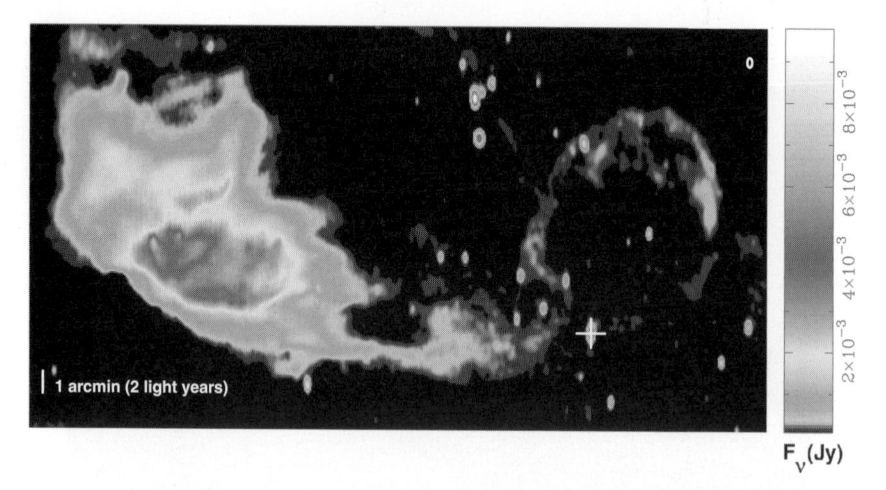


Fig. 7.17. The cross marks the location of the black hole Cygnus–X1 in this radio image. The bright region to the left (east) of the black hole is a dense cloud of gas existing in the space between the stars, the interstellar medium. The action of the jet from Cygnus–X1 has 'blown a bubble' in this gas cloud, extending to the north and west (right) of the black hole. Reproduced from Gallo *et al.* (2005).

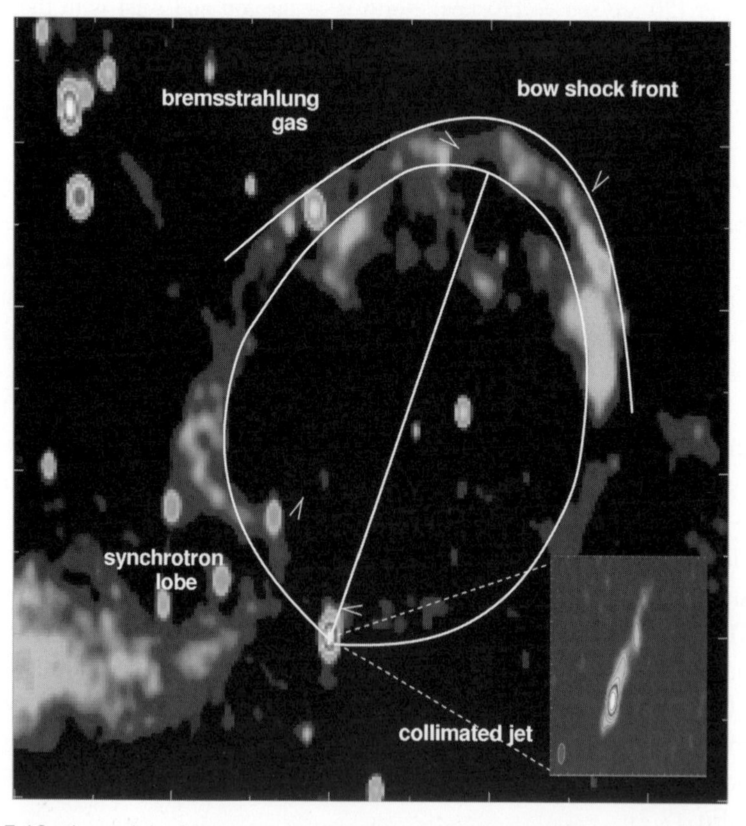


Fig. 7.18. A model of how the black hole created the bubble. The black hole's powerful jet (seen separately in the inset) has been pushing on interstellar gas for about a million years. At the edges of the shell the interstellar medium is heated as the bubble rapidly expands. Reproduced from Gallo *et al.* (2005).

Fig. 7.19. After approaching Pope John Paul II with an unidentified object concealed beneath his jacket, Zel'dovich produced a book of his collected papers, which he donated to the Pope. "Thanks" the Pope replied, to which Zel'dovich loudly responded "Not just 'thanks'! These are fifty years of my work!". The Pope kept Zel'dovich's collected papers (Zel'dovich, 1985) under his arm during the entire rest of the audience.

(see Eqs. (7.3) and (7.5)). What is even more important is that this electromagnetic energy can be released in a very short time, of the order of a second. This clearly contrasts with the rotational energy extraction process which, as we have shown in the jets in galactic and extragalactic sources, systematically occurs on time scales on the order of millions of years. The electromagnetic blackholic energy release leads to unprecedented power and luminosity in astrophysics, second only to the Big Bang. As we show in the next section, this is fundamental for the explanation of GRBs.

## 7.10 Gamma-ray bursts

It is very likely that GRBs would never have been detected without an outrageous idea put forward by Yakov Borisovich Zel'dovich. We had been close friends since September 1968, and I knew his very important contributions as inventor of the *Katiuscia* rockets and his later work with Andrei Sakharov on the Soviet A and H Bomb projects. I enjoyed many interesting and provocative discussions on relativistic astrophysics with Lifshitz, Ginzburg and him while visiting Moscow. His unpredictability and even irrationality had often surprised me (see Fig. 7.19). However, his proposal in the late fifties (see e.g. Foresta Martin, 1999) to show the clear dominance of the Soviet Union in space by having a rocket carry an atomic

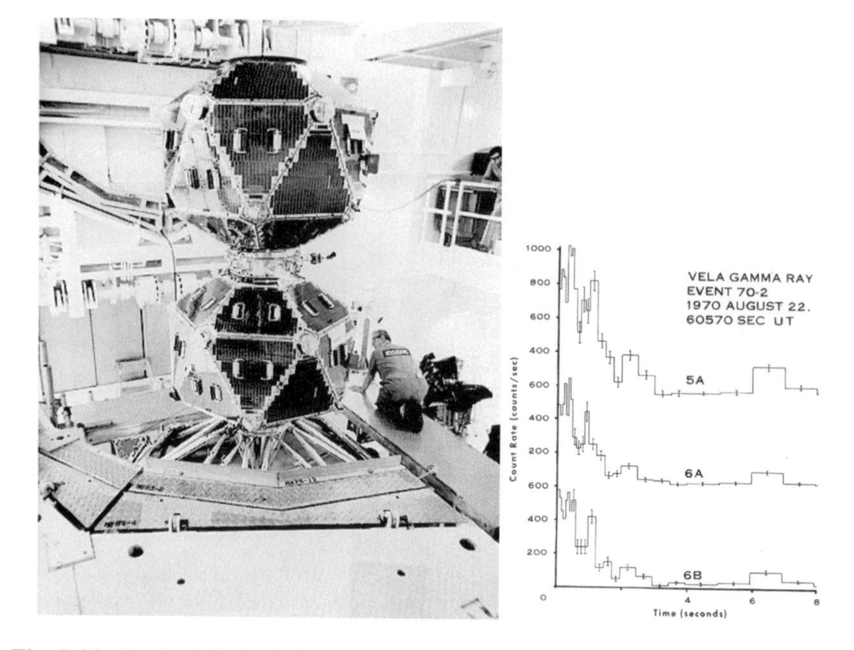


Fig. 7.20. On the left the *Vela* 5A and 5B satellites and, on the right side, a typical event as recorded by three of the *Vela* satellites. Details in I. Strong in Gursky & Ruffini (1975).

bomb to the moon and explode it on the lunar surface, was beyond belief! This would have been visible to a very large part of the world's population, all those facing the moon when the bomb went off. Fortunately, the proposal was not accepted, but it is very likely that it served as additional motivation for the United States of America to put a set of four *Vela* satellites into orbit, 150 000 miles above the Earth. They were top-secret omnidirectional detectors using atomic clocks to precisely record the arrival times of both X-rays and $\gamma$-rays (see Fig. 7.20). The direction of the source of the signals could then be calculated by triangulation. When they were made operational they immediately produced results. It was thought at first that the signals originated from nuclear bomb explosions on the earth but they were much too frequent, one per day! A systematic analysis by the military showed that they had not originated on the earth, nor even the solar system. These *Vela* satellites had discovered GRBs! The first public announcement of this came at the AAAS meeting in San Francisco in a special session on neutron stars, black holes and binary X-ray sources, organized by Herb Gursky and myself (Gursky & Ruffini, 1975).

A few months later, Thibault Damour and I published a theoretical framework for GRBs based on the vacuum polarization process in the field of a Kerr–Newman black hole (Damour & Ruffini, 1975). We showed how the pair creation predicted by the Heisenberg–Euler–Schwinger theory (Heisenberg & Euler, 1935; Schwinger, 1951) would lead to a transformation of the black hole, asymptotically close to reversibility. The electron–positron pairs created by this process were generated by what we now call the blackholic energy. In that paper we concluded that this "naturally leads to a very simple model for the explanation of the recently discovered GRBs". Our theory had two very clear signatures. It could only operate for black holes with mass $M_{BH}$ in the range 3.2–$10^6$ $M_\odot$ and the energy released had a characteristic value of

$$E = 1.8 \times 10^{54} \frac{M_{BH}}{M_\odot} \text{ ergs.} \tag{7.7}$$

Since nothing was then known about the location and the energetics of these sources we stopped working in the field, waiting for a clarification of the astrophysical scenario. As Rashid Sunyaev mentioned to me at that time, "There are too many models for $\gamma$-ray bursts". I reproduced a limited list of them later (see Fig. 11, p. 787 in Ruffini, 2001b).

The mystery of these sources became even more profound as the observations of the BATSE instrument on board the Compton Gamma Ray Observatory Satellite (CGRO) proved the isotropy of these sources in the sky (see Fig. 7.21). In addition to this data the CGRO satellite found an unprecedented number of GRBs and provided detailed information on their temporal structure and spectral properties (see Fig. 7.22). All this was encoded in the fourth BATSE catalog (Paciesas *et al.*, 1999). From the analysis there it soon became clear that there were two distinct families of GRBs; the short bursts lasting less than one second and harder in spectra, and the long bursts lasting more than one second and softer in spectra (see Fig. 7.23).

The situation changed drastically with the discovery of the "afterglow" of GRBs (Costa *et al.*, 1997) by the joint Italian–Dutch satellite *BeppoSAX* (see Fig. 7.24). This X-ray emission lasted for months after the "prompt" emission of a few seconds duration and allowed the GRB sources to be identified much more accurately. This then led to the optical identification of the GRBs by the largest telescopes in the world. These had just become operative and included the Hubble Space Telescope, the *Keck* telescope in Hawaii and the VLT in Chile. Also, the very large array in Socorro made radio identification possible.

We have recalled how the interplay between the X- and $\gamma$-ray satellites in space and the optical and radio observatories on the ground had been a major factor in the study of binary X-ray sources. This collaboration occurs now on a much larger scale for GRBs, thanks to the use of Space observatories like *Chandra* and *XMM-Newton*, dedicated space missions such as HETE and *Swift* and the unprecedented facilities on the ground. The first distance measurement for a GRB was made in

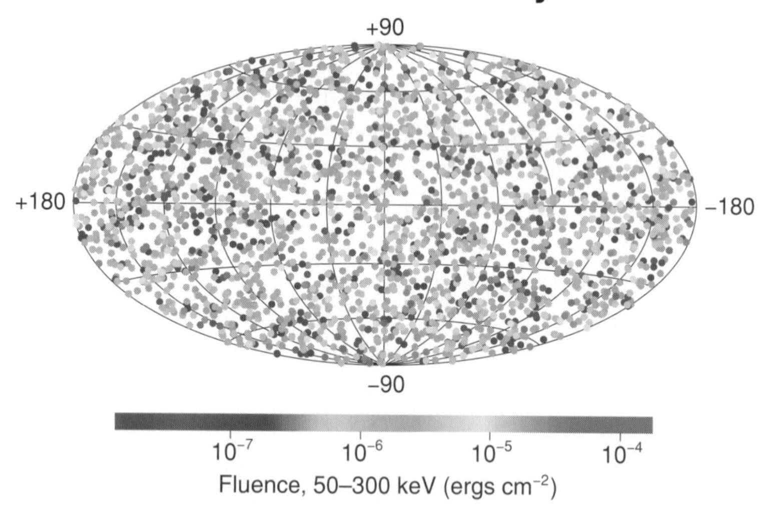


Fig. 7.21. The CGRO satellite and the position in the sky of the observed GRBs in galactic coordinates. Different shades correspond to different intensities at the detector. There is almost perfect isotropy, both in the spatial and in the energetic distribution.

1997 for GRB970228 and the truly enormous energy of this was determined to be $10^{54}$ ergs per burst. This proved the existence of a single astrophysical system emitting as much energy during its short lifetime as that emitted in the same time by all other stars of all galaxies in the Universe![13] It is interesting that this "quantum" of astrophysical energy coincided with the one Thibault Damour and I had already

[13] Luminosity of average star = $10^{33}$ erg $s^{-1}$, Stars per galaxy = $10^{12}$, Number of galaxies = $10^{9}$. Finally, $33 + 12 + 9 = 54$!

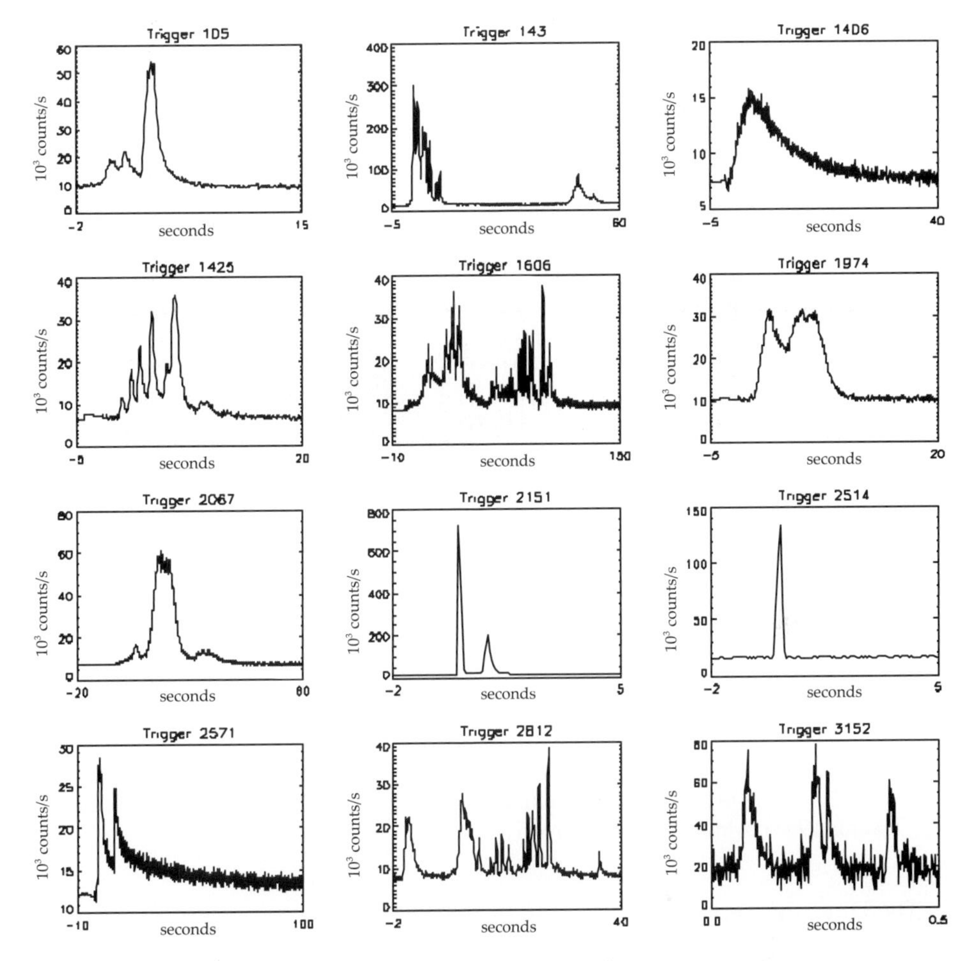


Fig. 7.22. Some GRB light curves observed by the BATSE instrument on board of the CGRO satellite.

predicted, see Eq. (7.7). We clearly imagined much stronger opposition to the concepts of this model from the establishment, possibly even stronger than that already encountered for the identification of Cygnus–X1 as a black hole. Once again, our imaginations have been too conservative.

## 7.11 Physics versus astrophysics

Among the many crucial advances made in physics, relativistic field theories and astrophysics that the 20th century has been rich with, two fruitful contributions will be remembered for their manifold consequences: the Dirac theory of the electron and the Kerr–Newman geometry. There are some analogies between these two discoveries.

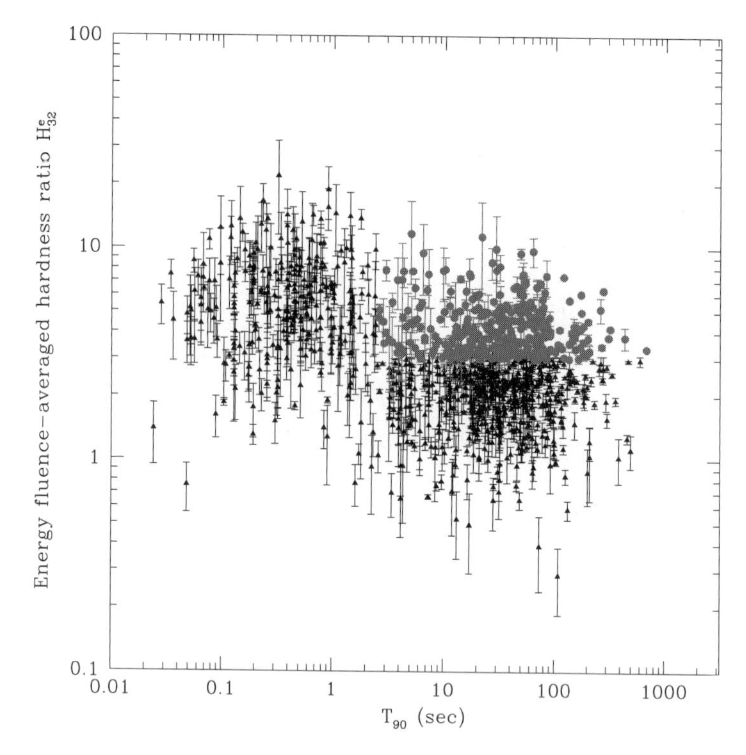


Fig. 7.23. The energy fluence-averaged hardness ratio for short ($T < 1$ s) and long ($T > 1$ s) GRBs are represented. Reproduced, by his kind permission, from Tavani (1998), where the details are given.

The Dirac theory:

(i) introduced for the first time the concept of the spin of an elementary particle in Minkowski space, gave the mathematical tools for developing such a study, and produced an enormous set of predictions and experimental verifications in the field of atomic physics;
(ii) predicted the matter–antimatter solutions which were splendidly confirmed by the discovery of the positron (Anderson, 1933); and
(iii) together with the works of Heisenberg & Euler (1935) and Schwinger (1951), lead to the concepts of vacuum polarization and the creation of electron–positron pairs in extreme electromagnetic fields. These have still not been observed, in spite of more than fifty years of repeated attempts in the leading high energy physics laboratories worldwide (Ruffini, Vitagliano & Xue, 2007).

The Kerr–Newman solution:

(i) is the exact mathematical solution corresponding to a spinning black hole satisfying the Einstein-Maxwell equations, generalizes to curved spacetimes the concepts of

Fig. 7.24. The Italian–Dutch satellite *BeppoSAX* (here represented) encountered enormous financial difficulties. In spite of this, *BeppoSAX* was ultimately a success, and achieved one of the most important discoveries ever in the field of astrophysics: the discovery of the GRB afterglow.

matter–antimatter solutions (see e.g. Damour, 1977; Deruelle, 1977) and has the same gyromagnetic ratio as the electron (Carter, 1968);

(ii) has been observationally verified with the discovery of Cygnus–X1 (see e.g. Giacconi & Ruffini, 1978); and

(iii) provides a field around an incipient black hole in which the vacuum polarization process generates electron–positron pairs. This extracts the blackholic energy, thereby emitting an enormous number of electron–positron pairs during the final collapse to the black hole. Such emission appears to be the natural explanation for GRBs (Damour & Ruffini, 1975).

Physicists had previously explored the new physical regimes predicted by the Dirac theory and compared their conclusions with the results of especially conceived experiments in high-energy laboratories on the ground. In such an approach, when the theoretical predictions are confirmed by a comprehensive set of experiments the great moment of discovery has occurred.

In astrophysics the situation is apparently different from that in particle physics. We cannot reproduce the experimental conditions in a laboratory because of the size of the systems involved. We have to use the entire universe instead by selecting among the billions of events those specific ones where the processes we are interested in occur, and then compare these with the theoretical predictions. When these agree that most exciting moment of discovery occurs again. All this is no different from an experiment done in the laboratory. In both cases there is a selection of the natural phenomena to be observed.

There is another difference between the approaches of physicists and some astrophysicists. The latter have purported to examine and justify the origins of the theory itself, instead of looking at the predictions of the theory and their verification by experiment or observations. Such an epistemological approach may appear to be tautological. If one were to take seriously such an approach by asking how an electron is born or why the Dirac equations apply one would need a unified field theory and possibly all its fundamental interactions. Such a theory, in turn, would be inconceivable without the knowledge gained by the theoretical studies of the Dirac equation and its experimental verification. Fortunately such an approach was never considered by those pragmatic physicists who have received the deserved rewards of their many fundamental discoveries (Dirac himself, Dyson, Feynman, Fermi, Lamb, Segré, etc.).

Paradoxically such an approach is gaining some proselytes in the astrophysical community. Our group, which initially included Natalie Deruelle and Thibault Damour, studied the Kerr–Newman metric using an approach similar to that in theoretical physics. We trusted the predictions of the equations more than the qualitative feelings of imagination. We studied the theoretical consequences of the Kerr–Newman black hole, a solution originating from the coupling of the Einstein and Maxwell equations, by far the most successfully tested theories in all of physics. We shall show in the following how we have been successful in applying these concepts to GRBs. Pressure has been mounting on us to provide a detailed model for the black hole formation. We have taken such a request positively, as a stimulus to reach a deeper understanding of the process of gravitational collapse leading both to the formation of black holes and to the quite different phenomenon of supernovae explosions, which we recall are also still far from being understood (Mezzacappa & Fuller, 2006). Just as for Dirac electrons, it is harder to explain the birth of black holes than it is to explain their observed activities by the theory. Only through an understanding of the GRB phenomenon will we gain information on possible precursors to black holes and how they form. Specifically, in the case of GRBs we believe that there is theoretical evidence not for just one but for a variety of GRB precursors. From the recent progress in understanding GRBs, we are confident that we are close to understanding the formation of Kerr–Newman black holes (see e.g. Ruffini, 2008). This progress will also help to clarify some crucial unexplained features of the gravitational collapse of stars in the range between 3.2 and $100 M_\odot$.

It is interesting that if one turns for a moment to larger black holes, originating in active galactic nuclei and expected to be at least a million solar masses, the situation is equally disappointing from the point of the above mentioned epistemological approach. As of today, there is no explanation for the birth of these maxi black holes. A likely possibility is that the "inos" explaining the dark matter of the host galaxy may form a relativistic cluster in the galactic core, leading to the formation of the black hole (Arbolino & Ruffini, 1988; Merafina & Ruffini, 1989, 1990, 1997; Bisnovatyi-Kogan *et al.*, 1998). The absence of a generally accepted mechanism for their formation should certainly not preclude the study of black holes, possibly charged, to help us understand active galactic nuclei. It would be scientifically unreasonable to stop this work until these somewhat epistemological demands can be answered.[14]

Having said this let us go to a quick outline of the work we did on Kerr–Newman geometries and their dyadospheres.

## 7.12 The concept of the dyadosphere

The evidence for the existence of GRBs with energies predicted by Eq. (7.7) convinced us to carefully analyze the vacuum polarization process leading to the creation of an electron–positron pair plasma in the field of a black hole. This pair creation process occurs in an electric field close to the critical value,

$$E_c = \frac{m^2 c^3}{e\hbar} = 1.32 \times 10^{16}\,\mathrm{V\,cm^{-1}}, \tag{7.8}$$

where $m$ and $e$ are the electron mass and charge. In Minkowski space, tunneling occurs between the matter–antimatter solutions of the Dirac equation (see Heisenberg & Euler (1935); Schwinger (1951)) leading to a pair creation rate that can be expressed in analytic form. Such a treatment has been generalized to the curved spacetime of a Kerr–Newman solution by Damour & Ruffini (1975). The concept of the "dyadosphere", which comes from the Greek "$\delta\upsilon\acute{\alpha}\varsigma, \delta\upsilon\acute{\alpha}\delta o\varsigma$" for "pairs", was initially introduced in Ruffini (1998) (see also Preparata, Ruffini & Xue, 1998). For simplicity, and yet to illustrate the basic gravitational and electrodynamical processes, a Reissner–Nordström black hole was assumed. The region outside the horizon where the electric field strength is larger than the critical value was given by Eq. (7.8). Clearly, this work excluded neither a more general metric nor the pair creation process in under-critical fields, *a priori*.

[14] There has been a recent attempt to deny the astrophysical relevance of vacuum polarization processes around a Kerr–Newman black hole by "proving" a no-go theorem for them. This proof consists of setting up a particular "straw man" and then demolishing it, hardly a proof of anything! No-go theorems are often used in physics but for them to be useful they must be proven results following from a clearly stated set of assumptions which then limit their domain of application. In this specific case the no-go theorem does not appear to have any validity since the counterexample created contradicts known physical facts and violates both energy conservation and causality (see Page, 2006 and the reply in Ruffini, Bianco & Xue, 2007).

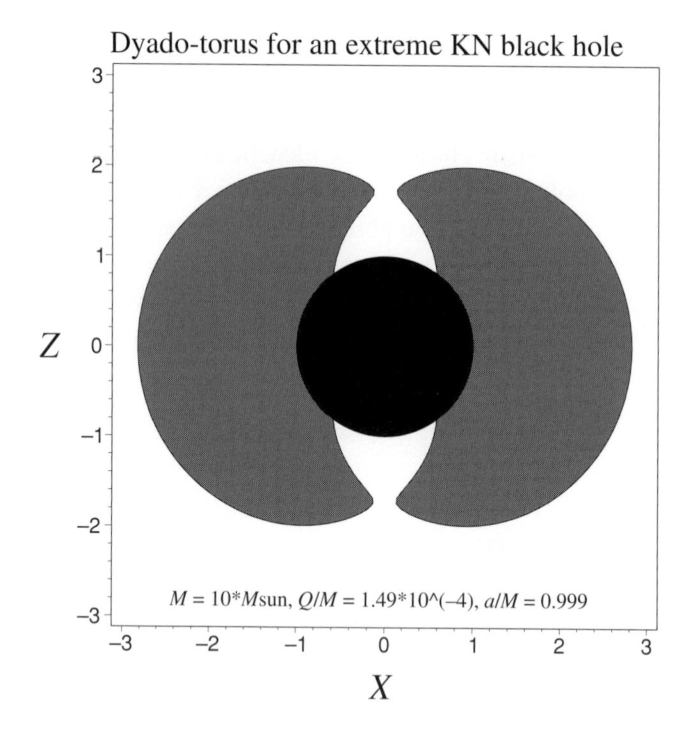


Fig. 7.25. The "dyado-torus" is the region outside the horizon of a Kerr–Newman black hole, where the electrodynamical processes generate electron–positron pairs by vacuum polarization processes. Details in Cherubini *et al.* (2007).

In the meantime, the dyadosphere concept has evolved in three major ways. They can occur in a variety of conditions, such as neutron star collapses, highly rotating and magnetized neutron stars and in Kerr–Newman spacetimes. The concept of the "dyado-torus" was also introduced to take into account the presence of angular momentum (see Fig. 7.25 and, e.g. Ruffini, Vitagliano & Xue, 2007, and references therein).

Since the work of Khriplovich (2000), dyadospheres have been considered with an electric field $E < E_c$. These turned out to be optically thin and characterized by the emission of $\sim 10^{21}$ eV particles. They are possibly relevant to Ultra High Energy Cosmic Rays (UHECRs, see e.g. Damour & Ruffini, 1975; Chardonnet *et al.*, 2003). These "under-critical" dyadospheres became an interesting complement to the "over-critical" ones, where $E > E_c$. The latter are initially optically thick and lead to an ultra-relativistic GRB afterglow, emitting radiation in both X- and $\gamma$-rays (Ruffini & Vitagliano, 2002; Ruffini, 2008).

Since 2002, we have considered the dynamical formation of a dyadosphere during gravitational collapse to a black hole. To generate the accelerations required by GRBs we restricted ourselves to optically thick dyadospheres. We found an explicit analytic treatment for an over-critical field self-sustained over the macroscopic astrophysical time scale of the gravitational collapse and creating an over-critical dyadosphere. To obtain such, we used a charged collapsing shell as our model. We only considered the regime in which the electric field is larger than the critical value given by Eq. (7.8). Starting from this assumed initial condition, we took into proper account:

(i) the dynamics of the shell (Cherubini, Ruffini & Vitagliano, 2002);
(ii) the electromagnetic blackholic energy extraction processes (Ruffini & Vitagliano, 2002, 2003);
(iii) some of the collective effects of the plasma formed by the electron–positron pairs created by the vacuum polarization process, including their feedback on the electromagnetic field and corresponding polarization effects (Ruffini, Vitagliano & Xue, 2003a,b); and
(iv) especially the electron–positron plasma oscillations, described by the Vlasov–Boltzmann equation (Gatoff, Kerman & Matsui, 1987).

Consistent initial conditions were obviously necessary when solving the field equations. It was also necessary to identify the physical processes that occur in the progenitor star and lead to its gravitational collapse to a black hole. We have identified an appropriate set of initial conditions for the electrodynamical structure of neutron stars, offering a natural explanation for the initial existence of an over-critical field. We have therefore given both the field equations and the initial conditions that describe the formation of an over-critical and optically thick astrophysical dyadosphere (see e.g. Ruffini, 2008; Ruffini, Rotondo & Xue, 2006, 2008).

## 7.13 The dynamics of the electron–positron plasma

Our GRB model, like all prevailing models in the existing literature (see e.g. Piran, 1999; Mészáros, 2002, 2006, and references therein), is based on the acceleration of an optically thick electron–positron plasma (EPP). The specific issue of the origin and energetics of such an EPP, either in relation to black hole physics or to other physical processes, has often been discussed qualitatively in the GRB scientific literature but never quantitatively with explicit equations. The concept of the dyadosphere is the only attempt, as far as we know, to do this. This relates such an electron–positron plasma to black hole physics and to the characteristics of the GRB progenitor star, using explicit equations that satisfy the existing physical laws. Far from being just a formal theoretical work, this is essential to show that the physical origin and energetics of GRBs are the blackholic energy of the Kerr–Newman metric.

If we turn now to the accelerating phase of the electron–positron plasma, our analysis differs from the other ones in the current literature, in both the dynamics and evolution of such a plasma and the details of the transparency condition.

The dynamics of the EPP was considered by Piran, Shemi & Narayan (1993) using a numerical approach, by Bisnovatyi-Kogan & Murzina (1995) using an analytic one and by Mészáros, Laguna & Rees (1993) using one that was both numerical and semi-analytic. We studied it in collaboration with Jim Wilson and Jay Salmonson at Livermore. Numerical simulations were developed at Livermore and a semi-analytic approach was developed in Rome (Ruffini *et al.*, 1999).

A conclusion common to all the treatments is that the EPP is initially optically thick and expands to very high values of the Lorentz gamma factor. A second common result is that the plasma shell expands in its co-moving frame and the Lorentz contraction is such that its width in the laboratory frame appears to be constant: the "Pair-Electro-Magnetic (PEM) Pulse".

In all treatments the EPP is assumed to have a baryon loading. This is acquired in our model when the pure EPP created in the dyadosphere expands to engulf the progenitor remnants. A new pulse is then formed with electron–positron–photons and baryons (PEMB Pulse, see Ruffini *et al.* (2000)), expanding until transparency is reached. At this point the emitted photons form what we define as the "Proper-GRB" (see Ruffini *et al.* (2001b)). The baryon loading is defined by a dimensionless quantity

$$B = \frac{M_B c^2}{E_{dya}}, \tag{7.9}$$

where $E_{dya}$ is the energy of the pairs created in the dyadosphere, and $M_B$ is the mass of the remnant. $B$ and $E_{dya}$ are the only free parameters characterizing the source in our theory.

Differences exist between our description of the rate equation for the electron–positron pairs and the ones by the other authors. The analogies and differences have been given in Ruffini *et al.* (2007); Bianco *et al.* (2006). From our analysis (Ruffini *et al.*, 2000) it became clear that such expanding dynamical evolution can only occur (see Fig. 7.26) for values of

$$B < 10^{-2}. \tag{7.10}$$

It follows that the collapse to a neutron star is drastically different from the collapse to a black hole leading to a GRB. Whilst in the former a very large amount of matter is expelled, in the latter the collapse process is smoother than any other one considered until today: almost 99.9% of the star has to be collapsing simultaneously.

We summarize in Fig. 7.27 some qualitative aspects of our model as well as the corresponding values of the Lorentz gamma factor as a function of the radial

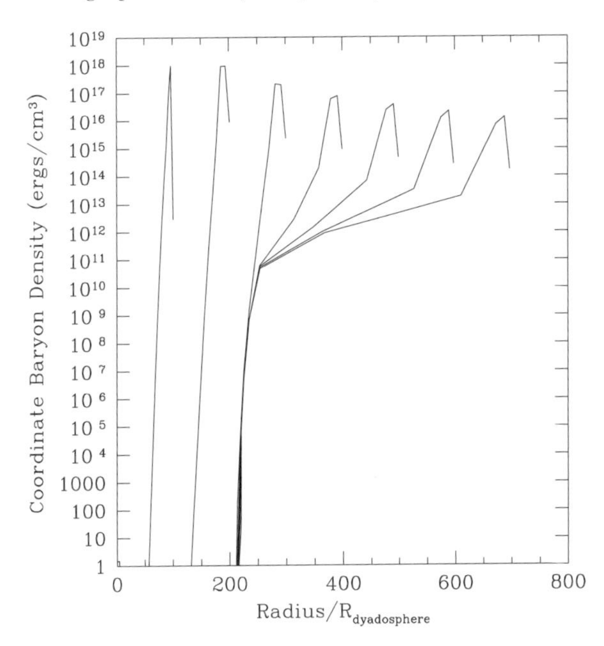


Fig. 7.26. The expansion of the PEMB pulse corresponding to a baryon loading $B = 10^{-2}$ is represented as a function of the radial coordinate. The instability following the encounter of the baryonic component is manifest. Details in Ruffini *et al.* (2000).

coordinate in the typical case of GRB991216 (Ruffini *et al.*, 2003, 2005a). The self-acceleration phase ends at point 4 where the Proper-GRB (P-GRB) is emitted.

## 7.14 The interaction of the accelerated baryonic matter (ABM Pulse) with the interstellar medium (ISM): the afterglow

After the plasma becomes transparent and the P-GRB is emitted, the accelerated baryonic matter (the ABM pulse) interacts with the interstellar medium (ISM). This creates the afterglow (see Fig. 7.27). I shall first summarize the commonalities between our approach and the ones in the current literature. A thin shell approximation is used in both (see Piran, 1999; Chiang & Dermer, 1999; Ruffini *et al.*, 2003, 2005a; Bianco & Ruffini, 2005b) to describe the collision between the ABM pulse and the ISM:

$$dE_{\text{int}} = (\gamma - 1)\, dM_{\text{ism}} c^2, \tag{7.11a}$$

$$d\gamma = -\frac{\gamma^2 - 1}{M} dM_{\text{ism}}, \tag{7.11b}$$

$$dM = \frac{1 - \varepsilon}{c^2} dE_{\text{int}} + dM_{\text{ism}}, \tag{7.11c}$$

$$dM_{\text{ism}} = 4\pi m_p n_{\text{ism}} r^2 dr, \tag{7.11d}$$

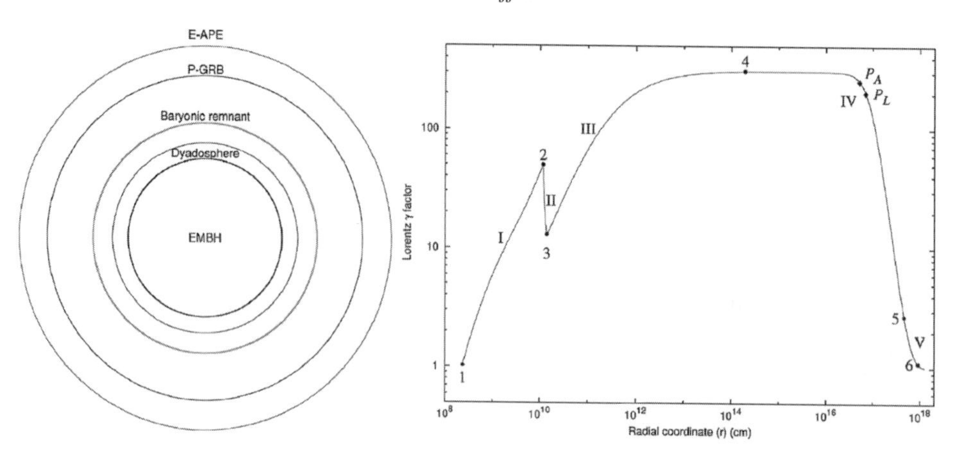


Fig. 7.27. The GRB afterglow phase is here represented together with the optically thick phase for GRB991216. The value of the Lorentz gamma factor is given from the transparency point all the way to the ultrarelativisitc, relativistic and non-relativistic regimes. Details in Ruffini *et al.* (2003).

where $E_{\rm int}$, $\gamma$ and $M$ are respectively the internal energy, the Lorentz factor and the mass–energy of the expanding baryonic shell, $n_{\rm ism}$ is the ISM number density which is assumed to be constant, $m_p$ is the proton mass, $\varepsilon$ is the emitted fraction of the energy developed in the collision with the ISM and $M_{\rm ism}$ is the amount of ISM mass swept up within the radius $r$:

$$M_{\rm ism} = (4/3)\pi(r^3 - r_\circ{}^3)m_p n_{\rm ism}.$$

Here $r_\circ$ is the starting radius of the baryonic shell, i.e. the plasma transparency radius. $\varepsilon = 0$ ($\varepsilon = 1$) corresponds to the "adiabatic" ("fully radiative") condition (see e.g. Bianco & Ruffini, 2005b).

In the current literature, following Blandford & McKee (1976), a so-called "ultra-relativistic" approximation $\gamma_\circ \gg \gamma \gg 1$ has been widely adopted to solve Eq. (7.11) (see e.g. Sari, 1997, 1998; Waxman, 1997; Panaitescu & Mészáros, 1998; Rees & Mészáros, 1998; Granot, Piran & Sari, 1999; Gruzinov & Waxman, 1999; Piran, 1999; van Paradijs, Koubeliotou & Wijers, 2000; Mészáros, 2002, and references therein). This leads to a simple constant-index power-law relation,

$$\gamma \propto r^{-a}, \tag{7.12}$$

with $a = 3$ for the fully radiative case and $a = 3/2$ for the fully adiabatic one. We have, instead, obtained the explicit analytic solution of the equations of motion for a shell with constant ISM density, in the entire range from ultra-relativistic to non-relativistic velocities. The resulting expressions, very different from the above power law, can be found in Bianco & Ruffini (2005b).

Knowledge of the equations of motion is essential for calculating the loci of source points of the signals arriving at the observer at the same time (Chandrasekhar, 1939), the "equitemporal surfaces" (EQTSs). When these are compared to the approximate ones obtained in the current literature, a remarkable difference is found (see Fig. 7.28 and Bianco & Ruffini, 2004, 2005a).

The most striking aspect of GRB theory is that these systems are among the very few in physics and astrophysics for which a completely detailed model can be computed in all its essential steps. The final result, however, depends crucially on the correctness of each theoretical step. All the GRB's observational properties are a function of the EQTSs, and all the observables must be calculated correctly.

I shall now turn to the last distinguishing feature between our theoretical model and the other ones in the current literature. We have proposed that the X- and $\gamma$-ray radiation has a thermal spectrum in the co-moving frame during the entire afterglow phase (Ruffini *et al.*, 2004). This follows an idea of Fermi, used to calculate a possible thermodynamic limit for high energy collisions between elementary particles. This thermalization procedure is justified for GRBs by recognizing that the ISM density is inhomogeneous. It has a filamentary structure with a density contrast $\Delta\rho/\rho$ as large as $10^9$ (Ruffini *et al.*, 2005b). The temperature is given by:

$$T_s = [\Delta E_{\rm int}/(4\pi r^2 \Delta\tau\sigma\mathcal{R})]^{1/4}, \tag{7.13}$$

where $\Delta E_{\rm int}$ is the internal energy developed in the collision with the ISM in time interval $\Delta\tau$ in the co-moving frame, $\sigma$ is the Stefan–Boltzmann constant and

$$\mathcal{R} = A_{\rm eff}/A, \tag{7.14}$$

is the ratio between the "effective emitting area" of the afterglow and the surface area of radius $r$. This factor $\mathcal{R}$ has to take into account both the ISM filamentary structure and any possible effect of fragmentation of the baryonic shell. These crucial steps lead to an evaluation of the source luminosity in a given energy band, essential for any comparison with the observational data. The source luminosity at detector arrival time $t_a^d$, per unit solid angle $d\Omega$ and in the energy band $[\nu_1, \nu_2]$, is given by (see Ruffini *et al.*, 2003, 2004):

$$\frac{dE_\gamma^{[\nu_1,\nu_2]}}{dt_a^d\, d\Omega} = \int_{EQTS} \frac{\Delta\varepsilon}{4\pi}\, v\, \cos\vartheta\, \Lambda^{-4}\, \frac{dt}{dt_a^d} W(\nu_1, \nu_2, T_{\rm arr})\, d\Sigma. \tag{7.15}$$

Here $\Delta\varepsilon = \Delta E_{\rm int}/V$ is the energy density released in the interaction of the ABM pulse with the ISM inhomogeneities measured in the comoving frame, $\Lambda = \gamma(1 - (v/c)\cos\vartheta)$ is the Doppler factor, $W(\nu_1, \nu_2, T_{\rm arr})$ is an "effective weight" required to evaluate only the contributions in the energy band $[\nu_1, \nu_2]$,

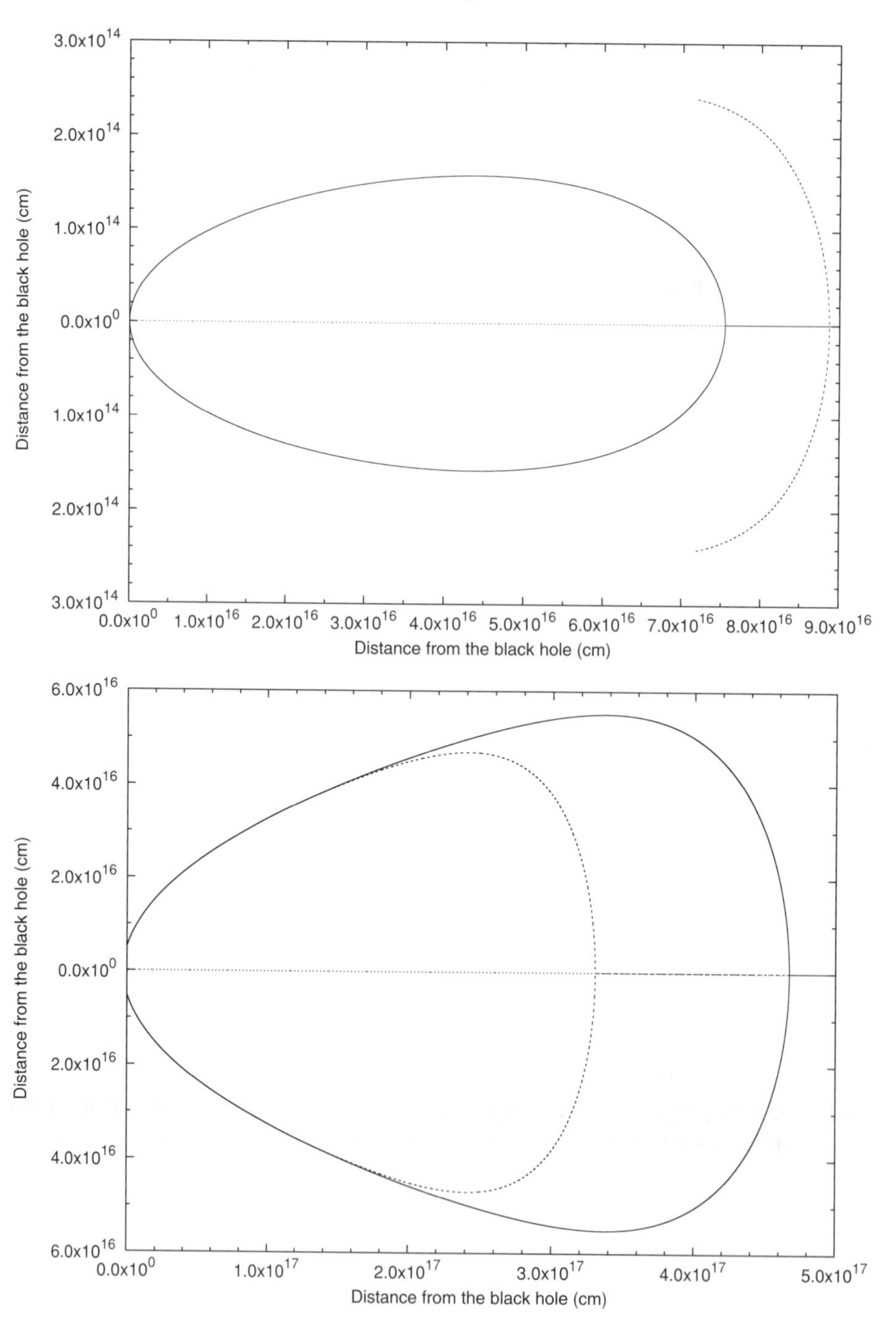


Fig. 7.28. Comparison between the EQTSs computed using the approximate formulas given by Panaitescu & Mészáros (1998) (dotted line) in the fully radiative case and the corresponding ones computed using our exact solution (solid line). The upper (lower) panel corresponds to $t_a^d = 35$ s ($t_a^d = 4$ day). Details in Bianco & Ruffini (2004).

$d\Sigma$ is the surface element of the EQTS at detector arrival time $t_a^d$ on which the integration is performed (see also Ruffini *et al.*, 2002) and $T_{\mathrm{arr}}$ is the observed temperature of the radiation emitted from $d\Sigma$:

$$T_{\mathrm{arr}} = T_s / \left[\gamma \left(1 - (v/c)\cos\vartheta\right)(1+z)\right]. \tag{7.16}$$

The "effective weight" $W(\nu_1, \nu_2, T_{\mathrm{arr}})$ is given by the ratio of the integral over the given energy band of a Planckian distribution at a temperature $T_{\mathrm{arr}}$ to the total integral $aT_{\mathrm{arr}}^4$:

$$W(\nu_1, \nu_2, T_{\mathrm{arr}}) = \frac{1}{aT_{\mathrm{arr}}^4} \int_{\nu_1}^{\nu_2} \rho(T_{\mathrm{arr}}, \nu)\, d\left(\frac{h\nu}{c}\right)^3, \tag{7.17}$$

where $\rho(T_{\mathrm{arr}}, \nu)$ is the Planckian distribution at temperature $T_{\mathrm{arr}}$:

$$\rho(T_{\mathrm{arr}}, \nu) = (2/h^3) h\nu / (e^{h\nu/(kT_{\mathrm{arr}})} - 1). \tag{7.18}$$

This apparently simple procedure needs a very complicated integration technique. Every value of the luminosity at any given detector arrival time is actually the outcome of an integration over the given EQTS of literally millions of different points, each one characterized by a different temperature and a different value of the Lorentz boost!

Historically, this procedure was used for GRB991216, where for the first time we recognized the existence of the P-GRB and its entire afterglow. Some much more detailed examples were made recently, using data obtained by the INTEGRAL and Swift satellites (Bernardini *et al.*, 2005; Ruffini *et al.*, 2006a).

## 7.15 The three paradigms for the interpretation of GRBs

Having outlined the main theoretical features of our model and some recent observational verifications, I would like to recall the three basic paradigms for understanding GRBs proposed already by us in 2001. These assume as a starting point that all GRBs, whether short or long, are characterized by the same basic process of gravitational collapse to a black hole.

The first paradigm, the relative spacetime transformation (RSTT) paradigm (Ruffini *et al.*, 2001a) emphasizes the importance of a global analysis of the GRB phenomenon encompassing both the optically thick and the afterglow phases. Since all the data are received at the detector arrival time it is essential to know the equations of motion for all relativistic phases with $\gamma > 1$ of the GRB sources in order to reconstruct the time coordinate in the laboratory frame. Contrary to other phenomena in non-relativistic physics or astrophysics where every phase can be examined separately from the others, for GRBs the phases are inter-related by their signals received in arrival time $t_a^d$. In order to describe the physics of the source at a given arrival time $t_a^d$, the laboratory time $t$ must be calculated taking necessarily into account the entire past worldline of the source.

The second paradigm, the interpretation of the burst structure (IBS) paradigm (Ruffini *et al.*, 2001b) covers three fundamental issues:

(i) the existence, in the canonical GRB, of two different components, the P-GRB and the afterglow related by precise equations determining their relative amplitude and temporal sequence (see Ruffini *et al.*, 2003);
(ii) the fact that the "prompt emission", usually considered as a burst in the literature, is not a burst at all in our model – it is just the emission from the peak of the afterglow (see Fig. 7.29);
(iii) the crucial role of the parameter $B$ in determining the relative amplitude of the P-GRB to the afterglow and discriminating between the short and the long bursts (see Fig. 7.30).

Both short and long bursts arise from the same physical phenomena, the dyadosphere. For values of the baryon loading $B < 10^{-4}$ (see Fig. 7.30) the P-GRB becomes prominent with respect to the afterglow. These correspond to the short bursts. In the limit of $B \to 0$, all the energy is emitted in the P-GRB and the afterglow goes to zero. The presence of baryonic matter in the range $10^{-4} < B < 10^{-2}$ leads to the prominence of the afterglow energy with respect to the P-GRB one. When the ISM density is large enough ($n_{\rm ism} \sim 1$ particle/cm$^3$), the afterglow peak emission is prominent with respect to the P-GRB and generates the so-called long bursts.

The third paradigm, the GRB-Supernova Time Sequence (GSTS) paradigm (Ruffini *et al.*, 2001c), deals with the relation between the GRB and the associated supernova process. Models of GRBs based on a single source (the "collapsar") generating both the supernova (SN) and the GRB abound in the literature (see e.g. Woosley & Bloom, 2006). In our approach we have emphasized the concept of induced gravitational collapse, which occurs strictly in a binary system. The SN originates from a normal star and the GRB from collapse to a black hole. The two phenomena are qualitatively very different. There is still much to be discovered about SNe due to their complexity. In contrast the GRB is much better known since its collapse to a black hole is now understood. The concept of induced collapse implies at least two alternative scenarios. In the first, the GRB triggers a SN explosion in the very last phase of the thermonuclear evolution of a companion star (Ruffini *et al.*, 2001c). In the second, the early phases of the SN induce gravitational collapse of a companion neutron star to a black hole (Ruffini, 2008). Of course, there is also the possibility that the collapse to a black hole that generates the GRB

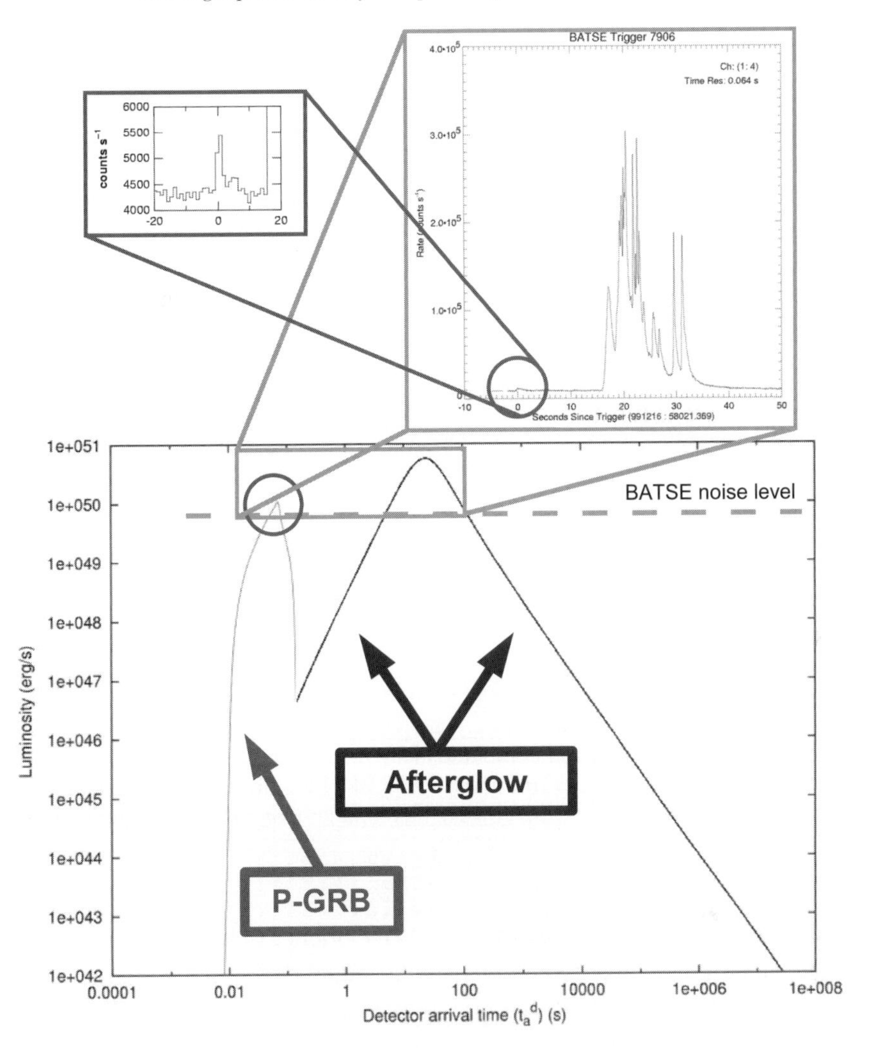


Fig. 7.29. GRB991216 within our theoretical framework. The prompt emission observed by BATSE is identified with the peak of the afterglow, while the small precursor is identified with the P-GRB. Details in Ruffini *et al.* (2001b, 2002, 2003, 2005a).

occurs in a single star system, clearly without any SN and fulfilling the very strong condition given by Eq. (7.10).

It is clear that GRBs are possibly the most important astrophysical systems ever discovered, both for physics and for astrophysics. Although enormously complex, they offer the possibility of being completely understood, allowing a detailed theoretical description. Through GRBs we are exploring some of the real frontiers of physics and astrophysics. These include the first precision analysis of the formation

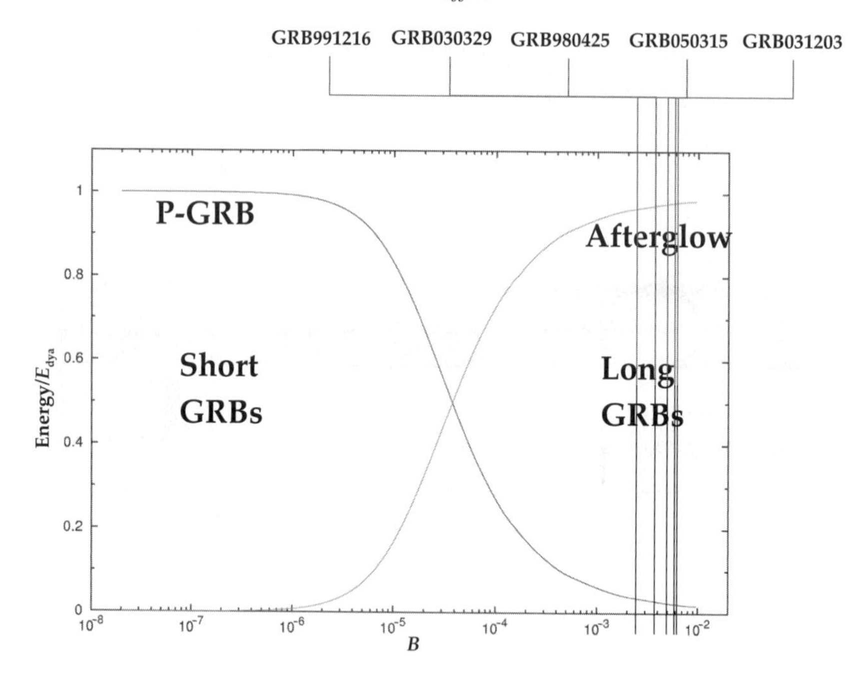


Fig. 7.30. The energy radiated in the P-GRB and in the afterglow, in units of the total energy of the dyadosphere ($E_{dya}$), are plotted as functions of the $B$ parameter. The values of the $B$ parameter computed in our theory for the sources GRB991216, GRB030329, GRB980425, GRB050315, GRB031203 are also represented. It is very remarkable that they are all consistently smaller than the absolute upper limit $B < 10^{-2}$ found in Ruffini *et al.* (2000).

of a Kerr–Newman geometry and the first clear evidence of the creation of electron–positron pairs by vacuum polarization using the blackholic energy. Also, there is the possibility that the early phases of the onset of SNe can be observed. This would provide essential data about this enormously complex and still not understood system.

## 7.16 The Kerr solution and art and science

The trajectories studied with Mark Johnston (see Johnston & Ruffini, 1974) were used in the design of a silver sculpture by Attilio Pierelli (see Fig. 7.31), the prize for the Marcel Grossmann Awards. We recall that these awards are traditionally given to a scientific institution or to scientists who have distinguished themselves in the field of relativistic astrophysics. In 2006 the Award was given to the Freie Universität Berlin, to Roy Kerr, to George Coyne and to Joachim Trumper. This same figure has become the logo of ICRA, the International Center for Relativistic Astrophysics, and its network ICRANet (see Fig. 7.32).

Fig. 7.31. The sculpture TEST (Traction of Events in Space and Time) by the Italian artist Attilio Pierelli as photographed by the Japanese artist Shu Takahashi (Imponente, 1985).

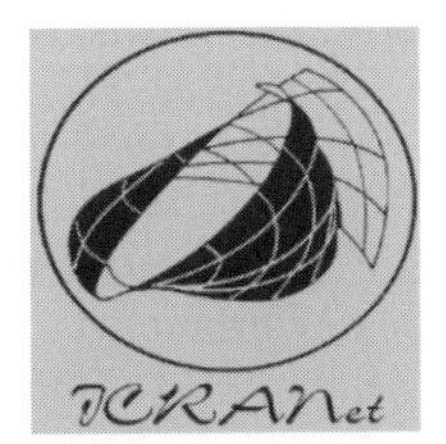


Fig. 7.32. The ICRA and ICRANet logos.

## Acknowledgments

Too many people have contributed through discussions to this manuscript to be mentioned. I would like to acknowledge specifically Roy Kerr, for having read the manuscript and making many important suggestions. I am especially thankful to José Funes, director of the Specola Vaticana, for offering the hospitality to Roy Kerr in the ICRANet office in Castel Gandolfo allowing us to finalize this manuscript. Finally, I like to thank all the members of the GRB group at ICRA and ICRANet, especially Carlo Luciano Bianco, for the preparation of the manuscript.

## References

Anderson, C. D. (1933), *Phys. Rev.* **43**, 491.
Arbolino, M. V. & Ruffini, R. (1988), *Astron. Astroph.* **192**, 107.

Atkinson, R. & Houtermans, F. G. (1929a), *Z. Phys.* **54**, 656.
Atkinson, R. & Houtermans, F. G. (1929b), *Z. Phys.* **58**, 478.
Bekenstein, J. D. (1973), *Phys. Rev.* **D 7**, 2333.
Bekenstein, J. D. (1974), *Phys. Rev.* **D 9**, 3292.
Bekenstein, J. D. (2006), *Of Gravity, Black Holes and Information* (Di Renzo Editore, Rome).
Bell, J. & Hewish, T. (1967), *Nature* **213**, 1214.
Bernardini, M. G., Bianco, C. L., Chardonnet, P., Fraschetti, F., Ruffini, R. & Xue, S.-S. (2005), *Astroph. J.* **634**, L29.
Bethe, H. (1939), *Phys. Rev.* **55**, 103.
Bianco, C. L. & Ruffini, R. (2004), *Astroph. J.* **605**, L1.
Bianco, C. L. & Ruffini, R. (2005a), *Astroph. J.* **620**, L23.
Bianco, C. L. & Ruffini, R. (2005b), *Astroph. J.* **633**, L13.
Bianco, C. L., Ruffini, R., Vereshchagin, G. V. & Xue, S.-S. (2006), *J. Kor. Phys. Soc.* **49**, 722.
Bisnovatyi-Kogan, G. S. & Murzina, M. V. A. (1995) *Phys. Rev.* **D 52**, 4380.
Bisnovatyi-Kogan, G. S., Merafina, M., Ruffini, R. & Vesperini, E. (1998), *Astroph. J.* **500**, 217.
Blandford, R. D. & McKee, C. F. (1976), *Phys. Fluids* **19**, 1130.
Blandford, R. D. & Znajek, R. L. (1977), *Mon. Not. Roy. Astron. Soc.* **179**, 433.
Bolton, C. T. (1972a), *Nature* **235**, 271.
Bolton, C. T. (1972b), *Nature Phys. Sci.* **240**, 124.
Burbidge, E. M., Burbidge, G., Fowler, W. A. & Hoyle, F. (1957), *Rev. Mod. Phys.* **29**, 547.
Carilli, C. L., Perley, R., Harris, D. E. & Barthel, P. D. (1998), *Phys. Plasmas* **5**, 1981.
Carter, B. (1968), *Phys. Rev.* **174**, 1559.
Chandrasekhar, S. (1930), *Mon. Not. Roy. Astron. Soc.* **91**, 4.
Chandrasekhar, S. (1935), *Mon. Not. Roy. Astron. Soc.* **95**, 207.
Chandrasekhar, S. (1939), *An Introduction to the Study of Stellar Structure* (University of Chicago Press, Chicago).
Chardonnet, P., Mattei, A., Ruffini, R. & Xue, S.-S. (2003), *Nuovo Cimento* **B 118**, 1063.
Cherubini, C., Geralico, A., Rueda Hernandez, J. A. & Ruffini, R. (2007), *Phys. Rev. D*, submitted.
Cherubini, C., Ruffini, R. & Vitagliano, L. (2002), *Phys. Lett.* **B 545**, 226.
Chiang, J. & Dermer, C. D. (1999), *Astroph. J.* **512**, 699.
Christodoulou, D. (1970), *Phys. Rev. Lett.* **25**, 1596.
Christodoulou, D. & Ruffini, R. (1971), *Phys. Rev.* **D 4**, 3552.
Costa, E., Frontera, F., Heise, J., *et al.* (1997), *Nature* **387**, 783.
Damour, T. (1975), *Ann. New York Acad. Sci.* **262**, 113.
Damour, T. (1977), *Proceedings of the 1st Marcel Grossmann Meeting*, ed. R. Ruffini (North-Holland, Amsterdam).
Damour, T. (1978), *Phys. Rev.* **D 18**, 3598.
Damour, T. (1982), *Proceedings of the 2nd Marcel Grossmann Meeting*, ed. R. Ruffini (North-Holland, Amsterdam).
Damour, T. & Ruffini, R. (1975), *Phys. Rev. Lett.* **35**, 463.
Damour, T., Ruffini, R., Hanni, R. S. & Wilson, J. (1978), *Phys. Rev.* **D 17**, 1518.
Deruelle, N. (1977), *Proceedings of the 1st Marcel Grossmann Meeting*, ed. R. Ruffini (North-Holland, Amsterdam).
Eddington, A. S. (1920), *Brit. Assoc. Rep.*, p. 45.
Eddington, A. S. (1935), *Mon. Not. Roy. Astron. Soc.* **95**, 194.
Einstein, E. (1939), *Ann. Math.* **40**, 922.

Fairbank, J. D., Deaver, B. S., Everitt, C. W. F. & Michelson, P. F. (1988), *Near Zero: New Frontiers of Physics* (W. H. Freeman, San Francisco).
Fermi, E., *et al.* (1942), December 2, as reproduced in *Am. J. Phys.* **20** (1952), 536.
Fermi, E. (1949), "Teoria sull'origine degli elementi", lecture held at the Physics Institute of the University of Rome "La Sapienza" on 7 October 1949, translated into English in: *Fermi and Astrophysics*, ed. V. Gurzadyan & R. Ruffini (World Scientific, Singapore, in press).
Finzi, A. & Wolf, R. A. (1968), *Astroph. J.* **153**, 865.
Floyd, R. M. & Penrose, R. (1971), *Nature Phys. Sci.* **229**, 177.
Foresta Martin, F. (1999), *Corriere della sera*, March 7, p. 27.
Gallo, E., Fender, R., Kaiser, C., Russell, D., Morganti, R., Oosterloo, T. & Heinz, S. (2005), *Nature* **436**, 819.
Gamow, G. (1938), *Nuclear Physics* (Oxford University Press, Oxford).
Gamow, G. & Houtermans, F. G. (1928), *Z. Phys.* **52**, 496.
Gatoff, G., Kerman, A. K. & Matsui, T. (1987), *Phys. Rev.* **D 36**, 114.
Giacconi, R. (2002), *Nobel Lecture*.
Giacconi, R. (2005), *Ann. Rev. Astron. Astroph.* **43**, 1.
Giacconi, R. & Ruffini, R. (1978), *Physics and Astrophysics of Neutron Stars and Black Holes* (North-Holland, Amsterdam).
Gold, T. (1968), *Nature* **218**, 731.
Gold, T. (1969), *Nature* **221**, 27.
Granot, J., Piran, T. & Sari, R. (1999), *Astroph. J.* **513**, 679.
Gruzinov, A. & Waxman, E. (1999), *Astroph. J.* **511**, 852.
Gursky, H. & Ruffini R. (1975), *Neutron Stars, Black Holes and Binary X-Ray Sources* (Springer, Berlin).
Harrison, B. K., Thorne, K. S., Wakano, M. & Wheeler, J. A. (1965), *Gravitation Theory and Gravitational Collapse* (University of Chicago Press, Chicago).
Hawking, S. W. (1971), *Phys. Rev. Lett.* **26**, 1344.
Hawking, S. W. (1974), *Nature* **238**, 30.
Hawking, S. W. (1975), *Comm. Math. Phys.* **43**, 199.
Hawking, S. W. (1976), *Phys. Rev.* **D 13**, 191.
Heisenberg, W. & Euler, H. (1935), *Zeit. Phys.* **98**, 714.
Imponente, A. (1985), Catalog presentation of the show of A. Pierelli, "TEST, Trascinamento di Eventi Spazio Temporali", Rome, Galleria MR, September–October 1985.
Institut International de Physique Solvay (1958), "La structure et l'évolution de l'univers – Rapports et discussions", XI Conseil de Physique, Université de Bruxelles, June 9–13, 1958, ed. R. Stoops, Bruxelles.
Jacobi, C. G. J. (1995), *C. G. J. Jacobi's Collected Works (Gesammelte Werke)*, (American Mathematical Society, Providence).
Johnston, M. & Ruffini, R. (1974), *Phys. Rev.* **D 10**, 2324.
Kerr, R. P. (1963), "Gravitational field of a spinning mass as an example of algebraically special metrics", *Phys. Rev. Lett.* **11**, 237.
Khriplovich, I. B. (2000), *Nuovo Cimento* **B 115**, 761.
Landau, L. D. (1932), *Phys. Ze. Sowjet.* **1**, 285.
Landau, L. D. & Lifshitz, E. M. (1981), *Quantum Mechanics*, 3rd edition (Butterworth-Heinemann, Oxford).
Leach, R. W. & Ruffini, R. (1973), *Astroph. J.* **180**, L15.
Maxwell, J. C. (1986), *Maxwell on Molecules and Gases*, ed. E. Garber, S. G. Brush & C. W. F. Everitt (MIT Press, Cambridge, MA).

Merafina, M. & Ruffini, R. (1989), *Astron. Astroph* **221**, 4.
Merafina, M. & Ruffini, R. (1990), *Astron. Astroph* **227**, 415.
Merafina, M. & Ruffini, R. (1997), *Int. J. Mod. Phys.* **D 6**, 785.
Mészáros, P. (2002), *Ann. Rev. Astron. Astroph.* **40**, 137.
Mészáros, P. (2006), *Rep. Prog. Phys.* **69**, 2259.
Mészáros, P., Laguna, P. & Rees, M. J. (1993), *Astroph. J.* **415**, 181.
Mezzacappa, A. & Fuller, G. M. (2006), *Open Issues in Core Collapse Supernova Theory* (World Scientific, Singapore).
Mirabel, F. & Rodrigues, I. (2003), *Science* **300**, 1119.
Misner, C. W., Thorne, K. S. & Wheeler, J. A. (1973), *Gravitation* (W. H. Freeman, San Francisco).
Ohanian, H. C. & Ruffini, R. (1974), *Gravitation and Spacetime, 2nd edition* (W. W. Norton, New York).
Oppenheimer, J. R. & Serber, R. (1938), *Phys. Rev.* **54**, 540.
Oppenheimer, J. R. & Snyder, H. (1939), *Phys. Rev.* **56**, 455.
Oppenheimer, J. R. & Volkoff, R. (1939), *Phys. Rev.* **55**, 374.
Paciesas, W. S., *et al.* (1999), *Astroph. J. Suppl. Ser.* **122**, 465.
Pacini, F. (1968), *Nature* **219**, 145.
Page, D. N. (2006), *Astroph. J.* **653**, 1400.
Panaitescu, A. & Mészáros, P. (1998), *Astroph. J.* **493**, L31.
Penrose, R. (1969), *Riv. Nuovo Cimento* **1**, 252.
Perrin, J. B. (1920), *Rev. Mois* **21**, 113.
Piran, T. (1999), *Phys. Rep.* **314**, 575.
Piran, T., Shemi, A. & Narayan, R. (1993), *Mon. Not. Roy. Astron. Soc.* **263**, 861.
Preparata, G., Ruffini, R. & Xue, S.-S. (1998), *Astron. Astroph.* **338**, L87.
Punsly, B. (2001), *Black Hole Gravitohydromagnetics* (Springer, Berlin).
Regge, T. & Wheeler, J. A. (1957), *Phys. Rev.* **108**, 1063.
Rees, M. J. & Mészáros, P. (1998), *Astroph. J.* **496**, L1.
Rees, M. J., Ruffini, R. & Wheeler, J. A. (1976), *Black holes, Gravitational Waves and Cosmology* (Gordon and Breach, New York).
Rhoades, C. E. & Ruffini, R. (1971), *Astroph. J.* **163**, L83.
Robinson, I., Schild, A. & Schücking, E. L. (1965), *Quasi-stellar Sources and Gravitational Collapse, including the Proceedings of the First Texas Symposium on Relativistic Astrophysics* (University of Chicago Press, Chicago).
Ruffini, P. (1803), *Memorie di matematica e fisica della Società Italiana delle Scienze* **10**, 410.
Ruffini, R. (1973), *Black Holes*, ed. C. DeWitt & B. S. DeWitt (Gordon and Breach, New York).
Ruffini, R. (1998), *Black Holes and High Energy Astrophysics*, ed. H. Sato & N. Sugiyama (Universal Academic Press, Tokyo).
Ruffini, R. (2001a), *Exploring the Universe*, ed. H. Gursky, R. Ruffini & L. Stella (World Scientific, Singapore).
Ruffini, R. (2001b), *Fluctuating Paths and Fields*, ed. W. Janke, A. Pelster, H.-J. Schmidt & M. Bachmann (World Scientific, Singapore).
Ruffini, R. (2008), *Proceedings of the 11th Marcel Grossmann Meeting*, ed. R. Jantzen, H. Kleinert & R. Ruffini (World Scientific, Singapore), pp. 1352–1358.
Ruffini, R. & Wheeler, J. A. (1971a), *Phys. Today* **24**, 30.
Ruffini, R. & Wheeler, J. A. (1971b), *The Significance of Space Research for Fundamental Physics*, ed. A. F. Moore & V. Hardy (European Space Research Organization (ESRO)), Book No. SP-52.

Ruffini, R. & Wilson, J. (1975), *Phys. Rev.* **D 12**, 2959.
Ruffini, R., Salmonson, J. D., Wilson, J. R. & Xue, S.-S. (1999), *Astron. Astroph.* **350**, 334.
Ruffini, R., Salmonson, J. D., Wilson, J. R. & Xue, S.-S. (2000), *Astron. Astroph.* **359**, 855.
Ruffini, R., Bianco, C. L., Chardonnet, P., Fraschetti, F. & Xue, S.-S. (2001a), *Astroph. J.* **555**, L107.
Ruffini, R., Bianco, C. L., Chardonnet, P., Fraschetti, F. & Xue, S.-S. (2001b), *Astroph. J.* **555**, L113.
Ruffini, R., Bianco, C. L., Chardonnet, P., Fraschetti, F. & Xue, S.-S. (2001c), *Astroph. J.* **555**, L117.
Ruffini, R., Bianco, C. L., Chardonnet, P., Fraschetti, F. & Xue, S.-S. (2002), *Astroph. J.* **581**, L19.
Ruffini, R. & Vitagliano, L. (2002), *Phys. Lett.* **B 545**, 233.
Ruffini, R., Bianco, C. L., Chardonnet, P., Fraschetti, F., Vitagliano, L. & Xue, S.-S. (2003), *Cosmology and Gravitation: Xth Brazilian School of Cosmology and Gravitation*, ed. M. Novello & S. E. Perez Bergliaffa, *AIP Conf. Proc.* **668**, 16.
Ruffini, R. & Sigismondi, C. (2003), *Nonlinear Gravitodynamics* (World Scientific, Singapore).
Ruffini, R. & Vitagliano, L. (2003), *Int. J. Mod. Phys.* **D 12**, 121.
Ruffini, R., Vitagliano, L. & Xue, S.-S. (2003a), *Phys. Lett.* **B 559**, 12.
Ruffini, R., Vitagliano, L. & Xue, S.-S. (2003b), *Phys. Lett.* **B 573**, 33.
Ruffini, R., Bianco, C. L., Chardonnet, P., Fraschetti, F., Gurzadyan, V. & Xue, S.-S. (2004), *Int. J. Mod. Phys.* **D 13**, 843.
Ruffini, R., Bernardini, M. G., Bianco, C. L., Chardonnet, P., Fraschetti, F., Gurzadyan, V., Vitagliano, L. & Xue, S.-S. (2005a), *Cosmology and Gravitation: XIth Brazilian School of Cosmology and Gravitation*, ed. M. Novello & S. E. Perez Bergliaffa, *AIP Conf. Proc.* **782**, 42.
Ruffini, R., Bianco, C. L., Chardonnet, P., Fraschetti, F., Gurzadyan, V. & Xue, S.-S. (2005b), *Int. J. Mod. Phys.* **D 14**, 97.
Ruffini, R., Bernardini, M. G., Bianco, C. L., Chardonnet, P., Fraschetti, F. & Xue, S.-S. (2006a), *Astroph. J.* **645**, L109.
Ruffini, R., Rotondo, M. & Xue, S.-S. (2006b), *Int. J. Mod. Phys.* **D 16**, 1.
Ruffini, R., Bianco, C. L., Vereshchagin, G. V. & Xue, S.-S. (2007a), *Relativistic Astrophysics and Cosmology – Einstein's Legacy*, ed. B. Aschenbach, V. Burwitz, G. Hasinger & B. Leibundgut (Springer, Berlin), pp. 402–406 [arxiv:astro-ph/0605385].
Ruffini, R., Bianco, C. L. & Xue, S.-S. (2007b), *Astroph. J.*, submitted.
Ruffini, R., Vitagliano, L. & Xue, S.-S. (2007c), *Phys. Rep.*, in press.
Ruffini, R., Rotondo, M. & Xue, S.-S. (2008), *Proceedings of the 11th Marcel Grossmann Meeting*, ed. R. Jantzen, H. Kleinert & R. Ruffini (World Scientific, Singapore), in press.
Sari, R. (1997), *Astroph. J.* **489**, L37.
Sari, R. (1998), *Astroph. J.* **494**, L49.
Schwinger, J. (1951), *Phys. Rev.* **82**, 664.
Shakura, N. I. (1972a), *Sov. Astron. AJ* **49**, 495.
Shakura, N. I. (1972b), *Sov. Astron. AJ* **49**, 642.
Shklovsky, I. S. (1967), *Astroph. J.* **148**, L1.
Shvartsman, V. F. (1971), *Sov. Astron. AJ* **15**, 377.
Størmer, C. (1934), *Astroph. Norvegica* **1**, 1.
Tavani, M. (1998), *Astroph. J.* **497**, L21.

van Paradijs, J., Kouveliotou, C. & Wijers, R. A. M. J. (2000), *Ann. Rev. Astron. Astroph.* **38**, 379.
von Weizsäckr, C. F. (1937), *Phys. Z.* **38**, 176.
von Weizsäcker, C. F. (1938), *Phys. Z.* **39**, 633.
Waxman, E. (1997), *Astroph. J.* **491**, L19.
Webster, B. & Murdin, P. (1972), *Nature* **235**, 37.
Woosley, S. E. & Bloom, J. S. (2006), *Ann. Rev. Astron. Astroph.* **44**, 507.
Zel'dovich, Ya. B. (1985), *Selected Works: Particles, Nuclei, Universe*, 2 volumes (Nauka, Moscow).
Zel'dovich, Ya. B. & Guseynov, O. K. (1965), *Dokl. Akad. Nauk. USSR* **162**, 791.
Zerilli, F. (1970), *Phys. Rev.* **D 2**, 2141.
Zerilli, F. (1974), *Phys. Rev.* **D 9**, 860.

# 8

# Supermassive black holes

Fulvio Melia

## 8.1 Introduction

Supermassive black holes have generally been recognized as the most destructive force in nature. But in recent years, they have undergone a dramatic shift in paradigm. These objects may have been critical to the formation of structure in the early universe, spawning bursts of star formation and nucleating proto-galactic condensations. Possibly half of all the radiation produced after the Big Bang may be attributed to them, whose number is now known to exceed 300 million. The most accessible among them is situated at the center of our galaxy. In the following pages, we will examine the evidence that has brought us to this point, and we will understand why many expect to actually image the event horizon of the galaxy's central black hole within this decade.

The supermassive black hole story begins in 1963, at the Mount Palomar observatory, where Schmidt (1963) was pondering over the nature of a star-like object with truly anomalous characteristics. Meanwhile, at the University of Texas, Kerr (1963) was making a breakthrough discovery of a solution to Einstein's field equations of general relativity. Kerr's work would eventually produce a description of space and time surrounding a spinning black hole, now thought to power the compact condensations of matter responsible for producing the mystery on Schmidt's desk that year.

The development and use of radio telescopes in the 1940s had led to the gradual realization that several regions of the sky are very bright emitters of centimeter-wavelength radiation. In the early 1960s, the British astronomer Cyril Hazard's idea of using lunar occultation to determine with which, if any, of the known visible astronomical objects the emitter of centimeter-wavelength radiation was associated, led to the successful identification of 3C 273 as a star-like object

The Kerr Spacetime: Rotating Black Holes in General Relativity. Ed. David L. Wiltshire, Matt Visser and Susan M. Scott. Published by Cambridge University Press. 

Fig. 8.1. The quasar 3C 273 (the bright object in the upper-left-hand corner) was one of the first objects to be recognized as a "quasi-stellar-radio-source" (quasar), due to its incredible optical and radio brightness. Insightful analysis led to the realization that 3C 273 is actually an incredibly powerful, distant object. This *Chandra* image has a size $\approx 22 \times 22$ square arcseconds, which at the distance to 3C 273, corresponds to about $2000 \times 2000$ square light-years. (Photograph courtesy of H. L. Marshall *et al.*, NASA, and MIT.)

in Virgo. Its redshifted lines, however, indicated that this was not a star at all, but rather an object lying at great cosmological distances. A recent image of this historic source was made with the *Chandra* X-ray telescope, and is shown in Fig. 8.1.

3C 273's total optical output varies significantly in only ten months or so, implying that its size cannot exceed a few light-years – basically the distance between the Sun and its nearest stellar neighbor. So it was clear right from the beginning that the quasar phenomenon must be associated with highly compact objects. Even more impressively, their X-ray output has now been seen to vary in a matter of only hours, corresponding to a source size smaller than Neptune's orbit. Each quasar typically releases far more energy than an entire galaxy, yet the central

Fig. 8.2. This Hubble Space Telescope (HST) image reveals the faint host galaxy within which dwells the bright quasar known as QSO 1229+204. The quasar is seen to lie in the core of an ordinary-looking galaxy with two spiral arms of stars connected by a bar-like feature. (Photograph courtesy of John Hutchings, Dominion Astrophysical Observatory, and NASA.)

engine that drives this powerful activity occupies a region smaller than our solar system.

The idea that such small volumes could be producing the power of a hundred billion Suns led to their early identification as radiative manifestations of supermassive black holes (see, e.g. Salpeter, 1964; Zel'dovich and Novikov, 1967; and Lynden-Bell, 1969). But are they "naked" – deep, dark pits of matter floating aimlessly across the primeval cosmic soup – or are they attached to more recognizable structures in the early universe?

In recent years, the task of source identification has been made easier using the Hubble Space Telescope (see Fig. 8.2). The most widely accepted view now is that quasars are found in galaxies with active, supermassive black holes at their centers. Because of their enormous distance from Earth, the "host" galaxies appear very

Fig. 8.3. The collision between two galaxies begins with the unraveling of the spiral disks. This HST image shows the interacting pair of galaxies NGC 2207 (the larger, more massive object on the left) and IC 2163 (the smaller one on the right), located some 114 million light-years from Earth. By this time, IC 2163s stars have begun to surf outward to the right on a tidal tail created by NGC 2207s strong gravity. (Image courtesy of Debra M. Elmegreen at Vassar College, Bruce G. Elmegreen at the IBM Research Division, NASA, M. Kaufman at Ohio State U., E. Brinks at U. de Guanajuato, C. Struck at Iowa State U., M. Klaric at Bell South, M. Thomasson at Onsala Space Observatory, and the Hubble Heritage Team based at the Space Telescope Space Institute and AURA.)

small and faint, and are very hard to see against the much brighter quasar light at their center.

Quasars actually reside in the nuclei of many different types of galaxy, from the normal to those highly disturbed by collisions or mergers. A supermassive black hole at the nucleus of one of these distant galaxies "turns on" when it begins to accrete stars and gas from its nearby environment; the rate at which matter is converted into energy can be as high as 10 solar masses per year. So the character and power of a quasar depend in part on how much matter is available for consumption. Disturbances induced by gravitational interactions with neighboring galaxies can trigger the infall of material toward the center of the quasar host galaxy (see Fig. 8.3). However, many quasars reside in apparently undisturbed galaxies, and this may be an indication that mechanisms other than a disruptive collision may also be able to effectively fuel the supermassive black hole residing at the core.

Some supermassive black holes may not be visible as quasars at all, but rather just sputter enough to become the fainter galactic nuclei in our galactic neighborhood. Our Milky Way galaxy and our neighbor, the Andromeda galaxy, harbor supermassive black holes with very little nearby plasma to absorb. The question concerning how the undisturbed galaxies spawn a quasar is still not fully answered. Perhaps the Next Generation Space Telescope, now under development and expected to fly soon after 2010, will be able to probe even deeper than the Hubble Space Telescope has done, and expose the additional clues we need to resolve this puzzle.

## 8.2 The quasar/supermassive black hole census

By now, some 15 000 distant quasars have been found, though the actual number of supermassive black holes discovered thus far is much greater. Because of their intrinsic brightness, the most distant quasars are seen at a time when the universe was a mere fraction of its present age, roughly one billion years after the Big Bang. The current distance record is held by an object found with the Sloan Digital Sky Survey (SDSS), with a redshift of $\sim$6.3, corresponding to a time roughly 700 million years after the Big Bang.

The SDSS has shown that the number of quasars rose dramatically from a billion years after the Big Bang to a peak around 2.5 billion years later, falling off sharply at later times toward the present. Quasars turn on when fresh matter is brought into their vicinity, and then fade into a barely perceptible glimmer not long thereafter.

However, not all the supermassive black holes in our midst have necessarily grown through the quasar phase. Quasars typically have masses $\sim 10^9\ M_\odot$. Yet the black hole at the center of our galaxy is barely $3.4 \times 10^6\ M_\odot$. In other words, not all the supermassive black holes in our vicinity are dormant quasars.

A recent discovery suggests how some of these "smaller" black holes may have gotten their start. *Chandra* has identified what appears to be a mid-sized black hole 600 light-years from the center of M82 (Matsumoto *et al.* 2001). With a mass of $\sim 500\ M_\odot$, this object could conceivably sink to the center of M82, and then grow to become a supermassive black hole in its own right, without having passed through the rapid accretion phase of a quasar.

So the class of known quasars may be a good tracer of supermassive black holes, but it clearly does not encompass all of them. Taking advantage of two patches of sky relatively devoid of nearby objects, *Chandra* produced two of the deepest images ever made of the distant cosmos at X-ray energies, one in the southern hemisphere and the other in the north – the latter, called the *Chandra* Deep Field

Fig. 8.4. The *Chandra* Deep Field North X-ray image. The vast majority of the 500 or so sources in this view (spanning a region approximately 28 arcminutes wide) are supermassive black holes. Extrapolating this number to the whole sky, one infers $\sim$ 300 million such objects spread across the universe. (Image courtesy of D. M. Alexander, F. E. Bauer, W. N. Brandt *et al.*, NASA and PSU.)

North, is shown in Fig. 8.4. Based on the number of suspected supermassive black holes in these images, one infers an overall population of $\sim$ 300 million throughout the cosmos.

And yet, these X-ray detections speak only of those particular supermassive black holes whose orientation facilitates the transmission of their high-energy radiation toward Earth. Their actual number must be higher than this; indeed, there is now growing evidence that many – perhaps the majority – of the supermassive black holes in the universe are obscured from view. The faint X-ray background pervading the intergalactic medium has been a puzzle for many years. Unlike the cosmic microwave background radiation left over from the Big Bang, the photons in the X-ray haze are too energetic to have been produced at early times. Instead, this radiation field suggests a more recent provenance associated with a population of sources whose overall radiative output may actually dominate over everything else in the cosmos. Stars and ordinary galaxies simply do not radiate profusely at such high energy, and therefore cannot fit the suggested profile.

A simple census shows that in order to produce such an X-ray glow with quasars alone, for every known source there ought to be ten more obscured ones. This would also mean that the growth of most supermassive black holes by accretion is hidden from the view of optical, UV, and near infrared telescopes. Fabian *et al.* (2000) have reported the discovery of an object they call a type-2 quasar. Invisible to optical light telescopes, the nucleus of this otherwise normal looking galaxy betrayed its supermassive guest with a glimmer of X-rays. The implication is that many more quasars, and their supermassive black-hole power sources, may be hidden in otherwise innocuous-looking galaxies.

And so, the all-pervasive X-ray haze, in combination with the discovery of gas-obscured quasars, now point to supermassive black holes as the agents behind perhaps *half* of all the universe's radiation produced after the Big Bang. Ordinary stars no longer monopolize the power as they had for decades prior to the advent of space-based astronomy.

## 8.3 Black holes in the nuclei of normal galaxies

Much closer to Earth, within hundreds of thousands of light-years as opposed to the 11 billion-light-year distance to the farthest quasars, supermassive black holes accrete at a lower rate than their quasar brethren and are therefore much fainter. An archetype of this group, Centaurus A, graces the southern constellation of Centaurus at a distance of 11 million light-years (see Fig. 8.5). At the center of the dark bands of dust, HST recently uncovered a disk of glowing, high-speed gas, swirling about a concentration of matter with the mass of $\sim 2 \times 10^6\ M_\odot$. This enormous mass within the central cavity cannot be due to normal stars, since these objects would shine brightly, producing an intense optical spike toward the middle, unlike the rather tempered look of the infrared image shown here.

Centaurus A is apparently funneling highly energetic particles into beams perpendicular to the dark strands of dust. It may therefore have much in common with the X-ray jet-producing black hole in 3C 273 (see Fig. 8.1), and another well-known active galactic nucleus, Cygnus A, shown in Fig. 8.6. The luminous extensions in this figure project out from the nucleus of Cygnus A, an incredible distance three times the size of the Milky Way. Yet located 600 million light-years from Earth, they cast an aspect only one-tenth the diameter of the full moon. Radio and X-ray observations show that objects such as this accelerate plasma to relativistic speeds on scales of 10–100 Schwarzschild radii (see Fig. 8.7), and that these jets are not rare.

Fig. 8.5. This image of Centaurus A is a composite of three photographs taken by the European Southern Observatory. The dramatic dark band is thought to be the remnant of a smaller spiral galaxy that collided, and ultimately merged, with a large elliptical galaxy. Imaging this galaxy at radio wavelengths, we would see two jets of plasma spewing forth from the central region in a direction perpendicular to the dark dust lanes (see Fig. 8.6 for the corresponding configuration in Cygnus A). These relativistic expulsions of plasma share much in common with the X-ray glowing stream shown in Fig. 8.1. (Photograph courtesy of Richard M. West and the European Southern Observatory.)

An important conclusion to draw from the morphology of jets like those in Cygnus A is that the process responsible for their formation must be stable for at least as long as it takes the streaming particles to journey from the center of the galaxy to the extremities of the giant radio lobes. Evidently, these pencil-thin jets of relativistic plasma have retained their current configuration for over one million years. The most conservative view regarding the nature of these objects is that a spinning black hole is ultimately responsible for this activity. The axis of its spin functions as a gyroscope, whose direction determines the orientation of the jets. Although the definitive mechanism for how the ejection takes place is yet to be determined, almost certainly the twisting motion of magnetized plasma near the

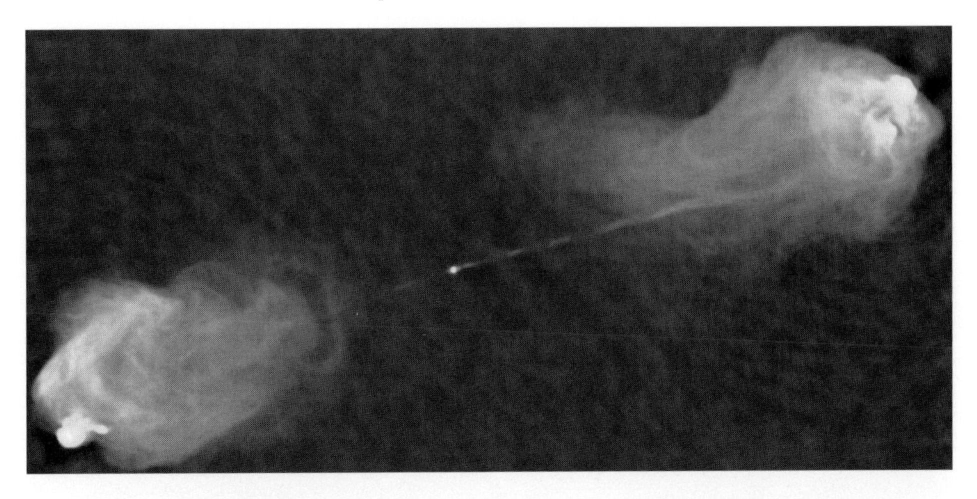

Fig. 8.6. A VLA image of the powerful central engine and its relativistic ejection of plasma in the nucleus of Cygnus A. Taken at 6 cm, this view reveals the highly ordered structure spanning over 500 000 light-years, fed by ultra-thin jets of energetic particles beamed from the compact radio core between them. The giant lobes are themselves formed when these jets plow into the tenuous gas that exists between galaxies. Despite its great distance from us (over 600 million light-years), it is still by far the closest powerful radio galaxy and one of the brightest radio sources in the sky. The fact that the jets must have been sustained in their tight configuration for over half a million (possibly as long as ten million) years means that a highly stable central object – probably a rapidly spinning supermassive black hole acting like an immovable gyroscope – must be the cause of all this activity. (Photograph courtesy of Chris Carilli and Rick Perley, NRAO, and AUI.)

black hole's event horizon is causing the expulsion. The Kerr spacetime, which describes the dragging of inertial frames about the black hole's spin axis, provides a natural setting for establishing the preferred direction for this ejective process.

## 8.4 Weighing supermassive black holes

Black hole masses are measured with a variety of techniques, though all have to do with the dynamics of matter within their gravitational influence. Knowledge of the radiating plasma's distance from the central object, and the force required to sustain its motion at that distance, is sufficient for us to extract the central mass.

One of the more compelling applications of this technique has been made to the spiral galaxy NGC 4258. Using global radio interferometry, Miyoshi *et al.* (1995) observed a disk of dense molecular material orbiting within the galaxy's nucleus at speeds of up to 650 miles per second. This disk produces sufficient radiation to excite condensations of water molecules, leading to strong maser emission at radio

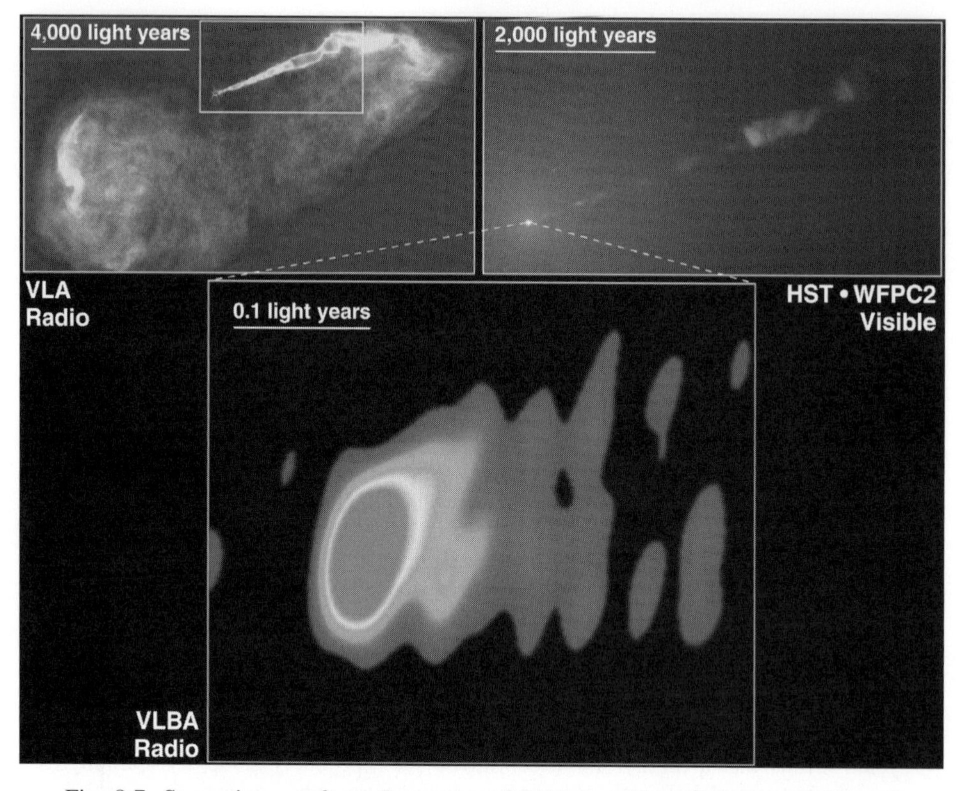


Fig. 8.7. Streaming out from the center of M87 is a black-hole-powered jet of plasma traveling at near lightspeed. Its source is a powerful central object with a mass of $\sim 3 \times 10^9\ M_\odot$. This sequence of photographs shows progressively magnified views: *top left:* a VLA image showing the full extent of the jets and the blobs at the termination points; *top right:* a visible light image of the giant elliptical galaxy M87, taken with NASA's Hubble Space Telescope; *bottom:* a Very Long Baseline Array (VLBA) radio image of the region surrounding the black hole. This view shows how the extragalactic jet is formed into a narrow beam within a few tenths of a light-year of the nucleus (the central core region is only a tenth of a light-year across), corresponding to only 100 Schwarzschild radii for a black hole of this mass. (Photographs courtesy of the National Radio Astronomy Observatory and the National Science Foundation [top left and bottom], and John Biretta at the Space Telescope Science Institute, and NASA [top right].)

wavelengths. The disk within which these water molecules are trapped is small compared to the galaxy itself, but it happens to be oriented fortuitously so that beams of microwaves are directed along our line-of-sight.

The maser clouds appear to trace a very thin disk, with a motion that follows Kepler's laws to within one part in 100, reaching a velocity (inferred from the Doppler shift of the lines) of about 1040 km per second at a distance of 0.5 light-years from the center. The implied central mass is $\sim (35 - 40) \times 10^6\ M_\odot$, concentrated within 0.5 light-years of the center in NGC 4258. This points to a matter density of at least $10^8$ $M_\odot$ per cubic light-year. If this mass were simply a highly concentrated star cluster, the stars would be separated by an average distance only somewhat greater than the diameter of the solar system, and with such proximity, they would not be able to survive the inevitable catastrophic collisions and mergers with each other. Because of the precision with which we can measure this concentration of dark mass, we regard the object in the nucleus of NGC 4258 as one of the two most compelling supermassive black holes now known, the other being the object sitting at the center of our own galaxy, about which we will have more to say shortly.

The fortuitous arrangement of factors that permits the use of this particular technique does not occur often, however, so other methods must be used. Often, clouds of gas orbiting the nucleus are irradiated by the central engine, and they in turn produce a spectrum with emission lines indicative of the plasma's ionized state. The method used to determine the distance of these ionized clouds from the black hole is known as reverberation. By monitoring the light emitted by the supermassive black hole and, independently, the radiation from its halo of irradiated clouds of gas, one can determine when a variation in the radiative output has occurred. When the quasar varies its brightness, so does the surrounding matter – but only after a certain time delay. The lag is clearly due to the time it took the irradiating light from the center to reach the clouds, and using the speed of light, this delay provides a measure of the distance between the nucleus and the orbiting plasma. Again, this procedure provides the speed of matter and its distance from the center, which together yield a determination of the gravitating mass.

Having said this, the best mass determination one can make is still based on kinematic studies of stars orbiting the central object. The center of our galaxy is close enough for this method to work spectacularly. Known as Sagittarius A*, the black hole at the center of the Milky Way may not be the most massive, nor the most energetic, but it is by far the closest, only 28 000 light-years away. Figure 8.8 shows an infrared image of the galactic center produced recently with the 8.2-meter VLT *Yepun* telescope at the European Southern Observatory in Paranal, Chile. The image we see here is sharp because of the use of adaptive optics, in which a mirror in the telescope moves constantly to correct for the effects of turbulence in the Earth's atmosphere. Sagittarius A* is so close to us compared to other supermassive black holes, that on an image such as this, we can identify individual stars orbiting a mere seven to 10 light-days from the source of gravity. In the nucleus of Andromeda, the nearest major galaxy to the Milky Way, the best we could do right now is about two light-years.

In the galactic center, stars orbit Sagittarius A* at speeds of up to five million kilometers per hour. This motion is so rapid, in fact, that we can easily detect their

Fig. 8.8. This (1.6–3.5 micron) image, taken by the European Southern Observatory's 8.2-meter telescope atop Paranal, Chile, provides one of the sharpest views of the stars surrounding the supermassive black hole at the galactic center. The location of the black hole itself is indicated by the two central arrows. This view represents a scale of approximately $2 \times 2$ square light-years. (Photograph courtesy of R. Genzel *et al.* at the Max-Planck-Institut für Extraterrestrische Physik, and the European Southern Observatory.)

proper motion on photographic plates taken only several years apart. Some of them complete an orbit about the center in only 15 years (Schödel, Ott, Genzel, *et al.*, 2002). In the middle of the photograph in Fig. 8.8, it appears that one of the fainter stars – designated as S2 – lies right on top of the position where the black hole is inferred to be. S2 is an otherwise "normal" star, though some 15 times more massive and seven times larger than the Sun. This star, S2, has now been tracked for over ten years and the loci defining its path trace a perfect ellipse with one focus at the position of the supermassive black hole. This photograph, taken near the middle of 2002, just happens to have caught S2 at the point of closest approach (known as the perenigricon), making it look like it was sitting right on top of the nucleus.

At perenigricon, S2 was a mere 17 light hours away from the black hole – roughly three times the distance between the Sun and Pluto, while traveling with a speed in excess of 5000 kilometers per second, by far the most extreme measurements ever made for such an orbit and velocity. We infer from these data that the mass of Sagittarius A* is $\sim 3.4 \times 10^6\ M_\odot$, compressed into a region no bigger than $\sim 17$ light hours. For this reason, Sagittarius A*, and the central object in NGC 4258, are considered to be the most precisely "weighed" supermassive black holes yet discovered.

## 8.5 The formation of supermassive black holes

An increasingly important question being asked in the context of supermassive black holes is which came first, the central black hole, or the surrounding galaxy? Quasars seem to have peaked 10 billion years ago, early in the universe's existence. The light from galaxies, on the other hand, originated much later – after the cosmos had aged another 2–4 billion years. Unfortunately, both measurements are subject to uncertainty, and no one can be sure we are measuring *all* of the light from quasars and galaxies, so this argument is not quite compelling. But we do see quasars as far out as we can look, and the most distant among them tend to be the most energetic objects in the universe, so at least *some* supermassive black holes must have existed near the very beginning. At the same time, images such as Fig. 8.3 provide evidence of mergers of smaller structures into bigger aggregates, but without a quasar. Perhaps not every collision feeds a black hole or, what is more likely, at least some galaxies must have formed first. Several scenarios for the formation of supermassive black holes are currently being examined. In every case, growth occurs when matter condenses following either the collapse of massive gas clouds, or the catabolism of smaller black holes in collisions and mergers.

All of the structure in the universe traces its beginnings to a brief era shortly after the Big Bang. Very few "fossils" remain from this period; one of the most important is the cosmic microwave background radiation. The rapid expansion that ensued lowered the matter density and temperature, and about one month after the Big Bang, the rate at which photons were created and annihilated could no longer keep up with the thinning plasma. The radiation and matter began to fall out of equilibrium with each other, forever imprinting the conditions of that era onto the radiation that reaches us to this day from all directions in space.

We now know that the temperature anisotropies are smaller than one part in a thousand, a limit below which density perturbations associated with ordinary matter would not have had sufficient time to evolve freely into the nonlinear structures we see today. Only a gravitationally dominant dark-matter component could then account for the strong condensation of mass into galaxies and supermassive black holes. The thinking behind this is that whereas the cosmic microwave background radiation interacted with ordinary matter, it would retain no imprint at all of the dark matter constituents in the universe. The nonluminous material could therefore be condensed unevenly (sometimes said to be 'clumped') all the way back to the Big Bang and we simply wouldn't know it.

The first billion years of evolution following the Big Bang must have been quite dramatic in terms of which constituents in the universe would eventually gain primacy and lasting influence on the structure we see today. The issue of how the fluctuations in density, mirrored by the uneven cosmic microwave background radiation, eventually condensed into supermassive black holes and galaxies is currently a topic of ongoing work. This question deals with the fundamental contents of the universe, and possibly what produced the Big Bang and what came before it. The evidence now seems to be pointing to a coeval history for these two dominant classes of objects – supermassive black holes and galaxies – though as we have already noted, at least some of the former must have existed quite early.

One possibility proposed by Balberg and Shapiro (2002) is that the first supermassive objects formed from the condensation of dark matter alone; only later would these seed black holes have imposed their influence on the latter. But this dark matter has to be somewhat peculiar, in the sense that its constituents must be able to exchange heat with each other. As long as this happens, a fraction of its elements evaporate away from the condensation, carrying with them the bulk of the energy, and the rest collapse and create an event horizon. The net result is that the inner core of such a clump forms a black hole, leaving the outer region and the extended halo in equilibrium about the central object. Over time, ordinary matter gathers around it, eventually forming stars, and planets.

Ordinary matter could not have achieved this early condensation because it simply wasn't sufficiently clumped initially. Perhaps this material formed the first stars, followed by more stars, eventually assembling a cluster of colliding objects. Over time, the inner core of such an assembly would have collapsed due to the evaporation of some of its members and the ensuing loss of energy into the extended halo, just as the dark matter did (see, e.g., Haehnelt and Rees 1993). A seed black hole might have formed in the cluster's core. Estimates show that once formed, such an object could have doubled its mass every 40 million years, so over the age of the universe, even a modestly appointed black hole could have grown into a billion-solar-mass object. The problem is that this could not have happened in only 700 million years, when the first supermassive black holes appeared.

Yet another method leading to black hole growth results from ongoing collisions between galaxies, which eventually lead to the merger of the black holes themselves. An example of such a process occurring right now is shown in Fig. 8.9.

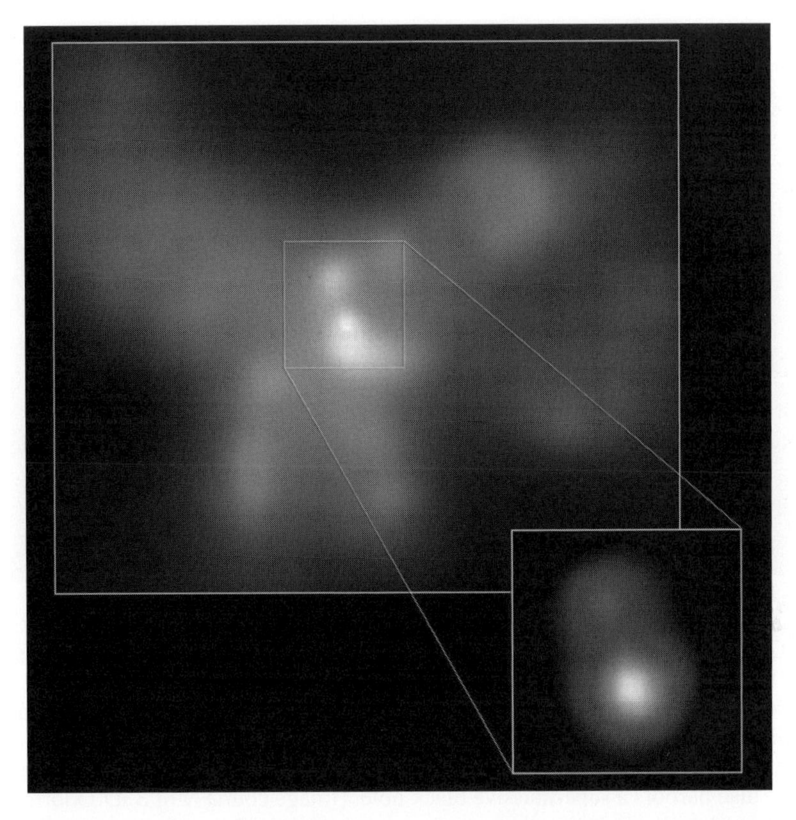

Fig. 8.9. NGC 6240 is a butterfly-shaped galaxy, believed to be the product of a collision between two smaller galaxies some 30 million years ago. This *Chandra* X-ray image (34000 light-years across) shows the heat generated by the merger activity, which created the extensive multimillion degree Celsius gas. We see here for the first time direct evidence that the nucleus of such a structure contains not one, but two active supermassive black holes, drifting toward each other across their 3 000-light-year separation; they're expected to merge into a bigger object several hundred million years hence. (Image courtesy of Stephanie Komossa, Günther Hasinger, and Joan Centrella, and the Max-Planck-Institut für Extraterrestrische Physik and NASA.)

Almost every large, normal galaxy harbors a supermassive black hole at its center. Some evidence for this has been provided by a recently completed survey of 100 nearby galaxies using the VLA, followed by closer scrutiny with the Very Long Baseline Interferometry array. At least 30 percent of this sample showed tiny, compact central radio sources bearing the unique signature of the quasar phenomenon. Of the hundreds of millions of supermassive black holes seen to pervade the cosmos, none of them appear to be isolated. And even more compelling is the work of Kormendy and Richstone (1995), who set about the task of systematically

Fig. 8.10. This image of the spiral galaxy NGC 2613 was captured on February 26, 2002, by the Very Large Telescope in Paranal, Chile. A search for a supermassive black hole in its core produced a null result (Bower *et al.* 1993). By now the preponderance of evidence suggests that highly flattened disk galaxies lacking a significant central hub, or spheroidal component, also lack a supermassive object in the nucleus. On the other hand, every galaxy that does possess a central bulge also harbors a supermassive black hole. (Image courtesy of S. D'Odorico *et al.*, who obtained it during the test phase of the VIMOS instrument at the European Southern Observatory's 8-meter *Melipal* telescope, and courtesy also of the European Southern Observatory.)

measuring as many black hole masses as is currently feasible. Direct measurements of supermassive black holes have been made in over 38 galaxies, based on the large rotation and random velocities of stars and gas near their centers. These objects are all relatively nearby because these direct methods don't work unless we can see the individual stars in motion about the central source of gravity. Curiously, none of the supermassive black holes have been found in galaxies that lack a central bulge (Fig. 8.10). Galaxies with a central bulge may have undergone one or more mergers in their past. Thus, a collision like that seen in Figs. 8.3 and 8.9 may have been required to create a central supermassive object.

Another recently inferred correspondence between supermassive black holes and their host galaxies seems to have clinched the case for a coeval growth (Ferrarese and Merritt 2000; Gebhardt *et al.* 2000). The data displayed in Fig. 8.11 show the relationship between the mass of the black hole and the velocity dispersion of stars within the spheroidal (or bulge) component of stars in the host galaxy. Based on the very tight correlation evident in this graph, it appears that the mass of the

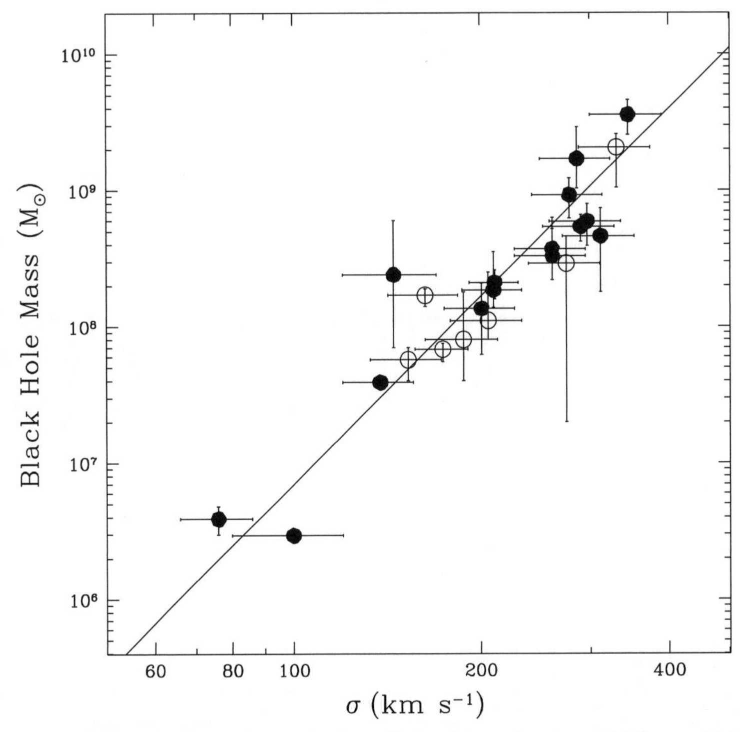


Fig. 8.11. Plotted against the velocity dispersion of stars orbiting within the spheroidal component of the host galaxy, the mass of a supermassive black hole is tightly correlated with the former, hinting at a coeval history. The stars are simply too far away from the nucleus to be directly affected by the black hole's gravity at the present time (Ferrarese and Merritt 2000).

central black hole can be predicted with remarkable accuracy simply by knowing the velocity of stars orbiting so far from it that its gravitational influence could not possibly be affecting their motion at the present time. Pinpointing the trajectory of a single object in the nuclei of the host galaxies is difficult and impracticable. Instead, these measurements are based on the accumulated light from a limited region of the galaxy's central hub, from which one may extract the collective Doppler shift, and thereby an average speed of the stars as a group. What these studies reveal is that the ratio of the black hole's mass to this average speed is constant across the whole sample of galaxies surveyed.

This result is one of the most surprising correlations yet discovered in the study of how the universe acquired its structure. Supermassive black holes, it seems, "know" about the motion of stars that are too distant to directly feel their gravity. This tight connection suggests an entangled history between a central black hole and the stellar activity in its halo. Although they may not be causally bonded today, they must have had an overlapping genesis in the past.

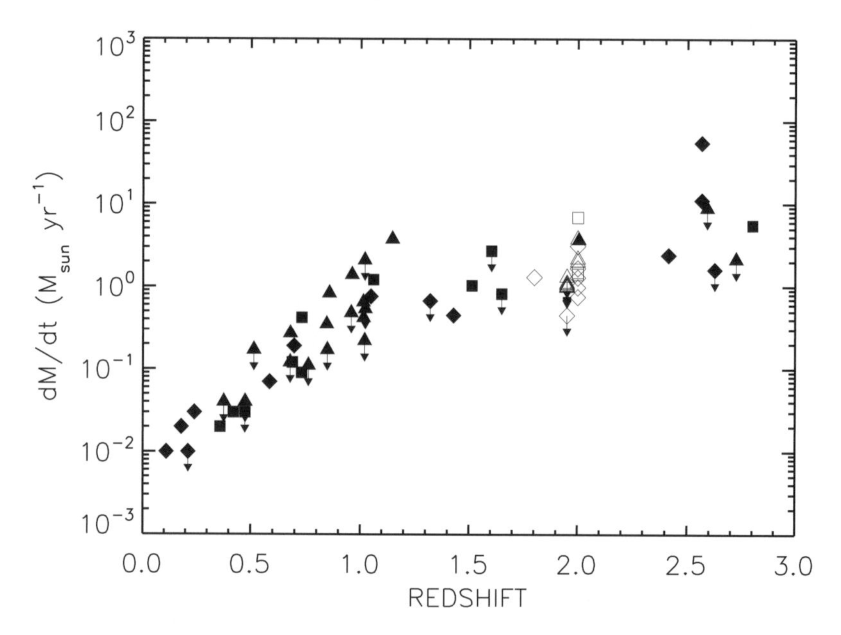


Fig. 8.12. The black hole accretion rate, in units of solar masses per year, versus redshift, assuming a canonical efficiency rate of 10 % conversion of rest mass energy into radiation. To generate a $10^9$ $M_\odot$ black hole over an accretion period of order 0.5 Gyr, a rate of order 2 $M_\odot$ yr$^{-1}$ is required. (From Barger *et al.* 2001.)

Additional information on the possible coeval history of supermassive black holes and their host galaxies is provided by images such as Fig. 8.4. If we adopt a canonical efficiency of 10 % for the conversion of rest mass energy into radiation, we obtain the mass inflow rate versus redshift relation shown in Fig. 8.12. This rate was as high as 10 solar masses per year at a redshift of 3 or so, and has declined monotonically to its much lower value of around 0.01 solar masses per year in the current epoch. An even more interesting evolutionary trait is that obtained by integrating the data in Fig. 8.12 over the black hole spatial distribution, which produces the redshift dependence of the so-called accretion rate density (Fig. 8.13). This quantity effectively gives us the rate at which black hole mass is increasing with time. The apparent $(1 + z)^3$ dependence of this quantity (exhibited in Fig. 8.13) is identical to the universe's star formation history, as deduced from the comoving UV intensity as a function of redshift.

Integrating this curve over time, we infer that about 0.006 of all the mass contained within the bulge of any given galaxy should be in the form of a central supermassive black hole. This is close to the value of $\sim$ 0.002 found in the local neighborhood of our galaxy. With these facts in hand, and satisfied that most supermassive black holes and their host galaxies grew interdependently, one may hypothesize that once created, a primordial condensation of matter continues to

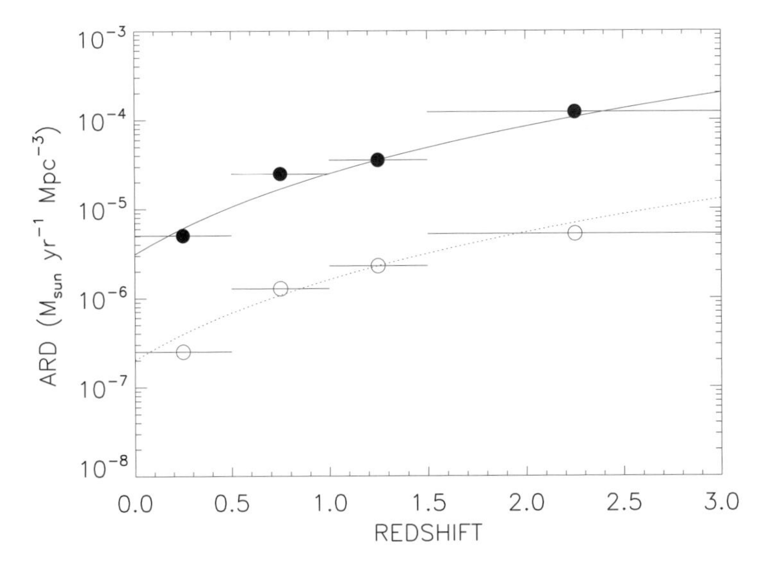


Fig. 8.13. The time history of the accretion rate density (effectively, the rate at which black hole matter is increasing, in units of solar masses per year per megaparsec cubed), integrated from the data exhibited in Fig. 8.12. The upper (solid) and lower (open) bounds in this figure correspond to the accretion rate density calculated from the bolometric and X-ray luminosities, respectively. The curves illustrate a $(1 + z)^3$ dependence, normalized to the $1 < z < 1.5$ redshift bin. (From Barger *et al.* 2001.)

grow with a direct feedback on its surroundings. This may happen either because the quasar heats up its environment and controls the rate at which additional matter can fall in from its cosmic neighborhood, or because mergers between galaxies affect the growth of colliding black holes in the same way that they determine the energy (and therefore the average speed) of the surrounding stellar distribution. In support of the idea that massive black holes may have formed prior to the epoch of galaxy definition, Silk and Rees (1998) suggest further that protogalactic star formation would have been influenced significantly by the quasar's extensive energy outflows. The ensuing feedback on the galaxy's spheroidal component could be the reason we now see such a tight correlation between the mass of the central object and the stellar velocities much farther out.

Another possible reason why black hole masses are correlated with the velocity dispersion of the host galaxy bulges may be the importance of galaxy mergers to the hierarchical construction of elaborate elliptical, or disk-plus-bulge profiles. Most large galaxies have experienced at least one major merger during their lifetime. Larger galaxies have grown with a succession of collisions and mergers, a process contributing significantly to the variety of shapes encompassed by the Hubble sequence. Meanwhile, the turbulence generated in the core of the collision drives

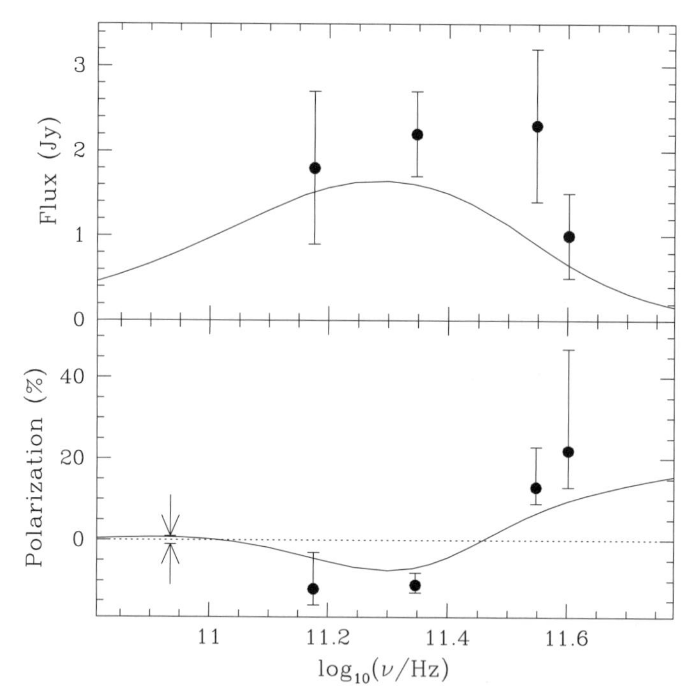


Fig. 8.14. Spectra showing total flux density and polarization from Sagittarius A*. The high-frequency data points are from Aitken *et al.* (2000) and the curves are from the magnetohydrodynamic model discussed in Bromley, Melia, and Liu (2001). The limit at 84 GHz is from Bower *et al.* (1999). At even lower frequencies, the best fit polarization is consistent with 0. In this figure, 'negative' polarization corresponds to the polarization vector being aligned with the spin axis of the black hole, whereas positive is for a perpendicular configuration. The polarization crosses 0 when this vector flips by 90° (Bromley, Melia, and Liu 2001).

most of the gas into the middle, where it forms new stars and feeds a central black hole, or a pair of black holes. The tight correlation between the black-hole mass and the halo velocity dispersion may be a direct consequence of this cosmic cascade.

## 8.6 Imaging a supermassive black hole

Of course, the most exciting development in black hole research would be the actual imaging of its event horizon (Melia 2003b). This may be feasible in the next several years, with the development of a global mm-VLBA. The spectrum of Sagittarius A* at the galactic center displays the well-known profile of a self-absorbed emitter longward of 3–7 mm, and a transparent medium at higher frequencies (see Fig. 8.14). The fact that Sagittarius A* becomes transparent across the mm/sub-mm bump in its spectrum means that a depression in intensity (literally, a 'black hole') should arise shortward of $\sim 1$ mm, due to the effects of strong light bending near the event horizon. The 'shadow' cast by the black hole in this fashion should have an apparent diameter of about 5 Schwarzschild radii (Falcke, Melia, and Agol 2000), nearly independent of the black hole spin or its orientation. The distance to Sagittarius A* being roughly 8 kpc, this diameter corresponds to an angular size of $\sim 30\,\mu$as, which approaches the resolution of current radio interferometers.

The fortunate aspect of this transition from optically-thick to optically-thin emission across the mm/sub-mm bump is that it happens to occur at roughly the same frequency where scatter broadening in the interstellar medium decreases below the intrinsic source size. Taking this effect into account, and the finite telescope resolution, one infers (Falcke, Melia, and Agol 2000) that the shadow of Sagittarius A* should be observable below $\sim 1$ mm.

Most recently, ray-tracing simulations have taken into account the magnetized Keplerian flow and the effects of light-bending and area amplification to produce the most detailed theoretical predictions of the images that mm-astronomy will produce in the near future. The polarimetric images shown in Fig. 8.15 demonstrate that future mm/sub-mm interferometry can directly reveal material flowing near the horizon in Sagittarius A*.

The blurring in these images is modeled by convolution with two Gaussians: the first is an approximate model of the scattering effects with an ellipsoidal filter whose major- and minor-axis FWHM values are 24.2 $\mu$as $\times$ $(\lambda/1.3\text{ mm})^2$ and 12.8 $\mu$as $\times$ $(\lambda/1.3\text{ mm})^2$, respectively, for emission at wavelength $\lambda$. In the simulations shown in Fig. 8.15, the spin axis of the disk was chosen to lie along the minor axis of the scattering ellipsoid, and a global interferometer was assumed with a 8000 km baseline. The second is a spherically symmetric filter with a FWHM of 33.5 $\mu$as $\times$ $(\lambda/1.3\text{ mm})$ to account for the resolution effects of an ideal interferometer. These images demonstrate clearly the viability of conducting polarimetric imaging of the black hole at the galactic center with upcoming VLBI techniques.

## 8.7 Conclusion

We have come a long way since 1963, when Roy Kerr discovered the spacetime that is now synonymous with black hole research. The past decade, in particular, has seen a dramatic shift in our view of the role played by supermassive black holes in the formation of structure in the universe. We now appear to be on the verge of finally "seeing" some of these objects, providing more direct evidence for their existence, and a more satisfying confirmation of their properties predicted by general relativity. Within the framework of the Kerr metric, Sagittarius A* at the galactic center, and possibly even the central black hole in the relatively nearby

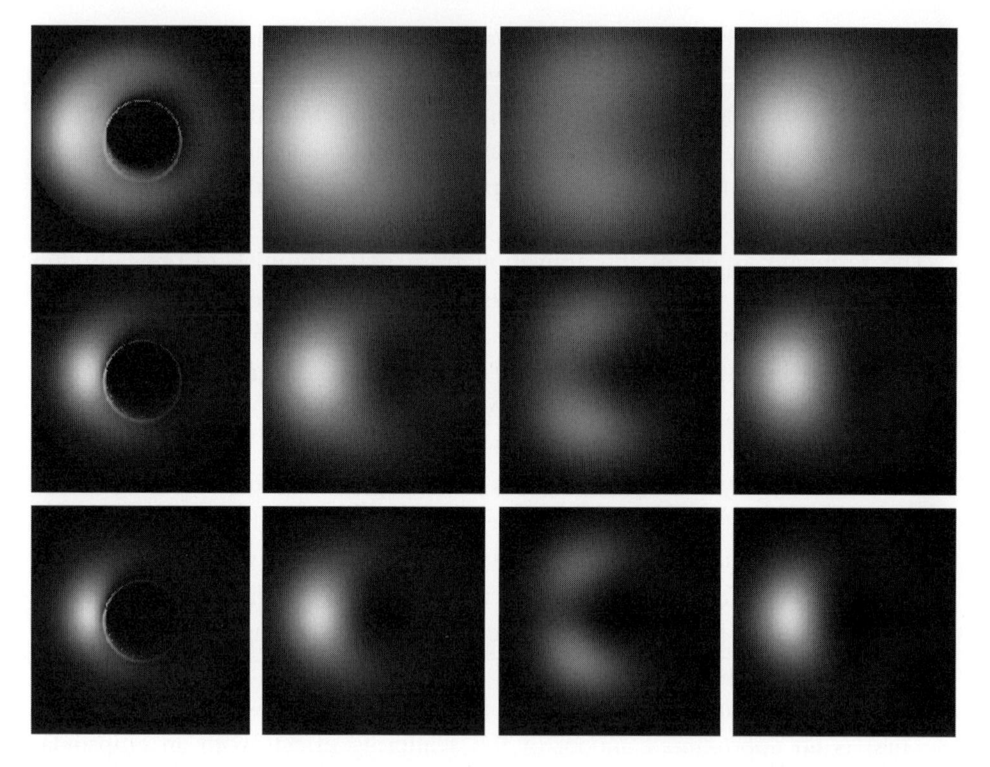

Fig. 8.15. Polarization maps at three wavelengths near the peak of the mm to sub-mm emission from Sagittarius A*. The top row shows emission at 1.5 mm, the middle row is at 1 mm, and the bottom row corresponds to 0.67 mm. The images in each row show the raw ray-tracing output (first column on the left), and an image blurred to account for finite VLBI resolution and interstellar scattering (second column). The two rightmost columns give the vertical and horizontal components of the polarized emission. The pixel brightness in all images scales linearly with flux (Bromley, Melia, and Liu 2001).

M81 galaxy, are expected to produce distinct shadows with a precise shape and size of the region where light bending and capture are important. This coming decade may finally give us a view into one of the most important and intriguing predictions of general relativity.

## References

Aitken, D. K. *et al.* (2000), "Detection of polarized millimeter and submillimeter emission from Sagittarius A*", *Astroph. J.* **534**, L173–L176.

Balberg, S. & Shapiro, S. L. (2002), "Gravothermal collapse of self-interacting dark matter halos and the origin of massive black holes", *Phys. Rev. Lett.* **88**, 101301.

Barger, A. J. *et al.* (2001), "Supermassive black hole accretion history inferred from a large sample of Chandra hard X-ray sources", *Astron. J.* **122**, 2177–2194.

Bower, G. A., Richstone, D. O., Bothun, G. D. & Heckman, T. M. (1993), "A search for dead quasars among nearby luminous galaxies. I. The stellar kinematics in the nuclei NGC 2613, NGC 4699, NGC 5746, and NGC 7331", *Astroph. J.* **402**, 76–94.
Bower, G. C. *et al.* (1999), "The linear polarization of Sagittarius A*. I. VLA spectropolarimetry at 4.8 and 8.4 GHz", *Astroph. J.* **521**, 582–586.
Bromley, B., Melia, F. & Liu, S. (2001), "Polarimetric imaging of the massive black hole at the Galactic Center", *Astroph. J.* **555**, L83–L86.
Fabian, A. C. *et al.* (2000), "Testing the connection between the X-ray and submillimeter source populations using Chandra", *Mon. Not. Roy. Astron. Soc.* **315**, L8–L12.
Ferrarese, L. & Merritt, D. (2000), "A fundamental relation between supermassive black holes and their host galaxies", *Astrophy. J.* **539**, L9–L12.
Gebhardt, K. *et al.* (2000), "A relationship between nuclear black hole mass and galaxy velocity dispersion", *Astroph. J.* **539**, L13–L16.
Falcke, H., Melia, F. & Agol, E. (2000), "Viewing the shadow of the black hole at the Galactic Center", *Astroph. J.* **528**, L13–L16.
Haehnelt, M. G. & Rees, M. J. (1993), "The formation of nuclei in newly formed galaxies and the evolution of the quasar population", *Mon. Not. Roy. Astron. Soc.* **263**, 168–178.
Kerr, R. P. (1963), "Gravitational field of a spinning mass as an example of algebraically special metrics", *Phys. Rev. Lett.* **11**, 237–238.
Kormendy, J. & Richstone, D. (1995), "Inward bound – the search for supermassive black holes in galactic nuclei", *Ann. Rev. Astron. Astrophys.* **33**, 581–624.
Lynden-Bell, D. (1969), "Galactic nuclei as collapsed old quasars", *Nature* **223**, 690.
Matsumoto, H. *et al.* (2001), "Discovery of a luminous, variable, off-center source in the nucleus of M82 with the Chandra high-resolution camera", *Astroph. J.* **547**, L25–L28.
Melia, F. (2003b). *The Black Hole at the Center of Our Galaxy* (Princeton University Press, Princeton).
Melia, F., Liu, S. & Coker, R. (2001), "A magnetic dynamo origin for the submillimeter excess in Sagittarius A*", *Astroph. J.* **553**, 146–157.
Miyoshi, M. *et al.* (1995), "Evidence for a black hole from high rotation velocities in a sub-parsec region of NGC 4258", *Nature* **373**, 127.
Salpeter, E. E. (1964), "Accretion of interstellar matter by massive objects", *Astroph. J.* **140**, 796–800.
Schmidt, M. (1963), "3C 273: a star-like object with large red-shift", *Nature* **197**, 1040.
Schödel, R. *et al.* (2002), "A star in a 15.2-year orbit around the supermassive black hole at the centre of the Milky Way", *Nature* **419**, 694–696.
Silk, J. & Rees, M. J. (1998), "Quasars and galaxy formation", *Astron. Astroph.* **331**, L1–L4.
Zel'dovich, Ya. B. & Novikov, I. D. (1967), "The hypothesis of cores retarded during expansion and the hot cosmological model", *Sov. Astron.* **10**, 602.

# 9

# The X-ray spectra of accreting Kerr black holes

Andrew C. Fabian and Giovanni Miniutti

## 9.1 Introduction

The form of the spacetime geometry around an astrophysical black hole (BH) is due to Roy Kerr (1963) who found the exact solution to the general relativistic Einstein's equations for the spacetime outside the horizon of a rotating BH. Since then, rotating BHs are known as Kerr BHs, the spacetime geometry of which depends on two parameters only, the BH mass and spin. In the limit of no rotation, Kerr's solution coincides with that for a non-rotating BH found earlier by Karl Schwarzschild (1916), which obviously depends on the BH mass only. The complete solution of the equations of motion in the Kerr spacetime is due to Brandon Carter (1968). The interested reader finds an excellent treatment of Einstein's theory of General Relativity and of BH spacetimes in the fundamental book *Gravitation* by Misner, Thorne & Wheeler (1973).

A large fraction of the accretion energy in luminous BH systems is dissipated in the innermost regions of the accretion flow. Most of the power is radiated from close to the smallest accretion disc radii in the relativistic region close to the BH (Shakura & Sunyaev 1973; Pringle 1981). Relativistic effects then affect the appearance of the X-ray spectrum from the disc through Doppler, aberration, gravitational redshift and light bending effects (Page & Thorne 1974; Cunningham 1975, 1976). The dominant feature in the 2–10 keV X-ray spectrum seen by a distant observer is an iron line with a broad skewed profile which carries unique information on the structure, geometry, and dynamics of the accretion flow in the immediate vicinity of the central BH, providing a tool to investigate the nature of the spacetime there (Fabian *et al.* 1989; Laor 1991).

Relativistic broad iron lines seen in the spectrum of several active galaxies and Galactic black hole binaries are reviewed here (see also Fabian *et al.* 2000;

The Kerr Spacetime: Rotating Black Holes in General Relativity. Ed. David L. Wiltshire, Matt Visser and Susan M. Scott. Published by Cambridge University Press. 

Reynolds & Nowak 2003 for previous reviews). Among others, the cases for relativistic lines in the Seyfert galaxies MCG–6-30-15 and IRAS 18325–5926, and the X-ray binaries XTE J1650–500 and GX 339-4 are very strong (Tanaka *et al.* 1995; Wilms *et al.* 2001; Fabian *et al.* 2002a; Iwasawa *et al.* 1996a; Iwasawa *et al.* 2004; Miller *et al.* 2002a; Miniutti, Fabian & Miller 2004; Miller *et al.* 2004a,b). In three out of four objects, the X-ray data require emission from within the innermost stable circular orbit of a non-rotating BH, suggesting that the BHs in many objects are rapidly spinning. The spectra of many other objects, in particular Narrow Line Seyfert 1 (NLS1) galaxies, can be successfully described by a simple two-component model comprising a highly variable power law continuum and a much more constant reflection component from the accretion disc (e.g. Fabian *et al.* 2004; Fabian *et al.* 2005; Ponti *et al.* 2006). The puzzling spectral variability of such sources is now beginning to be understood within the context of emission from the strong gravity regime (Miniutti *et al.* 2003; Miniutti & Fabian 2004). Some active galactic nuclei (AGN) and X-ray black hole binaries show either no line or only a narrow one (e.g. Page *et al.* 2004; Yaqoob & Padmanabhan 2004). This is discussed within the context of both theoretical implications and present observational limitations. Finally, the short-timescale variability of the broad iron line is beginning to be unveiled, providing exciting results that are dramatically improving our understanding of the innermost regions of the accretion flow, where General Relativity is no longer a small correction and becomes the most relevant physical ingredient (e.g. Turner *et al.* 2002; Iwasawa, Miniutti & Fabian 2004).

The advent of future X-ray missions with higher energy resolution (e.g. *Suzaku* successfully launched in July 2005) and much larger collecting area in the relevant iron band (XEUS and Constellation-X in the next decade) will open up a new window on the innermost regions of the accretion flow in AGN and X-ray binaries. This will enable us to further test the general picture we propose here, with the potential of mapping with great accuracy the strong field regime of General Relativity in a manner which is still inaccessible at other wavelengths.

## 9.2 Main components of the X-ray spectrum

Radiatively-efficient accreting BHs are expected to be surrounded by a dense disc radiating quasi-blackbody thermal EUV and soft X-ray emission (see e.g. the seminal paper by Shakura & Sunyaev 1973). However, the X-ray spectra of accreting BHs also exhibit a power law component extending to hard X-ray energies up to 200 keV or more. The most promising physical mechanism to produce such hard power law components is Comptonization of the soft X-ray photons in a corona above the disc, possibly fed by magnetic fields from the body of the disc itself (e.g. Haardt & Marschi 1991 and 1993, Zdziarski *et al.*

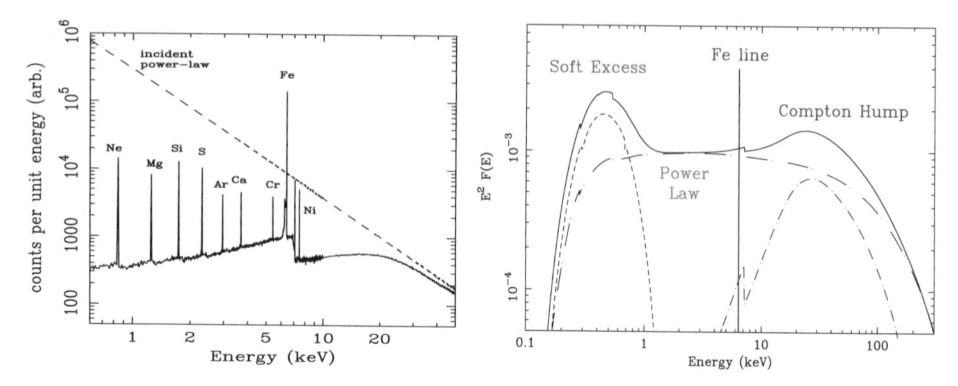


Fig. 9.1. *Left panel*: Monte Carlo simulations of the reflection spectrum from a slab of uniform density neutral matter with solar abundances. The incident power law continuum is also shown. Figure from Reynolds (1996). *Right panel*: The main components of the X-ray spectra of unobscured accreting BHs are shown: soft quasi-thermal X-ray emission from the accretion disc (dotted curve); power law from Comptonization of the soft X-rays in a corona above the disc (broken curve); reflection continuum and narrow Fe line due to reflection of the hard X-ray emission from dense gas (chain curve).

1994). Irradiation of the dense disc material by the hard X-rays then gives rise to a characteristic "reflection" spectrum which is the result of Compton scattering and photoelectric absorption followed either by Auger de-excitation or by fluorescent line emission (see e.g. Guilbert & Rees 1988; Lightman & White 1988; George & Fabian 1991, Matt, Perola & Piro 1991). This last process gives rise to an emission line spectrum where fluorescent narrow K$\alpha$ lines from the most abundant metals are seen. Thanks to a combination of large cosmic abundance and high fluorescent yield, the iron (Fe) K$\alpha$ line at 6.4 keV is the most prominent fluorescent line in the X-ray reflection spectrum. Photoelectric absorption is an energy-dependent process, so that incident soft X-rays are mostly absorbed, whereas hard photons tend to be Compton scattered back out of the disc. However, above a few tens of keV, Compton recoil reduces the backscattered photon flux and produces a broad hump-like structure around 20–30 keV (the so-called Compton hump). An example of the X-ray reflection spectrum from a neutral and uniform density semi-infinite slab of gas is shown in the left panel of Fig. 9.1 where fluorescent emission lines dominate below about 8 keV and the Compton hump is seen above 20 keV (from Reynolds 1996).

In the case of AGNs, the accretion disc is not the only reflector able to produce a X-ray reflection spectrum. The presence of a dusty molecular torus surrounding the accreting system is required by unification models at the parsec scale (Antonucci 1993). The torus provides the absorbing column (sometimes more than $10^{24}$ cm$^{-2}$) that is observed in Seyfert 2 galaxies which are thought to be distinguished from unobscured Seyfert 1 galaxies only because of a higher observer inclination. The torus itself is Compton-thick and therefore provides an additional reflector producing a spectrum very similar to that seen in the left panel of Fig. 9.1. As will be explained in detail in the next section, special and general relativistic effects shape the reflection spectrum from the centre of the accretion disc and not that from regions far away from the central BH (such as the torus), helping us to disentangle the two contributions.

In the right panel of Fig 9.1, we show the main components of the X-ray spectrum of accreting black holes. The "soft excess" represents here the soft X-ray emission from the disc, although some other interpretations are possible and will be discussed throughout this contribution. We also show the power law (with a high-energy cut-off characteristic of the coronal temperature) and reflection components. Reflection is here represented by the reflection continuum (characterized by the "Compton hump") and the Fe K$\alpha$ line at 6.4 keV that we assume here to be narrow, i.e. emitted from a distant reflector such as the torus. The spectrum is absorbed by a column of $2 \times 10^{20}$ cm$^{-2}$, representing the typical value for absorption by our Galaxy in the line of sight.

Another important component in the X-ray spectrum of AGN is due to the ubiquitous presence of warm gas surrounding the central nucleus. This gas is photoionized by the primary X-ray continuum and contributes to the spectrum both in absorption and in re-emission (Halpern 1984; Reynolds 1997; Kaastra *et al.* 2000; Kaspi *et al.* 2001). In the case of Seyfert 1 galaxies, the continuum level is so high that the emission component is diluted and generally not observable, except in particularly low flux states of the X-ray source. One remarkable case in this sense is provided by the low flux state of NGC 4051 (see the analysis of the high resolution *XMM-Newton*–RGS data below 2 keV by Pounds *et al.* 2004b; also Ponti *et al.* 2006).

In Seyfert 2 galaxies, the primary continuum is heavily absorbed by large columns in the line of sight (the torus) and therefore the emission component is readily detectable. On the other hand, absorption by this same photoionized gas is obviously easier to detect in Seyfert 1 than Seyfert 2 galaxies due to the higher level of the continuum. In the left panel of Fig. 9.2 we show both absorption and emission components on a primary power law continuum (plus cold absorption from our own Galaxy in the direction of the source) in the most relevant soft X-ray band. The gas is assumed to have a column of $N_{\mathrm{H}} = 5 \times 10^{21}$ cm$^{-2}$ and ionization parameter of $\xi = 31.6$ erg cm s$^{-1}$ and the spectrum is modelled with the tXSTAR code (Kallman & Krolik 1986). In the right panel of the same figure, we show a qualitative view of the typical spectrum for a Seyfert 2 galaxy. The hard spectrum is dominated by reflection from the torus, whereas in the soft band emission lines from photoionized gas can be detected because of the low continuum level (e.g.

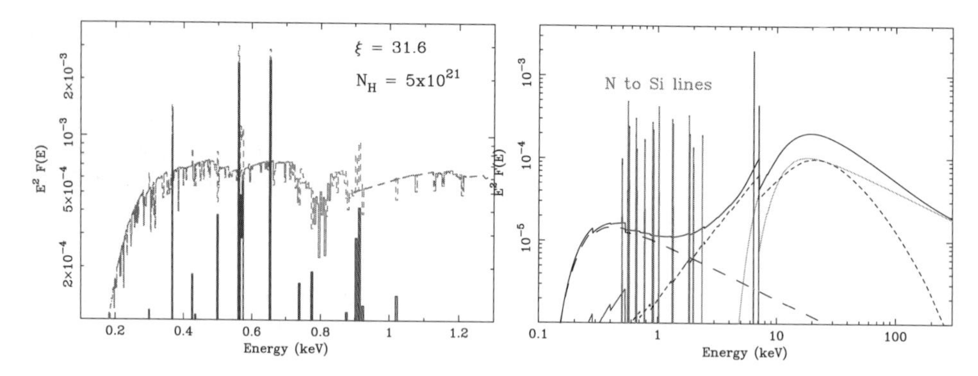


Fig. 9.2. *Left panel*: The effect of photoionized gas surrounding an AGN is shown both in absorption (broken curve) and emission (solid curve). The spectrum is from the tXSTAR code of Kallman & Krolik (1986). *Right panel*: The typical spectrum of a Seyfert 2 galaxy. The primary continuum (broken curve) is heavily absorbed by a large column (typically larger than $10^{24}$ cm$^{-2}$) which is identified as the torus, partially blocking the line of sight. The same column produces nearly neutral X-ray reflection (dotted curve) from the visible side of the torus. Emission lines due to photoionized gas (vertical lines) are detectable because of the reduced primary continuum flux.

Kinkhabwala *et al.* 2002; Bianchi *et al.* 2005 and many others). In the following we shall focus on unobscured accretion systems such as Seyfert 1 galaxies and BH binaries in which the innermost regions of the accretion flow are in principle observable.

## 9.3 The relativistic iron line

The Fe I K$\alpha$ emission line is a doublet comprising K$\alpha$1 and K$\alpha$2 lines at 6.404 and 6.391 keV, respectively. However, the energy separation has generally been too small to be detected and a weighted mean at 6.4 keV is generally assumed. The resulting line is symmetric and intrinsically much narrower than the spectral resolution commonly available today. Hence, the detailed line shape and broadening can be used to study the dynamics of the emitting region.

If the reflection spectrum, and therefore the Fe line, originates from the accretion disc, the line shape is distorted by Newtonian, special and general relativistic effects (see e.g. Fabian *et al.* 2000). This is illustrated schematically in Fig. 9.3. In the Newtonian case, each radius on the disc produces a symmetric double-peaked line profile with the peaks corresponding to emission from the approaching (blue) and receding (red) sides of the disc. Close to the BH, where orbital velocities become relativistic, relativistic beaming enhances the blue peak with respect to the red

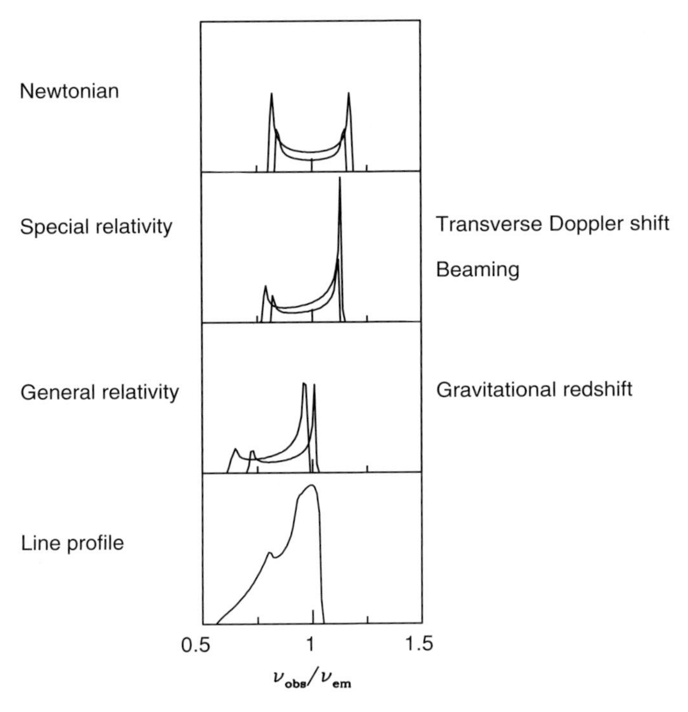


Fig. 9.3. The profile of an intrinsically narrow emission line is modified by the interplay of Doppler/gravitational energy shifts, relativistic beaming, and gravitational light bending occurring in the accretion disc (from Fabian *et al.* 2000). The upper panel shows the symmetric double-peaked profile from two annuli on a non-relativistic Newtonian disc. In the second panel, the effects of transverse Doppler shifts (making the profiles redder) and of relativistic beaming (enhancing the blue peak with respect to the red) are included. In the third panel, gravitational redshift is turned on, shifting the overall profile to the red side and reducing the blue peak strength. The disc inclination fixes the maximum energy at which the line can still be seen, mainly because of the angular dependence of relativistic beaming and of gravitational light bending effects. All these effects combined give rise to a broad, skewed line profile which is shown in the last panel, after integrating over the contributions from all the different annuli on the accretion disc.

one, and the transverse Doppler effect shifts the profile to lower energies. As we approach the central BH and gravity becomes strong enough, gravitational redshift becomes important with the effect that the overall line profile is shifted to lower energies. The disc inclination fixes the maximum energy at which the line can still be seen, mainly because of the angular dependence of relativistic beaming and of gravitational light bending effects. Integrating over all radii on the accretion disc, a broad and skewed line profile is produced, such as that shown in the bottom panel of Fig. 9.3. It is clear from the above discussion that the detailed profile of a broad relativistic line from the accretion disc has the extraordinary potential of revealing

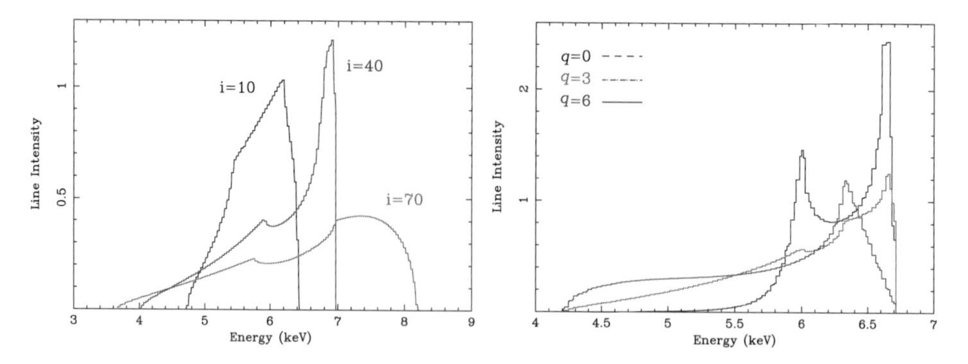


Fig. 9.4. *Left panel*: The dependence of the line profile on the observer inclination is shown. *Right panel*: The dependence of the line profile on the emissivity profile on the disc is shown. The disc emissivity is assumed to scale as $\epsilon = r^{-q}$. The steeper the emissivity, the broader and more redshifted the line profile, because more emphasis is given to the innermost radii where gravity dominates.

the dynamics of the innermost accretion flow in accreting BHs and even of testing Einstein's theory of General Relativity in a manner that is unaccessible to other wavelengths.

### *9.3.1 Dependence on disc inclination and emissivity*

The relativistic line profile exhibits a dependence for many physical parameters. The energy of the blue peak of the line is mainly dictated by the inclination of the observer line of sight with respect to the accretion disc axis. This is clear in the left panel of Fig. 9.4 where we show the result of fully relativistic computations (e.g. Fabian *et al.* 1989; Laor 1991; Dovčiak, Karas & Yaqoob 2004 among many others). The three profiles have all the same parameters but different observer inclination $i$. From Fig. 9.4, it is clear that the higher the inclination the bluer the line is, providing a quite robust tool to measure the inclination of the accretion disc.

Another important parameter is the form of the emissivity profile, i.e. the efficiency with which the line is emitted as a function of the radial position on the disc. This depends mainly on the illumination profile by the hard X-rays from the corona which is in turn determined by the energy dissipation on the disc and by the heating events in the corona (possibly associated with magnetic fields see e.g. Merloni & Fabian 2001a,b). The emissivity profile is generally assumed to be in the form of a simple power law $\epsilon(r) = r^{-q}$, where $q$ is the emissivity index (but see e.g. Beckwith & Done 2004). By assuming that the emissivity is a good tracer of the energy dissipation on the disc, the standard value for the emissivity index is $q = 3$ (e.g. Pringle 1981; also Reynolds & Nowak 2003 and Merloni & Fabian 2003 for a discussion on the dependence of the emissivity profile on boundary conditions). In the right panel of Fig. 9.4, we show the dependence of the line profile from this most important parameter. We show the cases of a uniform ($q = 0$), standard ($q = 3$), and steep ($q = 6$) emissivity profile. A steep emissivity profile indicates that the conversion of the X-ray photons from soft to hard in the corona is centrally concentrated thereby illuminating more efficiently the very inner regions of the accretion disc. As shown in Fig. 9.4, steeper emissivity profiles produce much broader and redshifted lines because more weight is given to the innermost disc, where gravitational redshift dominates.

### *9.3.2 Self-consistent ionized reflection models*

So far, we have assumed that the disc (or to be more precise its outer layers) is a slab of uniform density gas where hydrogen and helium are fully ionized, but all the metals are neutral. The real situation is likely to be much more complex. One first important step towards the accurate model of accretion disc atmospheres is made by considering thermal and ionization equilibrium. Results of such computations have been published over the last ten years or so with different degrees of complexity (e.g. Ross & Fabian 1993; Matt, Fabian & Ross 1993, 1996; Zycki *et al.* 1994; Nayakshin, Kazanas & Kallman 2000; Rózańska *et al.* 2002; Dumont *et al.* 2003). See Ballantyne, Ross & Fabian 2001 for a comparison between different hypotheses such as constant-density atmospheres and atmospheres in hydrostatic equilibrium. The recent work by Ross & Fabian (2005) extending and improving previous computations (e.g. Ross, Fabian & Young 1999; Ballantyne, Ross & Fabian 2001) is described here in some detail since it is used extensively in comparing X-ray data to theoretical models.

The illuminating radiation is assumed to have an exponential cut-off power law form with high-energy cut-off fixed at 300 keV and variable photon index $\Gamma$ between 1 and 3 roughly covering the observed range. The ionization parameter $\xi$ is defined as the ratio between the isotropic total illuminating flux and the comoving hydrogen number density of the gas; results are produced for $\xi$ ranging from 1 to $10^4$ erg cm s$^{-1}$. The local temperature and fractional ionization of the gas are computed self-consistently by solving the equations of thermal and ionization equilibrium and ions from C, N, O, Ne, Mg, Si, S, and Fe are treated. The available model grids allow for a variable Fe abundance.

In the left panel of Fig. 9.5 we show X-ray reflection spectra produced by the code for three different values of the ionization parameter (all other parameters being fixed). The ionization parameter has clearly a large effect on the resulting spectrum, most remarkably on the emission lines. For $\xi = 10^4$ erg cm s$^{-1}$ (top black) the surface layer is very highly ionized and the only noticeable line is a highly Compton-broadened Fe K$\alpha$ line peaking at 7 keV. The overall spectral

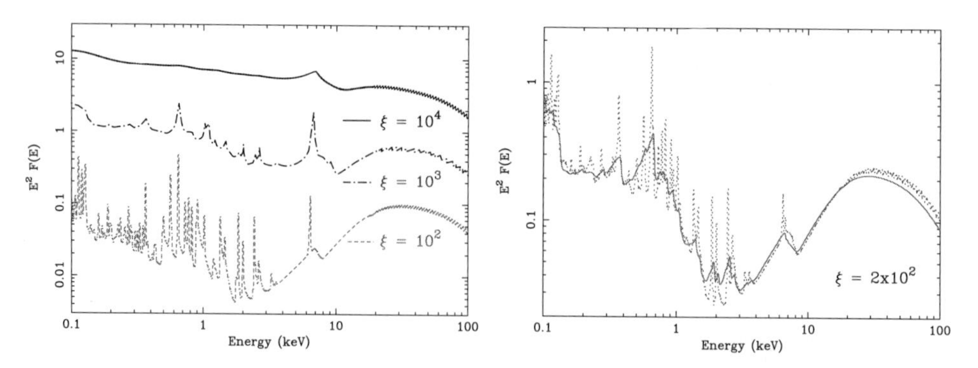


Fig. 9.5. *Left panel*: Computed X-ray reflection spectra as a function of the ionization parameter $\xi$ (from the code by Ross & Fabian 2005). The illuminating continuum has a photon index of $\Gamma = 2$ and the reflector is assumed to have cosmic (solar) abundances. *Right panel*: Relativistic effects on the X-ray reflection spectrum. We assume that the intrinsic rest-frame spectrum (dotted line) is emitted in an accretion disc and suffers all the relativistic effects discussed above (see text for details). The relativistically-blurred reflection spectrum is shown in red.

shape closely resembles that of the illuminating continuum (a cut-off power law with photon index $\Gamma = 2$). For $\xi = 10^3$ erg cm s$^{-1}$ (middle) the strong Fe K$\alpha$ line is dominated by the Fe XXV intercombination line, while K$\alpha$ lines from the lighter elements emerge in the 0.3–3 keV band. Further reducing the ionization parameter to $\xi = 10^2$ erg cm s$^{-1}$ gives rise to a spectrum dominated by emission features below 3 keV atop a deep absorption trough. The most prominent feature is the Fe K$\alpha$ one at 6.4 keV. No residual Compton broadening of the emission lines is visible.

In the right panel of Fig. 9.5 we show two versions of a model with an ionization parameter of $\xi = 2 \times 10^2$ erg cm s$^{-1}$. The dotted one is the X-ray reflection spectrum in the absence of any relativistic effect, whereas the solid curve shows the relativistically-blurred version of the same model, i.e. the spectrum that is observed if reflection occurs from the accretion disc. All sharp spectral features of the unblurred spectrum (dotted curve) are broadened by the relativistic effects explained above which makes it difficult to identify clear emission lines in the soft spectrum. Below about 2 keV the situation is often complicated by the presence of absorption/emission features due to photoionized gas complicating the soft part of the spectrum (see Fig. 9.2, left panel). Thanks to its strength, isolation, and to the fact that it occupies a region of the X-ray spectrum relatively free from absorption, the Fe line is however clearly seen. This is what makes this particular emission feature a remarkable and unique tool that allows us to investigate the dynamics of the innermost accretion flow via relativistic effects in accreting BH systems.

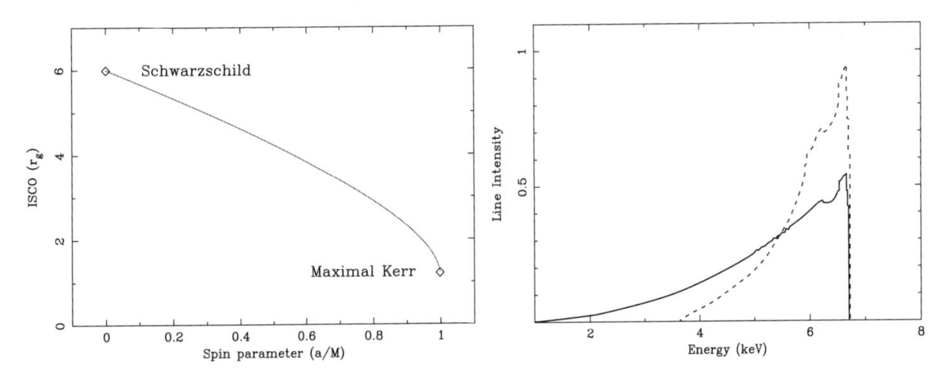


Fig. 9.6. *Left panel*: The dependence of the ISCO from the BH spin. We consider the allowed spin range from $a/M = 0$ (Schwarzschild solution) to $a/M = 0.998$ (Maximal Kerr solution). For these extremal cases, the ISCO is located at 6 $r_g$ ($a/M = 0$) and $\simeq 1.24\ r_g$ ($a/M = 0.998$). *Right panel*: The line profiles dependence from the inner disc radius is shown for the two extremal cases of a Schwarzschild BH (dotted curve, with inner disc radius at 6 $r_g$) and of a Maximal Kerr BH (solid curve, with inner disc radius at $\simeq 1.24\ r_g$).

### 9.3.3 Dependence on the inner disc radius

Einstein's equations imply the existence of an innermost radius within which the circular orbit of a test particle in the equatorial plane is no longer stable. This radius is known as the Innermost Stable Circular Orbit (ISCO), sometimes referred to as the marginally stable orbit (Bardeen, Press & Teukolsky 1972). Beyond the ISCO, test particles rapidly plunge into the BH on nearly geodesic orbits with constant energy and angular momentum. By making the standard assumption that the accretion disc is made of gas particles in circular, or nearly circular, orbital motion, the disc extends down to the ISCO, and emission from the plunging region is ignored. We shall discuss further this assumption in the next Section.

The actual radius of the ISCO depends on the BH spin parameter $a/M$ which can take any value from 0 (Schwarzschild BH) to 1 (maximally spinning Kerr BH). As pointed out by Thorne (1974), the maximal value for the spin parameter is likely to be about 0.998 and we shall refer to that case as "Maximal Kerr". The dependence of the ISCO from the BH spin parameter is illustrated in the left panel of Fig. 9.6. The ISCO lies at 6 $r_g$ from the centre for a Schwarzschild BH, and at $\simeq 1.24\ r_g$ for a Maximal Kerr one, where $r_g = GM/c^2$ is the gravitational radius. It should be stressed that the dependence is quite steep. If emission from say 3 $r_g$ can be detected the implied BH spin would be $a/M > 0.78$. Notice that accretion naturally causes a BH to spin up provided the disc angular momentum is oriented as that of the hole (the possible history of the spin of massive BHs is discussed e.g. by Volonteri *et al.* 2005 and references therein).

The inner boundary of the accretion disc, i.e. the location of the ISCO, has a large impact on the shape of the line profile, especially on its broad red wing. This is shown in the right panel of Fig. 9.6 where the cases of a Schwarzschild (dotted curve) and Maximal Kerr BH (solid curve) are computed. The line is much broader in the Kerr case because the smaller inner disc radius implies that the line photons are suffering stronger relativistic effects (such as gravitational redshift) which is visible in the resulting line profile. To summarize, the detection and modelling of a relativistic broad Fe line via X-ray observations of accreting BH potentially provides crucial information on the system inclination, the radial efficiency of the coronal hard X-ray emission, and also on one of the two parameters that characterize the Kerr solution, i.e. the BH spin.

### *9.3.4 The ISCO and its relation to black hole spin*

A key issue in using the profile of a broad iron line to determine BH spin is whether the derived inner radius equals the ISCO, or not. Reynolds & Begelman (1997) pointed out that matter in the plunge region within the ISCO can also contribute to reflection and so could confuse spin determination. The matter in the plunge region is however moving inward rapidly and so has a low density. This means that the ionization parameter is high, which changes the ability to produce a detectable iron line. Young, Ross & Fabian (1998) showed in the key case of MCG–6-30-15 that the emission (and absorption) produced by reflection in the plunge region does not resemble the observed line shape. The issue of matter falling within the ISCO was further considered by Gammie (1999) and Krolik (1999); see also Agol & Krolik (2000) with the inclusion of magnetic fields which can connect the outer part of the plunge region with the disc and slow matter down (also Krolik & Hawley 2002). In fact, Dovčiak, Karas & Yaqoob (2004) have shown that the difference in the flow direction has a negligible effect on the profile shape.

We suspect that the effect of magnetic fields on the iron-line appearance is small. The ionization parameter can be written as

$$\xi = 3 \times 10^7 f(v_{\rm rad}/c) \quad \text{erg cm s}^{-1}\,, \qquad (9.1)$$

where $f$ is the volume filling factor of the flow which is falling radially inward at velocity $v_{\rm rad}$ (Fabian & Miniutti, in preparation). Since $f$ and $v_{\rm rad}$ are unlikely to be much smaller than 0.1 over most of the plunge region, it is most unlikely that a strong iron line, requiring $\xi \sim$ 100–1000 erg cm s$^{-1}$, can be produced. We suspect that the inner radius determined from iron line studies cannot originate from far within the ISCO and that the inner disc radius $r_{\rm in}$ is never more than one $r_g$ or so within the ISCO.

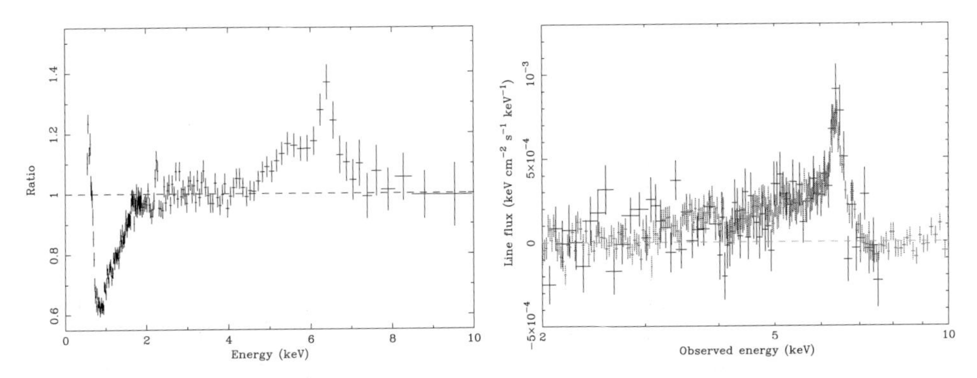


Fig. 9.7. *Left panel*: Ratio to a power law model of the 1994 ASCA S0 spectrum of MCG–6-30-15. The broad Fe line is clearly detected, together with absorption from photoionized gas below 2 keV. *Right panel*: Superposition of the broad Fe line profile of MCG–6-30-15 from the long *XMM-Newton* (grey) and *Chandra* (black) observations. The observations are non-simultaneous but the line profiles superimpose very well (Young *et al.* 2005).

## 9.4 The broad relativistic Fe line of MCG–6-30-15

The X-ray spectrum of the bright Seyfert 1 galaxy MCG–6-30-15 ($z = 0.007\,75$) has a broad emission feature stretching from below 4 keV to about 7 keV. The shape of this feature, first clearly resolved with *ASCA* by Tanaka *et al.* (1995), is skewed and peaks at about 6.4 keV. This profile is consistent with that predicted from iron fluorescence from an accretion disc inclined at 30 deg extending down to within about six gravitational radii ($6\,r_g = 6GM/c^2$) of a BH. During the *ASCA* observation the line appeared to be occasionally redshifted to lower energies, implying emission from even smaller radii and therefore strongly suggesting that the BH in this object is rapidly spinning (Iwasawa *et al.* 1996b). The broad iron line of MCG–6-30-15 has been detected by all major X-ray missions (*ASCA*, *RXTE*, *BeppoSAX*, *Chandra*, and *XMM-Newton*) with very similar results.

In Fig. 9.7 we show the line profile obtained at different times with three different detectors. In the right panel, we show the broadband spectrum (as a ratio with a power law model) of the S0 detector on board of *ASCA* during the 170 ks observation in 1994 (Iwasawa *et al.* 1996b). In the right panel, we show the line profile from non-simultaneous *XMM-Newton* (grey, 320 ks in 2001) and *Chandra* (black, 522 ks in 2004) observatories (Fabian *et al.* 2002a; Young *et al.* 2005). The large effective area provided by *XMM-Newton* confirmed earlier results by Iwasawa *et al.* (1996b; 1999) showing that the broad red wing of the line extends down to about 3 keV implying a spin parameter $a/M > 0.93$ (Dabrowski *et al.* 1997; Reynolds *et al.* 2004). Here we report in some detail results from the *XMM-Newton* observations of MCG–6-30-15 (Wilms *et al.* 2001; Fabian *et al.* 2002a; Fabian & Vaughan 2003;

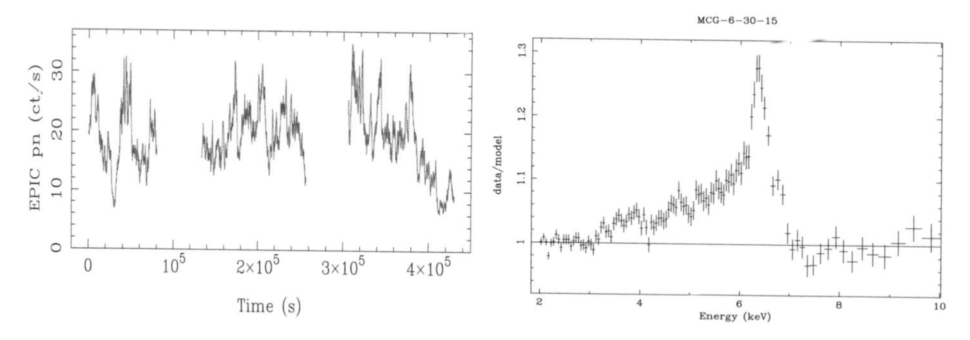


Fig. 9.8. *Left panel*: The broadband light curve of the long *XMM-Newton* observation in 2001. Three satellites' orbits have been devoted to study MCG–6-30-15 in 2001, for a total exposure of about 320 ks. *Right panel*: The broad iron line in MCG–6-30-15 from the *XMM-Newton* observation in 2001 (Fabian *et al.* 2002a) is shown as a ratio to the continuum model.

Vaughan & Fabian 2004; Reynolds *et al.* 2004) mentioning also the most important consequences of the recent long *Chandra* observation (Young *et al.* 2005).

The 2001 *XMM-Newton* observation was very long (about 320 ks corresponding to three full satellite orbits) and simultaneous with a *BeppoSAX* one, providing unprecedented spectral coverage with *XMM-Newton* being more sensitive than *BeppoSAX* in the 0.2–10 keV band, and *BeppoSAX* extending the data set up to 200 keV. The broadband *XMM-Newton* light curve for the three orbits is shown in the left panel of Fig. 9.8. In the right panel of Fig. 9.8 we show the observed broad Fe line profile from the time-averaged spectrum. We discuss here our best description of the X-ray spectrum of MCG–6-30-15, focusing on the broad Fe line, and later review alternatives to our modelling which, however, all prove largely unsatisfactory when tested against the data.

The broadband spectrum of MCG–6-30-15 is best described by a relativistically blurred reflection spectrum modified by absorption by photoionized gas below about 2 keV (Fabian *et al.* 2002a). For the X-ray reflection model, we used the code by Ross & Fabian (1993; 2005) allowing the photon index $\Gamma$, the ionization parameter $\xi$, and the relative strength of the reflection spectrum with respect to the power law continuum free to vary. To account for Doppler and gravitational effects, the reflection spectrum is convolved with a relativistic kernel computed using a modified version of the code by Ari Laor (1991) which is appropriate for the Maximal Kerr case. However, the inner disc radius is not fixed to the ISCO but can be larger. Therefore, if the inner disc radius can be inferred from the data, information on the spin can in principle be obtained by making use of the ISCO–spin relation (see Fig. 9.6). The emissivity profile has the general form of a broken power law with index $q_{\rm in}$ from the inner disc radius to a break radius $r_{\rm br}$ and $q_{\rm out}$ outwards.

The best-fitting parameters indicate a weakly ionized disc with ionization parameter $\xi < 30$ erg cm s$^{-1}$ and a strong reflection fraction[1] of about $R \simeq 2.2$, a result which is fully confirmed by the *BeppoSAX* data where the Compton hump around 20–30 keV is clearly seen. Fe is measured to be three times more abundant than standard (solar). The parameters of the relativistic blurring are the disc inclination $i = 33° \pm 1°$, the inner disc radius $r_{\rm in} = 1.8 \pm 0.1\ r_g$, and the emissivity parameters $q_{\rm in} = 6.9 \pm 0.6$, $q_{\rm out} = 3.0 \pm 0.1$, and $r_{\rm br} = 3.4 \pm 0.2$ (Vaughan & Fabian 2004). The measured value of the inner disc radius implies emission from far beyond the ISCO of a Schwarzschild BH ($6\ r_g$) and therefore suggests the BH in MCG–6-30-15 is rapidly spinning. If we take the measured value of $r_{\rm in}$ as a measure of the ISCO, we can constrain the BH spin to be $a/M = 0.96 \pm 0.01$, providing one of the most remarkable indications so far for the astrophysical relevance of the Kerr solution to the Einstein's equations. The inferred emissivity profile indicates a standard emissivity from about $3.4\ r_g$ to the outer disc boundary. However, a steep profile is required in the innermost few gravitational radii, indicating that most of the accretion power is released in the region where General Relativity is in the strong field regime. Tapping of black holes spin by magnetic fields in the disc is a strong possibility to account for the peaking of the power so close to the hole (Wilms *et al.* 2001; Reynolds *et al.* 2004; Garofalo & Reynolds 2005).

### 9.4.1 Alternatives to a relativistic line

The claim that iron line studies are probing the strong field regime of General Relativity just a few gravitational radii from the BH is a bold one, and should always be tested again alternative models. In this spirit, we discuss here alternatives to the broad relativistic Fe line in the X-ray spectrum of MCG–6-30-15 (see also Fabian *et al.* 1995; Fabian *et al.* 2000).

The broad Fe line profile seen in the right panel of Fig. 9.8 comprises two main components: a relatively narrow core peaking around 6.4 keV, and an extended broad red wing from about 3 to 6 keV. If the broad red wing is not due to a relativistic line but instead to some spectral curvature independent of any relativistic effect, we are left with the line core only. If this is emitted from far from the BH (such as from the torus), it has to be narrow and unresolved at the *XMM-Newton* resolution (about 100 eV). However, the line core itself is unambiguously resolved by *XMM-Newton*. When modelled by a Gaussian emission line, the line core width is $\sigma = 352^{+106}_{-50}$ eV, much larger than the spectral resolution (Vaughan & Fabian 2004). Considering a blend of lines from different Fe ionization states as responsible for the broadening

[1] The reflection fraction ($R$) is defined as the relative normalization of the reflection component with respect to the illuminating continuum. $R = 1$ means that the primary source subtends a solid angle $\Omega/2\pi = 1$ at the disc. Values larger than 1 imply that the disc is seeing more illuminating continuum than any observer at infinity and therefore that the primary source is anisotropic and preferentially shines towards the disc.

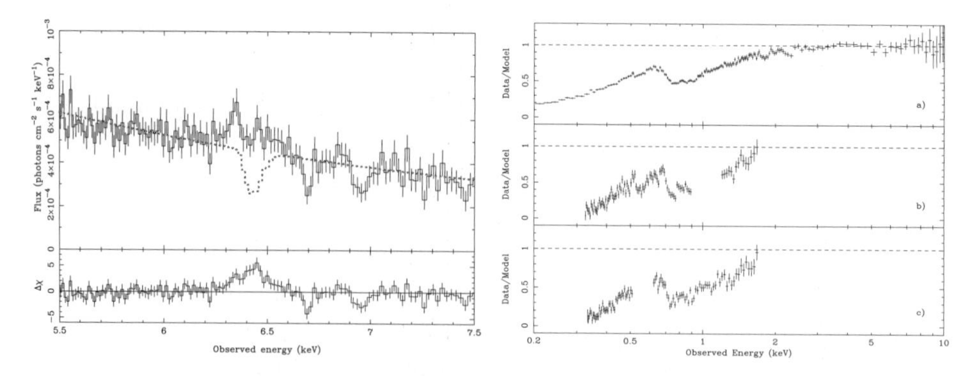


Fig. 9.9. *Left panel*: The *Chandra* HEG spectrum (solid curve) is overlaid to the best-fit ionized absorption model of the broad red wing of the iron line. The ionized absorber produces a curved spectral shape that approximates the broad red wing, but predicts iron absorption features in the 6.4–6.6 keV range (rest-frame) which are inconsistent with the data (Young *et al.* 2005). *Right panel*: The difference spectrum obtained by subtracting the low flux state spectrum from the high flux state one, plotted as a ratio to a power law in the 3–10 keV band. The top panel refers to the EPIC–pn data, the two lower panels to the RGS gratings on *XMM-Newton* (Turner *et al.* 2003).

produces an unacceptable fit. Therefore, even ignoring for a moment the broad red wing of the line, the relatively narrow line core itself already suggests that the line is emitted in a fast orbiting medium close to the central BH and excludes any strong contribution from a reflector located at the parsec (and even sub-parsec) scale such as the torus and/or the Broad Line Region. A small contribution is possible and likely, but at the level of 15 per cent maximum of the line core.

As for the broad red wing of the line below 6 keV (see right panel of Fig. 9.8) several attempts have been made to explain the spectral shape without invoking relativistic effects. The broad red wing of the line has been modelled by different emission components, i.e. a thermal component, a blend of broadened emission lines (other than Fe), and complex absorption. In the latter model, which is probably the only serious one, the continuum passes through a large column of moderately ionized gas which can cause significant curvature above the Fe L edge (about 0.7 keV) up to about 5–6 keV, therefore with the potential of reproducing the broad red wing of the line. The gas must be sufficiently highly ionized to avoid undetected excessive opacity in the soft band, but not so highly ionized to lose all of its L-shell electrons. The best-fit ionized absorption model to both *XMM-Newton* and *Chandra* data can reproduce the red wing of the line (though the fit statistic is much worse than when the data are fitted with a relativistic line). However, the model predicts a clear complex of narrow absorption lines between 6.4 and 6.7 keV. This is inconsistent with the high resolution *Chandra* data as shown in the left panel of Fig. 9.9 from Young *et al.* (2005). We conclude that ionized absorption cannot reproduce the broad red wing of the relativistic Fe line.

A neutral partial covering model (which would be consistent with *Chandra* data) is also ruled out because it is inconsistent with the high-energy data provided by *RXTE* and *BeppoSAX*, in particular the unambiguous requirement for a strong reflection component (Reynolds *et al.* 2004; Vaughan & Fabian 2004). As a further test, a partial covering model was added to the best-fit model with a relativistic line to see if the derived parameters (most importantly the inner disc radius providing information on the BH spin) are robust. No differences were noticed and $r_{\rm in}$ remains at $1.8 \pm 0.1\ r_g$ (Vaughan & Fabian 2004; Young *et al.* 2005). An ionized partial covering model can not be simply ruled out, but *Chandra* data constrain the covering fraction to be less than 5 per cent, insufficient to produce enough spectral curvature in the relevant energy band, therefore requiring the presence of a broad relativistic line as for the neutral partial covering model above. As discussed in the next Section, the presence of complex absorption above 3 keV is also strongly (and in our opinion, unambiguously) ruled out by spectral variability analysis.

### *9.4.2 A puzzling spectral variability*

The X-ray continuum emission of MCG–6-30-15 is known to be highly variable (see Vaughan, Fabian & Nandra 2003; Vaughan & Fabian 2004; Reynolds *et al.* 2004; McHardy *et al.* 2005 for some recent analyses) as is also clear from the light curve in Fig. 9.8. Several different methods are used to explore the spectral X-ray variability in accreting BH sources. One such a method is provided by the analysis of the so-called difference spectrum, i.e. the spectrum that can be obtained by subtracting a low flux state spectrum from a high flux one (Fabian & Vaughan 2003). In this way any component that remained constant within the two flux levels is effectively removed from the spectrum which therefore carries information on the variable components plus absorption only. The *XMM-Newton* 2001 observation was split into two different flux states according to the light curve (left panel of Fig. 9.8) and two spectra were extracted as representative of the high and low flux states. The low flux spectrum was then subtracted from the high flux one.

In the right panel of Fig. 9.9 we show the resulting difference spectrum as a ratio to a power law fitted in the 3–10 keV band (from Turner *et al.* 2003). The EPIC–pn data are shown in the upper panel, while the high resolution *XMM-Newton*–RGS gratings are plotted in the middle and lower panels. It is clear from Fig. 9.9, that the variable component is very well described by the power law in the 3–10 keV band and is affected by absorption from photoionized gas only below 3 keV. This demonstrates in a model-independent way that the X-ray variability in MCG–6-30-15 is due to a relatively steep ($\Gamma \simeq 2.2$) power law component, and that the amplitude changes of the continuum in MCG–6-30-15 are due to this power law.

The difference spectrum also clarifies in a model-independent way that there is no subtle additional absorption above 3 keV that might influence the broad red

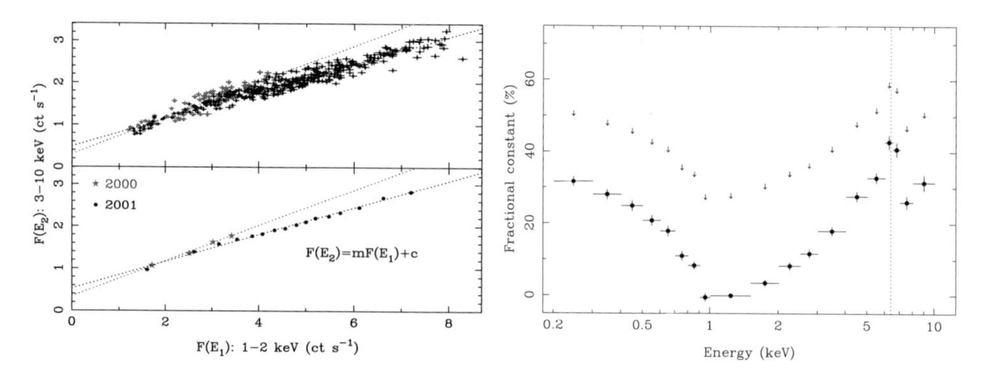


Fig. 9.10. *Left panel*: Flux–flux plot for the 2000 (grey) and 2001 (black) *XMM-Newton* observations. The 3–10 keV count rates are plotted versus the simultaneous 1–2 keV count rates. The upper panel shows the raw data, the lower is a binned version of the same plot. *Right panel*: The fractional spectrum of the constant component of MCG–6-30-15 constructed from the $y$-axis offset in flux–flux plots generated in different bands. The spectrum strongly resembles that of a reflection component, with a prominent contribution at 6.4 keV (vertical dotted line). Figures from Vaughan & Fabian (2004).

wing of the relativistic Fe line. Absorption by either neutral or ionized gas has to show up in the difference spectrum (as seen below 3 keV) but is not detected in the Fe band above 3 keV. This is in our opinion conclusive on the reality of the broad red wing of the Fe line. The case for the relativistic Fe line in MCG–6-30-15 therefore looks even more robust now than before.

It should be stressed that the difference spectrum, unlike the true spectrum of the source, does not exhibit any emission feature (neither narrow nor broad) in the Fe band. This implies that the relativistic Fe line in MCG–6-30-15 is much less variable than the power law continuum and has therefore been removed by the subtraction. As a test, the *XMM-Newton* 2001 exposure was divided into 10 ks long chunks and spectral fits performed. Direct spectral analysis on the 10 ks spectra confirms that the variability in MCG–6-30-15 closely follows a two-component model comprising a Power Law Component (hereafter PLC) with constant slope of about $\Gamma = 2.2$ and highly variable normalization (a factor 3–4) and a Reflection-Dominated Component (RDC) that carries the broad Fe line and whose variability is confined within 25 per cent (Fabian & Vaughan 2003; Vaughan & Fabian 2004) confirming previous results (Shih, Iwasawa & Fabian 2002; Matsumoto *et al.* 2003).

A further powerful model-independent tool to explore the spectral variability is provided by flux–flux plot analysis which consists of plotting the count rates in two different energy bands against each other (Taylor, Uttley & McHardy 2003). The largest variability in MCG–6-30-15 occurs in the 1–2 keV band which is therefore chosen as the reference energy band. In the left panel of Fig. 9.10 we show the count rates in the 3–10 keV band plotted against the simultaneous count rates in the 1–2 keV band (from Vaughan & Fabian 2004). Black points refer to the 2001 long *XMM-Newton* observation, grey ones to the previous shorter *XMM-Newton* observation in 2000 (Wilms *et al.* 2001). Figure 9.10 is divided into two panels, the lower one being simply a binned version. It is clear that a very significant correlation is found in both observations. The scatter in the unbinned version is a manifestation of the different shape of the power spectral density in the two bands and lack of coherence. The scatter is removed in the binned version which is most useful for fitting purposes.

The relationship is clearly linear which is an unambiguous sign that the variability in both bands is dominated by a spectral component which has the same spectral shape and varies in normalization only. However, extrapolating the linear relation to low count rates leaves a clear offset on the $y$-axis. This strongly suggests the presence of a second component that varies very little and contributes more to the 3–10 keV than to the 1–2 keV band. In other words, the flux in each band is the sum of a variable and a constant component, with the constant component contributing more to the harder than softer energies. From the previous discussion on the difference spectrum, the variable component is easily identified with a power law with constant slope $\Gamma = 2.2$ dominating the variability by variation in normalization only. Moreover, we already know that the reflection component varies little because no Fe line is seen in the difference spectrum and because of the result of direct time-resolved spectral analysis. It is therefore likely that the constant component is associated with the X-ray reflection spectrum, including the relativistic Fe line.

Flux–flux plots can be produced for any chosen band by plotting the relative count rates as a function of those in the reference band. Each plot will produce a different $y$-axis offset which represents the fractional contribution of the constant component to the spectrum in that particular energy band. Therefore, by recording the value of the offset in each band, the "fractional spectrum" of the constant component can be obtained and the hypothesis that this is similar to a X-ray reflection spectrum can be tested. The fractional spectrum of the constant component is shown in the right panel of Fig. 9.10 (from Vaughan & Fabian 2004; see Taylor, Uttley & McHardy for a similar result based on *RXTE* data). The black circles are the result obtained by assuming that the contribution of the constant component in the 1–1.5 keV band is null, the arrows indicate the upper limits (notice that the shape, which is relevant here, does not change). The fractional spectrum is soft below $\sim 1$ keV and hard at higher energies with a strong contribution at 6.4 keV. This strongly resembles a typical X-ray reflection spectrum, supporting our previous hypothesis that the constant component in MCG–6-30-15 is indeed reflection-dominated.

### 9.4.3 Interpretation: the light bending model

If the observed power law component (PLC) continuum drives the iron fluorescence then the line flux should respond to variations in the incident continuum on timescales comparable to the light-crossing, or hydrodynamical time of the inner accretion disc (Fabian *et al.* 1989; Stella 1990; Matt & Perola 1992; Reynolds *et al.* 1999). This timescale ($\sim 100 M_6$ s for reflection from within 10 $r_g$ around a black hole of mass $10^6 M_6$ $M_\odot$) is short enough that a single, long observation spans many light-crossing times. This has motivated observational efforts to find variations in the line flux (e.g. Iwasawa *et al.* 1996b, 1999; Reynolds 2000; Vaughan & Edelson 2001; Shih, Iwasawa & Fabian 2002; Matsumoto *et al.* 2003). However, the variability analysis discussed above implies that the reflection-dominated component (RDC) does not follow in general the PLC variations.

Explaining the relatively small variability of the RDC, compared with that of the PLC, provides a significant challenge. The RDC and PLC appear partially disconnected. Since however both show the effects of the warm absorber they must originate in a similar location. As the extensive red wing of the iron line in the RDC indicates emission peaking at only a few gravitational radii ($GM/c^2$) we must assume that this is indeed where that component originates. On the other hand, this is also the region where most of the accretion power is dissipated and therefore the PLC must originate in the corona above the innermost disc as well, so that light-travel-time effects cannot be invoked to explain the lack of response of the RDC to the PLC variation.

We have developed a model for the variability of the PLC and RDC components in accreting BH systems which is based on the idea that since both components originate in the immediate vicinity of the BH, they both suffer relativistic effects due to strong gravity (Fabian & Vaughan 2003; Miniutti *et al.* 2003; Miniutti & Fabian 2004). Some of the effects of the strong gravitational field on the line shape and intensity had already been explored by Martocchia *et al.* (2000, 2002a) and Dabrowski & Lasenby (2001). We assume a geometry in which a Maximal Kerr BH is surrounded by the accretion disc extending down to the ISCO. The PLC is emitted from a primary source above the accretion disc. Its location is defined by its radial distance from the BH axis and its height (h) above the accretion disc. Several geometries have been studied and the results we present are qualitatively maintained for any source location and geometry within the innermost 4–5 gravitational radii from the black hole axis. The main parameter in our model is the height of the primary source above the accretion disc and the main requirement is for the source to be compact. The primary PLC source could be physically realized by flares related to magnetic reconnection in the inner corona, emission from the base of a jet close to the BH, internal shocks in aborted jets, dissipation of the rotational BH energy via magnetic processes, etc. Any mechanism producing a compact PLC-emitting region above the innermost region of the accretion flow would be relevant for our model (e.g. Blandford & Znajek 1977; Markoff, Falcke & Fender 2001; Li 2003; Ghisellini, Haardt & Matt 2004 and many others).

A fraction of the radiation emitted by the primary source directly reaches the observer at infinity and constitutes the direct continuum which is observed as the PLC of the spectrum. The remaining radiation illuminates the accretion disc (or is lost into the hole event horizon). The radiation that illuminates the disc is reprocessed into the RDC. For simplicity, we assume that the intrinsic luminosity of the primary source is constant. The basic idea is that the relevant parameter for the variability of both the PLC and the illuminating continuum on the disc (which drives the RDC variability) is the height of the primary source above the accretion disc and that the variability is driven by general relativistic effects the most important of which is gravitational light bending.

As an example, if the source height is small (of the order of few gravitational radii) a large fraction of the emitted photons is bent towards the disc by the strong gravitational field of the BH enhancing the illuminating continuum and strongly reducing the PLC at infinity, so that the spectrum is reflection-dominated. If the source height increases, gravitational light bending is reduced so that the observed PLC increases. Finally, if the height is very large so that light bending has little effect, the standard picture of reflection models with approximately half of the emitted photons being intercepted by the disc and the remaining half reaching the observer as the PLC is recovered.

With this setup, we compute the PLC and Fe line flux (representative of the whole RDC) as a function of the primary source height above the disc (see Miniutti *et al.* 2003; Miniutti & Fabian 2004 for more details). Our results are presented in the left panel of Fig. 9.11 for the case of an accretion disc seen at an inclination of 30° (the relevant case for MCG–6-30-15). The black dots indicate different values for the source height ($h = 1, 5, 10, 20\ r_g$ from left to right). Figure 9.11 shows us that, once general relativistic effects are properly accounted for, the broad Fe line (and RDC) is not expected to respond simply to the observed PLC variation anymore. Three regimes can be identified (I, II, and III in Fig. 9.11). A correlation is seen only in regime I, where the primary source height is between about 1 and 3–4 $r_g$. On the other hand, if the primary source has a height between about 4 $r_g$ and 12 $r_g$ (regime II) the broad Fe line (and RDC) variability is confined within 10 per cent only, while the PLC can vary by a factor $\sim 4$. We recall here that the observed variability of the RDC in the 2001 *XMM-Newton* observation is confined within 25 per cent despite variations by a factor 3–4 in the PLC.

In the right panel of Fig. 9.11 we show the value of the reflection fraction predicted by the model as a function of the source height. When the source height

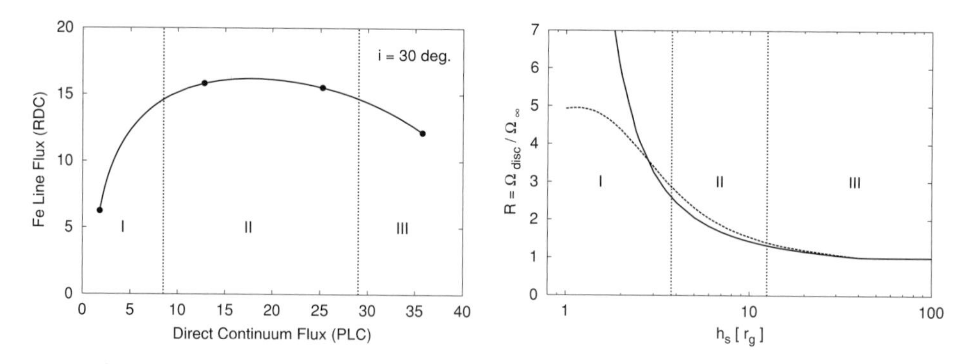


Fig. 9.11. *Left panel*: The Fe line flux (representative of the whole RDC) as a function of the PLC flux observed at infinity. The variability is only due to changes in the primary source height. Black dots represent heights of $h = 1,\ 5,\ 10,\ 20\ r_g$. Three regimes can be identified. In regime II the RDC variation is within 10 per cent while the PLC varies by about a factor 4. *Right panel*: The reflection fraction as a function of the source height (nearly the same if plotted versus the PLC flux). We show the cases of a source on the disc axis (solid) and one at 2 $r_g$ from the axis.

is large, the standard value $R = 1$ is recovered. However, as the height decreases, light bending comes into play increasing the number of photons bent towards the disc while simultaneously reducing the number of those able to escape from the gravitational attraction. Thus, the reflection fraction increases dramatically. Two cases are shown in Fig. 9.11, one for a source on the disc axis (solid), one for a ring-like configuration at 2 $r_g$ from the black hole axis. Regime II, which seems most relevant so far, is characterized by $R \simeq 2$. Quite remarkably this is precisely the measured value in MCG–6-30-15 during the simultaneous *XMM-Newton–BeppoSAX* 2001 observation we are discussing here.

Thus, if our model is qualitatively correct, it implies that MCG–6-30-15 was observed in regime II by *XMM-Newton* during the 2001 observation with most of the accretion power being dissipated within 5 $r_g$ in radius and in the range 4–12 $r_g$ in height above the disc. This is itself a remarkable result implying that we are looking down to the region of spacetime where General Relativity exhibits itself at its best. Since the model computes self-consistently the disc illumination, the emissivity profile can be computed. In regime II, it turns out to be best described by a broken power law, very steep in the innermost region of the disc and flattening outwards, as observed in MCG–6-30-15 (see left panel of Fig. 9.12). We computed fully relativistic line profiles with our model and compared them with the data (Fig. 9.12, right panel). The line profiles obtained from regime II provide a very good match to the *XMM-Newton* data (Miniutti *et al.* 2003).

As is clear from Fig. 9.11 (left panel), the model predicts a correlation between RDC and PLC at low flux levels. Fortunately, the *XMM-Newton* 2000 observation

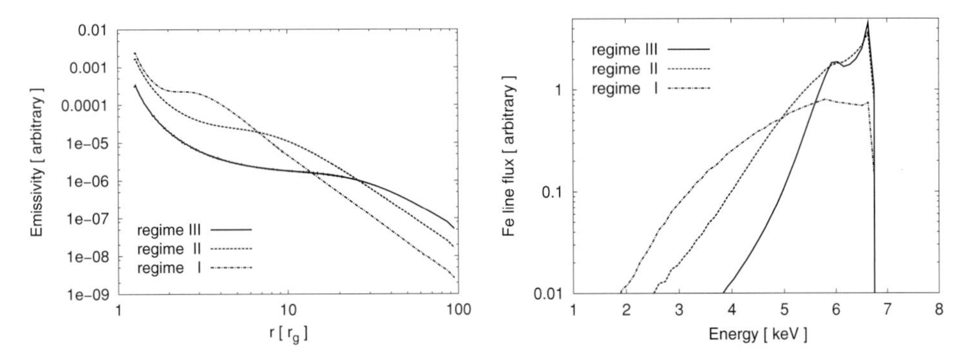


Fig. 9.12. *Left panel*: Typical emissivity profiles computed in the light bending model for the three relevant regimes. The emissivity is steeper in the inner part of the disc and flattens outwards. *Right panel*: Typical line profiles in the three regimes. A log scale is used to enhance the line profile differences.

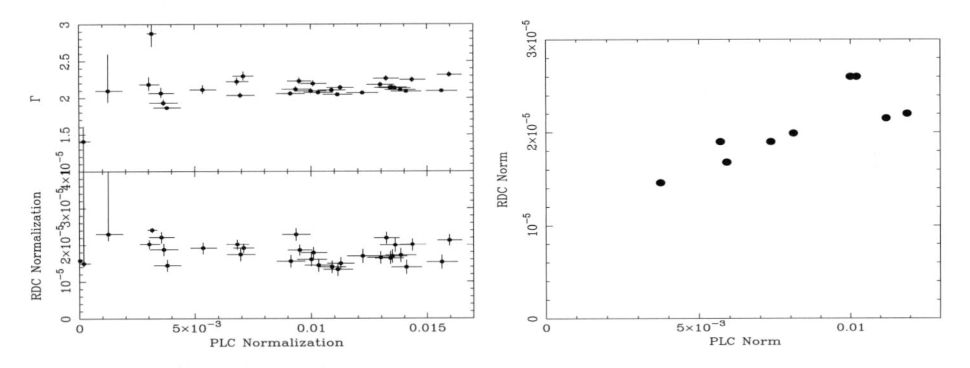


Fig. 9.13. *Left panel*: PLC photon index (top) and RDC normalization as a function of the PLC normalization for the 2001 observation. *Right panel*: RDC normalization as a function of the PLC normalization for the 2000 *XMM-Newton* observation.

(that we did not discuss so far) caught MCG–6-30-15 in a slightly lower flux state than the 2001 one. The flux state is not remarkably lower, but some indications of the general behaviour can still be inferred. The details of the spectral analysis of the 2000 observation can be found in Wilms *et al.* (2001) and Reynolds *et al.* (2004). Here we only focus on the variability of the two components. If the source did not change dramatically its properties between the two observations, we should observe a somewhat correlated variability between the RDC and the PLC. A comparison between the two observations in this sense is shown in Fig. 9.13 where results of spectral analysis performed on 10 ks chunks for both observations are presented. In the right panel, the PLC slope Γ and RDC normalization are shown as a function of the PLC normalization for the higher flux 2001 *XMM-Newton* observation. The relevant panel is the bottom one showing the constancy of the RDC despite the

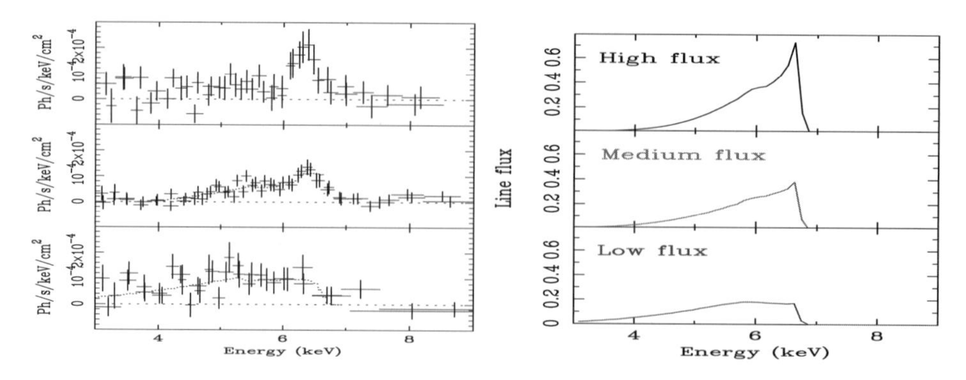


Fig. 9.14. *Left panel*: The observed variations of the line profile during the 1994 *ASCA* observation (Iwasawa *et al.* 1996). From top to bottom the high, medium, and low flux states profiles are shown. *Right panel*: The changes in the line profile predicted by the light bending model for the observed flux states.

large PLC variation. In the right panel of the same figure, we show the RDC versus PLC normalization for the lower flux 2000 *XMM-Newton* observation. As already reported by Reynolds *et al.* (2004), a clear correlation is seen between the two components in this lower flux observation. This result was not known yet when the light bending model was being developed and provides unexpected support for the overall picture, matching one of the main predictions of the model.

The puzzling variability of the RDC and PLC, the large value of the reflection fraction, and even the emissivity and line profiles all seem to indicate the relevance of the light bending model for MCG–6-30-15. Given the very strong case for Fe line emission from the relativistic region in MCG–6-30-15, it would be a rather surprising coincidence if strong relativistic effects were not responsible for some, if not all, of the behaviour discussed above. We conclude by stressing that in all cases in which broad Fe lines are detected strong relativistic effects on the X-ray emission and variability cannot be ignored and have to be included in theoretical models. The light bending model is probably crude and incomplete, but represents a first significant step in this direction.

### 9.4.4 Occasional Fe line variability

As mentioned, the RDC and broad Fe line in MCG–6-30-15 is almost constant despite large variation in the PLC. However, it exhibits occasionally marked variations in the line profile such as during the 1994 *ASCA* observation (Iwasawa *et al.* 1996b). Spectra were extracted in a high, medium, and very low flux state, the difference in count rate between the low and high flux being about a factor 4. The resulting line profiles are shown in the left panel of Fig. 9.14 where, from top to bottom, the line profile is for the high, medium and low flux state respectively.

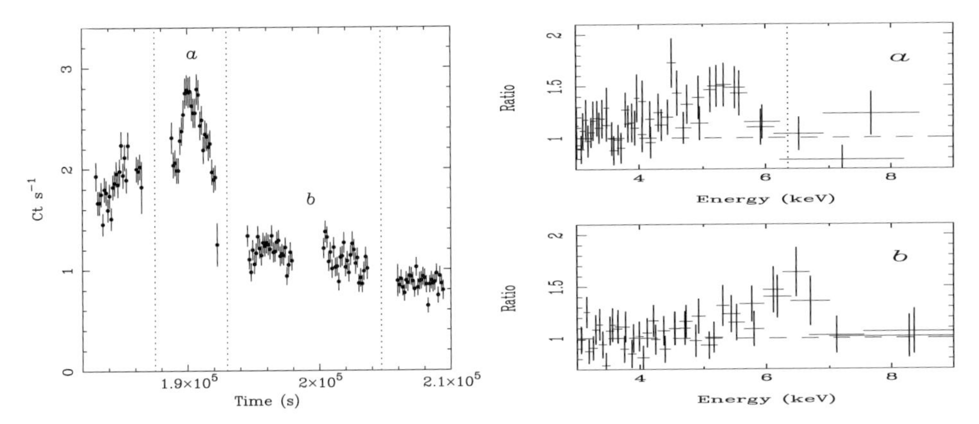


Fig. 9.15. *Left panel*: The *ASCA* light curve around the bright flare (Iwasawa *et al.* 1999). The two time-intervals relevant for the right panel of the figure are here defined. *Right panel*: During the flare (a) the broad line is redshifted well below the Fe K$\alpha$ rest-frame energy (6.4 keV, vertical line). When the flare ceased (b) the broad line recovered to the ordinary profile.

The general trend appears to be that the line is more peaked in high than low flux states and becomes much broader as the flux decreases. In particular, in the lowest flux state, no peak is visible at 6.4 keV and the line is very broad and redshifted. During the so-called deep minimum (bottom left panel in Fig. 9.14) the best-fitting parameters for the line profile indicate that emission is coming from within 6 $r_g$ providing the first evidence for the presence of a Kerr BH in MCG–6-30-15, later confirmed by higher quality *XMM-Newton* data.

In the framework of the light bending model, low flux states correspond to situations in which the primary source is very close to the BH thereby illuminating very efficiently the inner disc. In higher flux states, the source has a larger height, and the outer disc starts to be illuminated as well. Therefore, the model predicts very broad and redshifted lines in low flux states (see also Fig. 9.12, right panel) while a peak at 6.4 keV due to emission from the outer regions of the disc appears at higher flux levels. In the right panel of Fig. 9.14 we show three profiles obtained in the light bending model for three different heights of the primary source ($h = 6, 4, 2\ r_g$ from top to bottom). The profiles match the observed data rather well. It should be stressed that the PLC variation between $h = 2\ r_g$ (low flux) and $h = 6\ r_g$ (high flux) is predicted to be a factor 5, very similar to the factor 4 of the data.

Further remarkable evidence for variability of the Fe line on short timescales is provided by the 1997 *ASCA* observation (Iwasawa *et al.* 1999). During a bright flare, the Fe line profile changed dramatically. This is shown in Fig. 9.15 where the light curve and line profiles in two relevant time-intervals are shown. One possibility to explain such dramatic change is that a bright corotating flare appeared at about

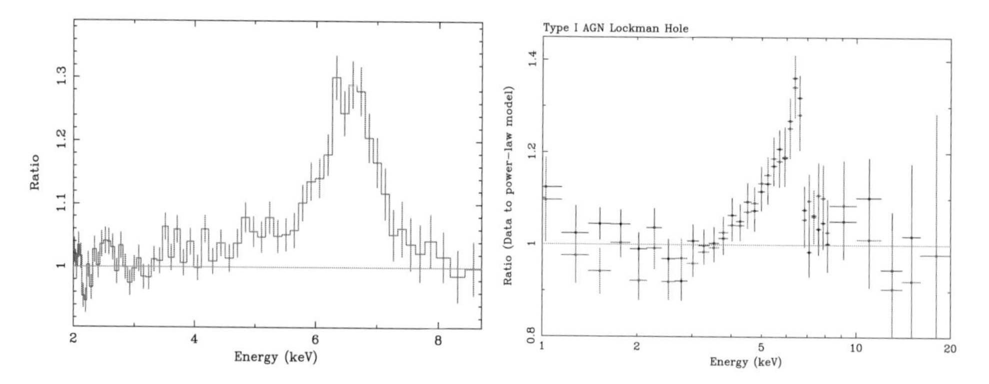


Fig. 9.16. *Left panel*: Ratio of the spectrum of IRAS 18325–5926 to a power law for *XMM-Newton* data. *Right panel*: Ratio plot of the mean unfolded spectrum for type-1 AGN in the Lockman Hole with respect to a power law (Streblyanska *et al.* 2005). *XMM-Newton* EPIC–pn and MOS detectors are used.

5 $r_g$ from the BH. Given the low BH mass in MCG–6-30-15 (about $10^6\ M_\odot$) the flare has enough time to complete more than one orbit during the time-interval 'a', therefore resulting effectively in a ring-like emitting structure which provides a good fit to the data (Iwasawa *et al.* 1999).

## 9.5 Other AGN: Seyfert1 and NLS1 galaxies

The case for a broad relativistic Fe line is very strong in MCG–6-30-15. In this source, the best-fitting parameters imply that Fe has about three times the solar abundance, the disc is lowly ionized, and the RDC is particularly strong with respect to the PLC ($R \simeq 2.2$). These conditions make it easier to detect the relativistic line in this AGN than in others. Indeed, it should be stressed that if these conditions were not met in MCG–6-30-15, the detection of a relativistic Fe line would be much more difficult, challenging present X-ray observatories. A discussion on the detectability of a broad Fe line in less favourable conditions is deferred to Section 9.7 below.

Here we just show another interesting case of a broad Fe line, presented in Fig. 9.16. In the left panel, we plot the ratio of the *XMM-Newton* data of IRAS 18325–5926 with respect to a power law model fitted in the 2–10 keV band. The Fe line in IRAS 18325–5926 is clearly broad and exhibits a strong red wing extending down to about 4 keV (Iwasawa *et al.* 1996a, Iwasawa *et al.* 2004; Iwasawa *et al.*, in preparation). Some other AGN in which resolved Fe lines have been reported in the past are discussed in Nandra *et al.* (1997a,b; 1999); Bianchi *et al.* (2001); Lamer *et al.* (2003); Longinotti *et al.* (2003); Balestra *et al.* (2004); Porquet *et al.* (2004b); Jiménez–Bailón *et al.* (2005), the list being non-exhaustive. Some of the Fe lines in the list are not extreme and do not imply that the reflector extends down below the ISCO of a non-rotating BH, thereby not constraining the BH spin. The line width in these objects is however inconsistent with emission from a very distant reflector such as the torus and/or the Broad Line Region and seems often inconsistent with being a blend of numerous narrower lines.

It should be stressed that not all broad lines are clear and unambiguous as in MCG–6-30-15, and other interpretations of the data are often possible. A possible scenario is that complex absorption by large columns of highly ionized matter produces a curvature in the spectrum that resembles that of a broad Fe line (e.g. Turner *et al.* 2005). As mentioned, this was conclusively excluded for MCG–6-30-15 but is a possible alternative to relativistic lines in other cases. The recently launched Astro–E2 mission has enough effective area and energy resolution in the relevant Fe band to test this scenario by detecting or not the absorption features associated with it. High-energy data will also be crucial because they can indicate the presence/lack of strong reflection components through the detection of the Compton hump therefore suggesting which interpretation is preferable.

The 770 ks *XMM-Newton* observation of the Lockman Hole field provides also some interesting information. The Lockman Hole is a special field in the sky because of extremely low Galactic absorption. It is therefore an ideal field to study faint AGN as a population and their contribution to the X-ray background. The field comprises 53 type-1 and 41 type-2 AGN with known redshifts. Streblyanska *et al.* (2005) have obtained the mean spectrum of type-1 AGN by stacking together the individual spectra and correcting for the source redshift (see also Brusa *et al.* 2005 for a spectral stacking results from the Chandra Deep Fields). The ratio of the data to a power law model for the mean type-1 AGN spectrum is shown in the right panel of Fig. 9.16 and exhibits a broad emission feature peaking at 6.4 keV. A prominent broad red wing is also visible. The observed equivalent width is larger than the average value for bright nearby Seyfert 1 galaxies but similar for example to MCG–6-30-15 and might indicate that Fe is overabundant with respect to solar values. Shemmer *et al.* (2004) have suggested that the metallicity in AGNs is correlated with the mass accretion rate (close to the Eddington limit), which is in turn related to the AGN luminosity. It is possible that the large equivalent widths measured here are due to high metallicity in distant and luminous objects. The average line profile is best-fitted with an inner disc radius of 3 $r_g$. This may be an indication that most of the AGN contain a Kerr BH at their very centre which would have serious implications for the dominant BH growth mechanisms and, more broadly, on cosmology and galaxy evolution.

### *9.5.1 1H 0707–495: a key to the nature of NLS1?*

Narrow Line Seyfert 1 galaxies tend to show steep soft X-ray spectra, extreme X-ray variability, and sometimes broad iron emission features. One such extreme

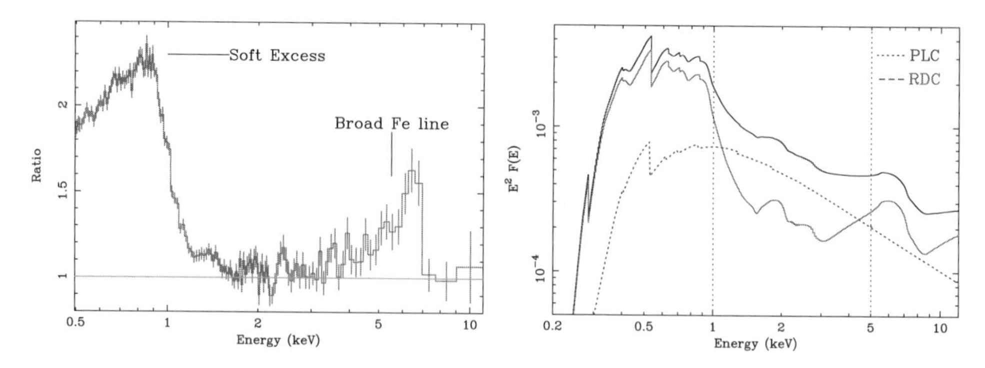


Fig. 9.17. *Left panel*: Ratio of the spectrum of the NLS1 1H0707 to a power-law fitted between 2 and 4 keV and above 7.5 keV. *Right panel*: Spectral decomposition of 1H0707–495 in terms of a variable power-law (dotted curve) and a blurred reflection component (broken). Vertical lines separate the regions in which one of the two components dominates the spectrum.

object is 1H 0707–495 which has a marked drop in its spectrum above 7 keV. This is either an absorption edge showing partial-covering in the source (Boller *et al.* 2002; Gallo *et al.* 2004; Tanaka *et al.* 2004) or the blue wing of a massive, very broad, iron line (Fabian *et al.* 2002b, 2004). Here we explore in some detail this latter possibility as an example of what could be a more general picture relevant to many other objects. 1H 0707–495 has been observed twice by *XMM-Newton* with a deep spectral drop at 7 keV discovered in the first observation. The drop had shifted to 7.5 keV in the second observation two years later.

In the left panel of Fig. 9.17 we show the ratio of the data (second observation) to a simple power law model fitted between 2 and 4 keV and above 7.5 keV. A large skewed emission feature is seen, very similar to that of MCG–6-30-15 (see the right panel of Fig. 9.8) together with a steep soft excess below 1–2 keV. We have fitted the broadband 0.5–10 keV spectrum with a simple two-component model comprising a PLC and a relativistically blurred RDC obtaining an excellent description of the entire data set. The data require Fe to have super-solar abundance (3 times the solar value) and the reflector to be ionized at the level of $\xi \simeq 650$ erg cm s$^{-1}$. The X-ray reflection spectrum comes from the innermost regions of the disc with inner disc radius measured at 2.3 $r_g$ and steep emissivity profile (Fabian *et al.* 2004). In comparison, the emissivity was even steeper during the first observation which explains why the marked spectral drop (that we interpret as the combination of the reflection edge and blue peak of the Fe line) was at lower energy (7 keV in the first and 7.5 keV in the second observation). The shift is due to gravitational redshift which is more effective in lowering the energy of sharp features when the emissivity is steeper (i.e. the first observation).

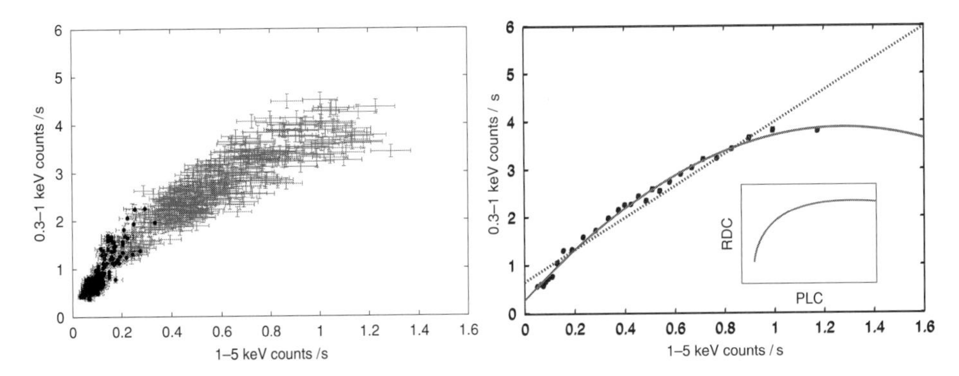


Fig. 9.18. *Left panel*: Flux–flux plot for the two observations of 1H 0707–495. The 0.3–1 keV band is representative of the RDC, whereas the 1–5 keV band of the PLC (see right panel of Fig. 9.17). *Right panel*: Binned version of the flux–flux plot. A linear relation is a much worse fit than a curved one. In the insert panel, we show the prediction of the light bending model.

The best-fit model is shown in the right panel of Fig. 9.17. The solution implies that the spectrum is almost completely reflection-dominated, i.e. the PLC illuminates much more efficiently the disc than the observer at infinity. This seems therefore to be a promising source in which to test further the light bending model that successfully reproduces the spectral shape and variability of the key source MCG–6-30-15. In fact the model predicts the existence of reflection-dominated states in regime I and part of regime II.

From the right panel of Fig. 9.17, the RDC dominates the spectrum below about 1 keV and above 5 keV, whereas the PLC dominates in the intermediate 1–5 keV band (the vertical lines in Fig. 9.17 separate the different energy bands). As a zeroth-order approximation, we can then consider the 1–5 keV flux as representative of the PLC, while the 0.3–1 keV flux is associated with the RDC. If so, by plotting the 0.3–1 keV count rate as a function of the 1–5 keV one, we could infer the RDC correlation with the PLC. This is shown in the left panel of Fig. 9.18 for both the first (black) and second (grey) observations. There is clearly a correlation that, however, is not linear. This is more clearly shown in the binned version of the same plot, shown in the right panel of Fig. 9.18. The correlation is non-linear, and a curved relationship is preferred by the data. In the insert, we show the prediction of the light bending model which seems to catch the overall behaviour surprisingly well. The insert is appropriate for an accretion disc seen at 30°, while the disc inclination in 1H 0707–495 is likely more 50°. As shown in Miniutti & Fabian (2004) the light bending model predicted behaviour for larger inclination is even more similar to that seen in the data.

We conclude by noting that, quite remarkably, a relativistically blurred reflection-dominated model describes well the *XMM-Newton* spectrum of 1H 0707–49 over the entire observed energy band (for both observations). Strong gravitational light bending around a Kerr BH seems to be the simplest explanation for the peculiar spectrum of this source and its remarkable variability. More detailed analysis of the data sets is given in Fabian *et al.* (2002b; 2004) where further support in favour of the light bending model is given. The same scenario described above is also relevant for 1H 0419–577 (Fabian *et al.* 2005) and NGC 4051 (Ponti *et al.* 2006). An alternative description in terms of partial covering is provided by Boller *et al.* (2002); Gallo *et al.* (2004); Tanaka *et al.* (2004). See also Pounds *et al.* (2004a) for a similar partial-covering interpretation of the low flux state of 1H 0419–577.

## 9.6 Galactic black hole candidates

Black hole binaries are accreting systems containing a dark compact primary with a mass larger than about 3 $M_{\odot}$ and a non-degenerate companion. These are generally referred to as confirmed BH binaries because the high mass of the primary excludes it is a neutron star. Only 18 such systems are known in our Galaxy, but they are thought to be representative of a large population of systems (possibly around 30 millions of sources). The first to be discovered was Cyg-X1, containing a BH of 6.9–13.2 $M_{\odot}$ with a O/B star as companion. In addition, many other X-ray sources in the Galaxy do exhibit all the observed characteristics of BH binaries. Many of these sources can not be claimed as secure BH binaries because the mass of the primary is not well constrained, and they are generally referred to as Galactic BH Candidates (GBHCs).

BH binaries and GBHCs exhibit a fascinating phenomenology which manifests itself in complex X-ray spectral and temporal properties. In many respects, these sources can be thought of as scaled-down versions of AGN both powered by accretion into the central BH. It is not our purpose here to explore the complexity of BH binaries and GBHCs, and we refer the interested reader to more specialist reviews such as, for example, the one by Remillard & McClintock (2006). Our interest here is to focus on the broad relativistic Fe lines that have been detected in many objects, sometimes suggesting that the central BH is rapidly spinning. One such a case, the GBHC XTE J1650–500, will be presented in some detail, mainly because its 2001 outburst was covered by three different X-ray detectors (*RXTE*, *BeppoSAX*, *XMM-Newton*) and is therefore particularly well characterized (though this is clearly not the only source which benefits from such an extensive X-ray coverage).

Examples of broad Fe lines detected with *ASCA* in the X-ray spectra of GBHCs are shown in the left panel of Fig. 9.19 from Miller *et al.* (2004). Other broad lines have been found in other GBHCs, especially in the high and very high/intermediate states where the accretion disc is believed to extend down to small radii (Martocchia *et al.* 2002b; Miller *et al.* 2002a,b,c; 2003). In the right panel of the same figure,

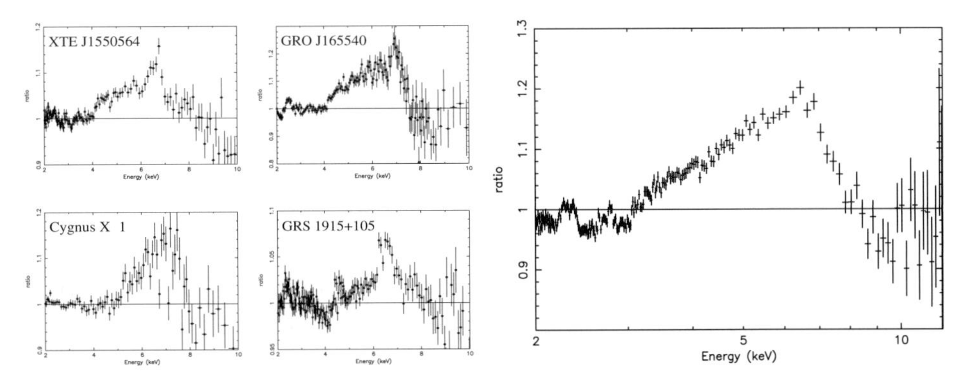


Fig. 9.19. *Left panel*: Prominent relativistic lines observed with *ASCA* in GBHCs. When fitted with relativistically blurred reflection models all sources but GRS 1915+105 require emission from within the ISCO of a non-rotating BH, suggesting that the BHs in most sources are rapidly spinning. *Right panel*: The broad iron line in GX 339–4 as observed by *XMM-Newton* (Miller *et al.* 2004b).

we show the *XMM-Newton* Fe line profile of GX 339–4 (Miller *et al.* 2004b). A similar profile is detected by *Chandra* as well (Miller *et al.* 2004a).

### 9.6.1 The case of XTE J1650–500

XTE J1650–500 is a X-ray binary with a period of about 7.6 hours and an optical mass function $f(M) = 2.73 \pm 0.56\ M_{\odot}$. If typical mass ratios with the companion and an inclination of about 50° are assumed (the latter in agreement with X-ray spectroscopy) the mass of the primary turns out to be $7.3 \pm 0.6\ M_{\odot}$ strongly suggesting the presence of a central BH (Orosz *et al.* 2004). The source outburst was followed up during 2001 by *BeppoSAX* (three observations), *XMM-Newton* (one observation) and *RXTE* itself (57 pointed observations). The *XMM-Newton* observation revealed the presence of a broad relativistic line (Miller *et al.* 2002a) which is shown in the top left panel of Fig. 9.20. The line profile extends down to about 4 keV and implies emission from within the ISCO of a non-rotating BH. An inner disc radius at the ISCO of a Maximal Kerr hole ($\simeq 1.24\ r_g$) is preferred over one at the ISCO of a Schwarzschild BH at the $6\sigma$ level, strongly suggesting that the BH is rapidly, possibly maximally, spinning. The emissivity profile is much steeper than standard and implies that energy dissipation preferentially occurs in the inner few $r_g$ from the centre.

The source was observed three times by *BeppoSAX* and in all cases a broad relativistic Fe line was detected confirming the main results of Miller *et al.* (2002a) with respect to both inner disc radius and steep emissivity (Miniutti, Fabian & Miller 2004). The broadband coverage provided by *BeppoSAX* (up to 200 keV)

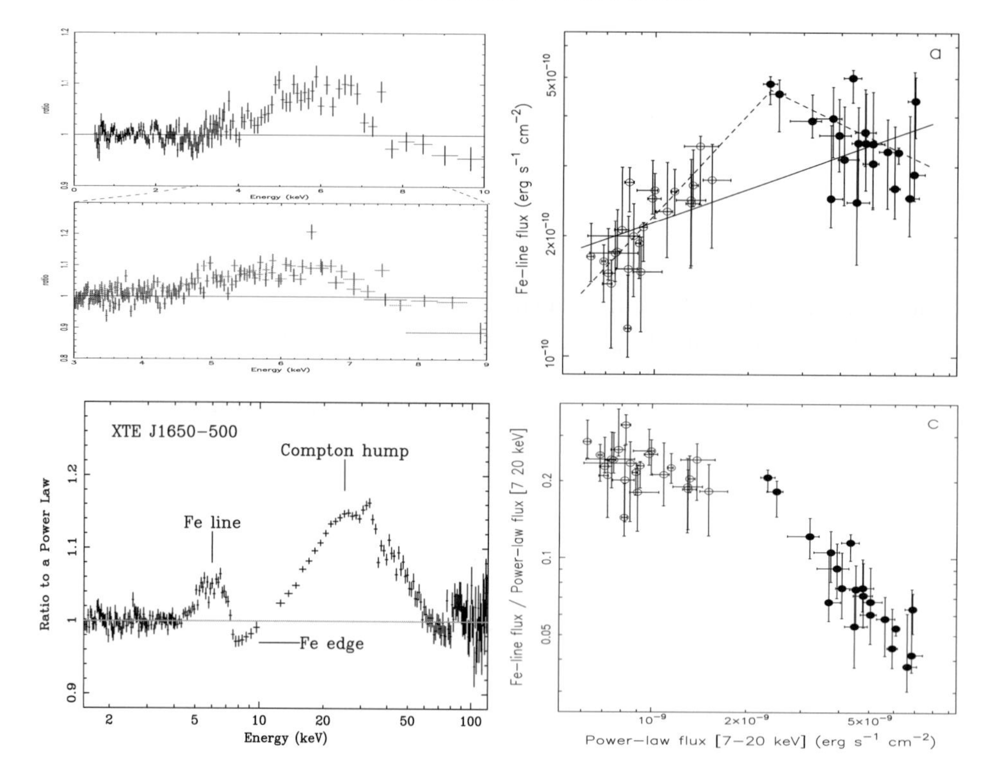


Fig. 9.20. *Left top panel*: The line profile of XTE J1650–500 as observed with *XMM-Newton* (Miller *et al.* 2002a). *Left middle panel*: Superposition of the broad lines in XTE J1650–500 (grey) and Cyg X–1 (black) from Miller *et al.* 2002a. *Left bottom panel*: Broadband *BeppoSAX* spectrum of XTE J1650–500 (as a ratio to the continuum). The signatures of relativistically-blurred reflection are clearly seen. *Right top panel*: The broad Fe line flux is plotted versus the ionizing continuum. *Right bottom panel*: The ratio between the line flux and the continuum (i.e. a measure of the reflection fraction) is plotted against the ionizing flux. Both right panels are from Rossi *et al.* (2005) and refer to *RXTE* data.

allowed us to detect the large reflection hump around 20–30 keV which is the unambiguous sign of the presence of a X-ray reflection spectrum, associated with the broad Fe line, giving more strength, if necessary, to the interpretation of the broad spectral feature as a relativistic Fe line. The broadband spectrum during one of the *BeppoSAX* observations is shown in the bottom-left panel of Fig. 9.20 as a ratio with a power law (and soft thermal emission) model.

The steep emissivity profile and small inner disc radius are very reminiscent of the case of MCG–6-30-15. In the latter case, the variability of the Fe line was succesfully explained by the light bending model. It is therefore an interesting exercise to study the Fe line variability in XTE J1650–500 as well and to put to test the light bending model not only in AGN, but in GBHCs as well. The three

*BeppoSAX* observations already gave some indication in that direction (Miniutti, Fabian & Miller 2004) suggesting the relevance of the light bending model for this object. Further study by Rossi *et al.* (2005) confirmed the first indication in a quite spectacular way. By using the 57 pointed observations by *RXTE*, Rossi *et al.* studied the broad Fe line variability in detail during the outburst. The results of their analysis is shown in the right panels of Fig. 9.20. The Fe line flux is correlated with the power law continuum at low fluxes and then saturates (or even shows some marginal evidence for anti-correlation) at higher flux levels. Its ratio with the power law continuum flux (a proxy for the reflection fraction) is generally anti-correlated with the continuum and saturates only at low fluxes. Those two results match very well the predictions of the light bending model (see e.g. Fig. 9.11). We therefore suggest that during the 2001 outburst of XTE J1650–500 most of the primary emission comes from a compact corona (or the base of a jet) located within the innermost few $r_g$ from the BH and that strong gravitational light bending is responsible for most of the observed variability, as in MCG–6-30-15.

## 9.7 Broad-line-free sources and nature of the soft excess

A broad relativistic Fe line is present in some but not all AGN spectra. On the other hand *XMM-Newton* and *Chandra* have shown that a narrow Fe line is almost ubiquitous in the X-ray spectrum of Seyfert galaxies (e.g. Page *et al.* 2004; Yaqoob & Padmanabhan 2004). The narrow line is generally interpreted as due to reflection from distant matter (at the parsec scale or so) such as the putative molecular torus, in good agreement with unification schemes. Here we explore in some detail some of the possible explanations for the non-ubiquitous detection of relativistic broad lines in sources that are expected to be radiatively efficient and, according to the standard model, should have an accretion disc extending down to small radii and therefore a broad Fe line as well.

### *9.7.1 Observational limitations*

The X-ray spectrum of Seyfert 1 and NLS1 galaxies and quasars is characterized by the presence of a narrow Fe line and by a soft excess, i.e. soft excess emission with respect to the 2–10 keV band spectral shape. The soft excess is often interpreted as thermal emission from the accretion disc. However, this interpretation is somewhat controversial. For example, Gierlinski & Done (2004) have selected a sample of 26 radio-quiet PG quasars for which good-quality *XMM-Newton* observations are available and tried to characterize their soft excesses. When interpreted as thermal emission, the soft excess in the sample has a mean temperature of 120 eV with a very small variance of 20 eV only. On the other hand, the maximum temperature of a standard Shakura–Sunyaev disc depends on the BH mass and on the mass accretion rate, which can in turn be related to the ratio between disc luminosity and Eddington luminosity as $T_{\max} \propto M_{\rm BH}^{-1/4}(L_{\rm disc}/L_{\rm Edd})^{1/4}$. For the BH masses and luminosities of the PG quasars in the Gierlinski & Done sample, one would expect temperatures in the range 3–70 eV. Thus two problems arise: i) the measured temperature of the soft excess is too uniform in sources that exhibit large differences in BH mass and luminosity; ii) the measured temperature is by far too high with respect to that predicted from standard accretion disc models. The uniformity of the soft excess spectral shape (and therefore of the inferred temperatures) was noted before by e.g. Walter & Fink (1993) and Czerny *et al.* (2003).

The uniformity and implausibly high temperature of the soft excess raises the question on the true nature of the soft excess. A uniform temperature could be easily understood if the soft spectral shape was due to atomic rather than truly thermal processes, the obvious candidate being ionized reflection from the disc. The uniformity and too large temperature of the soft excess may be then the result of applying the wrong spectral model (thermal emission) to the data. When a broad Fe line is present (e.g. MCG–6-30-15, 1H 0707–495) the parameters of the reflection and relativistic blurring are all constrained by the line shape and energy. It is then remarkable that the same model, once extrapolated in the soft band, provides an excellent description of the broadband spectrum and does not need any additional soft excess. We therefore suggest that relativistically blurred ionized reflection may well be responsible for the soft excess in many other objects as well (see also Ross & Fabian 2005; Crummy *et al.*, in preparation).

We then construct a theoretical model for what we believe could well be the typical X-ray spectrum of radiatively efficient X-ray sources. The model comprises a power law component and reflection from distant matter (represented by neutral reflection continuum and narrow Fe line). The soft excess is provided by ionized reflection from the accretion disc that also contributes to the hard band mainly with a broad relativistic line. We assume solar abundances and isotropic illumination of the accretion disc with reflection fraction $R \sim 1$. These conditions are less favourable for the detection of the relativistic Fe line than those found for example in MCG–6-30-15 and 1H 0707–495, in which the Fe abundance is super-solar and the RDC contribution particularly high, and should represent the typical and most common situation. The ionized reflection spectrum is convolved with a Laor model reproducing the Doppler and gravitational effects in an accretion disc around a Maximal Kerr BH (the other parameters being the disc inclination and the emissivity index, chosen to be $i = 30°$ and $q = 3$).

The model is shown in the top left panel of Fig. 9.21: a flux of $3 \times 10^{-12}$ erg cm$^{-2}$ s$^{-1}$ in the 2–10 keV band is assumed. We then simulated an *XMM-Newton* observation with a typical exposure time of 30 ks. The simulated

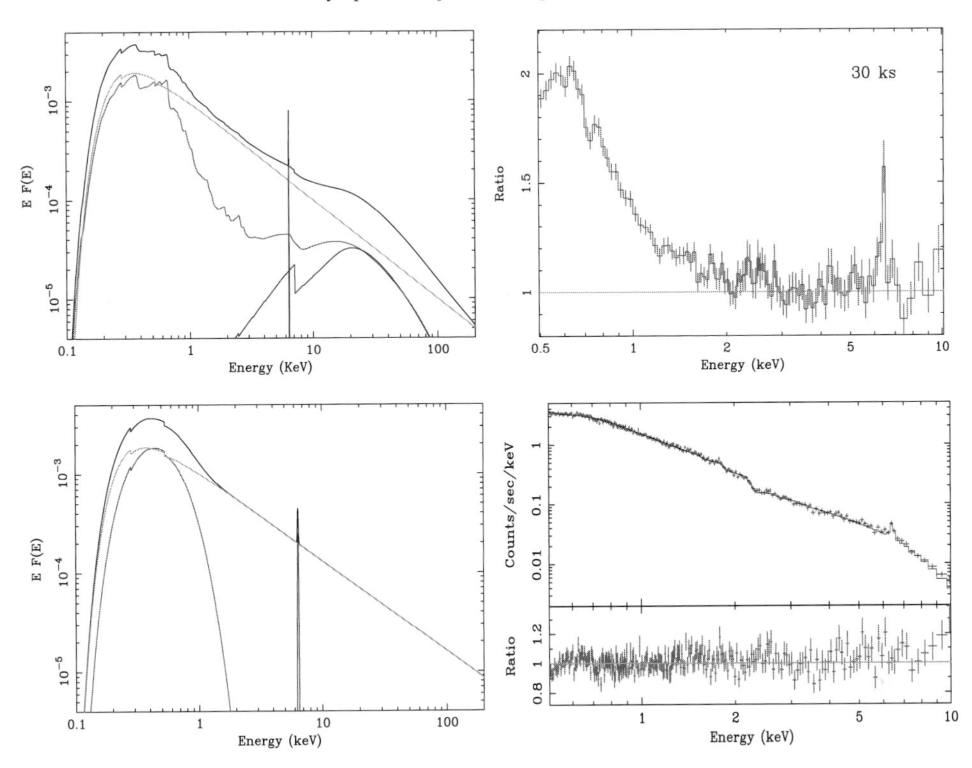


Fig. 9.21. *Top left panel*: The model we assumed for the simulations, based on our experience with X-ray data of Seyfert 1 and NLS1 galaxies. *Top right panel*: 30 ks *XMM-Newton* simulated spectrum with the model shown in the top left panel as a ratio with a power law fitted in the 2–10 keV band. The only prominent features are a narrow 6.4 keV Fe line and a steep soft-excess below about 1–2 keV, as generally observed. *Bottom left panel*: The "standard model" that is often used to fit the X-ray spectra of Seyfert 1 and NLS1 galaxies. *Bottom right panel*: The 30 ks *XMM-Newton* simulated spectrum is fitted with the "standard model" shown in the bottom left panel. The fit is excellent with reduced $\chi^2$ of 0.98.

spectrum is shown in the top right panel of Fig. 9.21 as a ratio with a power law in the 2–10 keV band. The main features in the simulated spectrum are a narrow 6.4 keV Fe K$\alpha$ line and a soft excess below 1–2 keV, as generally observed in Seyfert galaxies and quasars. The broad Fe line is completely lost into the continuum and no detection can be claimed with any significance. In fact, the spectrum can be described by the generally adopted model (hereafter "standard model") shown in the bottom left panel of Fig. 9.21 comprising a power law, narrow Fe line, and blackbody emission to model the soft excess. The fit with this model is shown in the bottom right panel of the same figure and is excellent. It should be noted that the measured temperature of the soft excess is $kT = 125 \pm 15$ eV, exactly in the range in which the "uniform temperature" of the soft excess is found.

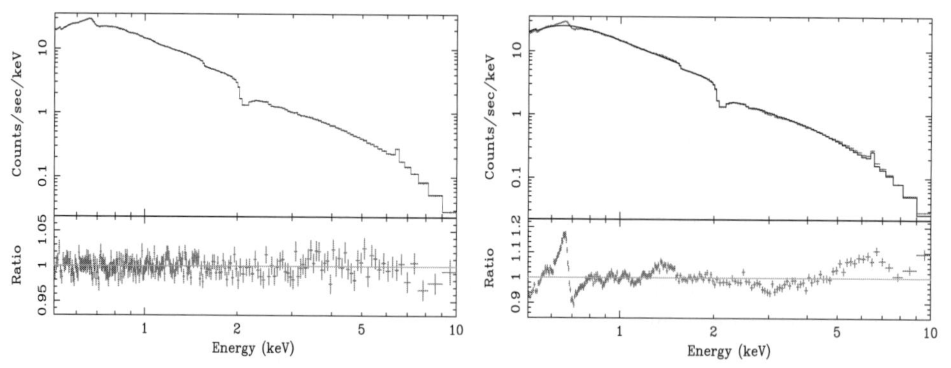


Fig. 9.22. *Left panel*: 100 ks Constellation-X (or XEUS with similar results) simulated spectrum with the same model as for the 30 ks *XMM-Newton* simulation. *Right panel*: The Constellation-X simulated spectrum is fitted with the "standard model". The fit is now totally unacceptable. The broad Fe line is now clearly seen in the 4–7 keV band, and residuals corresponding to reflection features are left at all energy of the spectrum. Future large collecting area missions such as XEUS and Constellation-X will therefore be able to distinguish easily between the two proposed models.

The non-detection of the broad relativistic line is also very significant. This means that if the soft excess is due to ionized reflection from the disc, the associated broad Fe line cannot be detected with present-quality data unless some specific conditions are met such as high Fe abundance, large RDC contribution, etc. This would naturally explain not only the nature of the soft excess in many sources but also the reason why the broad relativistic line is not ubiquitously detected in radiatively efficient sources. In the future, when X-ray missions with much larger effective area at 6 keV will be launched (XEUS/Constellation-X) our interpretation can be tested against data. In Fig. 9.22 we show the same spectrum simulated for a 100 ks Constellation-X observation. In the left panel, we show the spectrum and the ratio with the "real model" (top left panel of Fig. 9.21), while in the right panel, the fit is made by using the "standard model" (bottom left of Fig. 9.21). The "standard model" clearly is an unacceptable fit to the data and leaves residuals throughout the observed band, most remarkably the broad Fe line between 4 and 7 keV as well as soft residuals due to unmodelled soft features in the reflection spectrum.

### *9.7.2 Generalization of the light bending model*

Good examples of broad-line-free sources from long *XMM-Newton* exposures (among others) are Akn 120, which has no warm absorber (Vaughan *et al.* 2004), the Seyfert 1 galaxy NGC 3783 (Reeves *et al.* 2004), and the broad line radio galaxy 3C 120 (Ballantyne, Fabian & Iwasawa 2004). Various possibilities for the lack of any line have been proposed by the authors of those and other papers including:

(a) the central part of the disc is missing; (b) the disc surface is fully ionized (i.e. the Fe is); (c) the coronal emissivity function is flat, which could be due to (d) the primary X-ray sources being elevated well above the disc at say 100 $r_g$.

There are also intermediate sources where the data are either poor or there are complex absorption components so that one cannot argue conclusively that there is a relativistic line present. Some narrow line components are expected from outflow, warm absorbers and distant matter in the source. One common approach in complex cases, which is not recommended, is to continue adding absorption and emission components to the spectral model until the reduced $\chi^2$ of the fit is acceptable, and then claim that model as the solution. Very broad lines are difficult to establish conclusively unless there is something such as clear spectral variability indicating that the power-law is free of Fe–K features, as found for MCG–6-30-15, or for GX 339-4 and XTE J1650–500 where the complexities of an AGN are not expected.

Our interpretation of the spectral behaviour of MCG–6-30-15 and some other sources means that we are observing the effects of very strong gravitational light bending within a few gravitational radii of a rapidly spinning black hole. The short term (10–300 ks) behaviour is explained, without large intrinsic luminosity variability, through small variations in the position of the emitting region in a region where spacetime is strongly curved. This implies that some of the rapid variability is due to changes in the source position. Now BHCs in the (intermediate) high/soft state have high frequency breaks at higher frequency, for the same source, than when in the low/hard state (cf. Cyg X-1, Uttley & McHardy 2004). This additional variability when in the soft state could be identified with relativistic light-bending effects on the power-law continuum.

This picture suggests a possible generalization of the light-bending model to unify the AGN and GBHC in their different states. Note the work of Fender, Belloni & Gallo (2004) which emphasizes that jetted emission occurs commonly in the hard state of GBHCs. The key parameter may be the mean height of the main coronal activity above the black hole. Assume that much of the power of the inner disc passes into the corona (Merloni & Fabian 2001a,b) and that the coronal activity is magnetically focused close to the central axis. Then at low Eddington ratio the coronal height is large (say 100 $r_g$ or more), the corona is radiatively inefficient and most of the energy passes into an outflow; basically the power flows into a jet. Reflection is then appropriate for Euclidean geometry and a flat disc and there is only modest broadening to the lines. If the X-ray emission from the (relativistic) jet dominates then X-ray reflection is small (see e.g. Beloborodov 1999). The high frequency break to the power spectrum is low ($\sim 0.001 c/r_g$).

When the Eddington fraction rises above say ten per cent, the height of the activity drops below $\sim 20\ r_g$, the corona is more radiatively efficient and more high frequency variability occurs due to light bending and the turnover of the power spectrum rises above $0.01c/r_g$. The X-ray spectrum is dominated at low heights by reflection, including reflection-boosted thermal disc emission, and a broad iron line is seen. Any jet is weak. The objects with the highest spin and highest accretion rate give the most extreme behaviour. Observations suggest that these include NLS1 and some very high state, and intermediate state, GBHCs. Some broad-line-free sources do not however fit this model, so more work is required also to match the observed timing properties and not only the spectral variability.

## 9.8 Short-timescale variability in the Fe K band

In addition to the major line emission around 6.4 keV, transient emission features at energies lower than 6.4 keV are sometimes observed in X-ray spectra of AGN. As discussed, an early example was found in the ASCA observation of MCG–6-30-15 in 1997, which was interpreted as Fe K emission induced from a localized region of the disc, possibly due to illumination by a flare above it (Iwasawa *et al.* 1999, see Fig. 9.15). More examples followed in recent years with improved sensitivity provided by *XMM-Newton* and *Chandra* X-ray Observatory (Turner *et al.* 2002; Petrucci *et al.* 2002; Guainazzi *et al.* 2003; Yaqoob *et al.* 2003; Ponti *et al.* 2004; Dovčiak *et al.* 2004; Turner Kraemer & Reeves 2004; Longinotti *et al.* 2004; Porquet *et al.* 2004a; Miniutti & Fabian 2006). These features can be attributed to an Fe K$\alpha$ line arising from relatively localized reflecting spots on the accretion disc. If the spot is close to the central black hole, then the line emission is redshifted, depending on the location of the spot on the disc (Iwasawa *et al.* 1999; Ruszkowski 2000; Nayakshin & Kazanas 2001; Dovčiak *et al.* 2004). Here we present one case only, probably the most spectacular observed so far (NGC 3516, Iwasawa, Miniutti & Fabian 2004).

### *9.8.1 The remarkable case of NGC 3516*

Iwasawa, Miniutti & Fabian (2004) selected one of the *XMM-Newton* observations of the bright Seyfert galaxy NGC3516, for which Bianchi *et al.* (2004) reported excess emission at around 6 keV in addition to a stronger 6.4 keV Fe K$\alpha$ line in the time-averaged EPIC spectrum and studied the short-timescale variability of the 6 keV emission feature. We consider the 5–7.1 keV band, free from absorption which can affect energies above and below and fit an absorbed power law model to the data. The only residual emission features in the time-averaged spectrum are then two relatively narrow emission lines at 6.4 keV and 6 keV (hereafter the "line core" and "red feature" respectively). This is shown in the left panel of Fig. 9.23 where the line(s) profiles detected in the time-averaged spectrum are plotted.

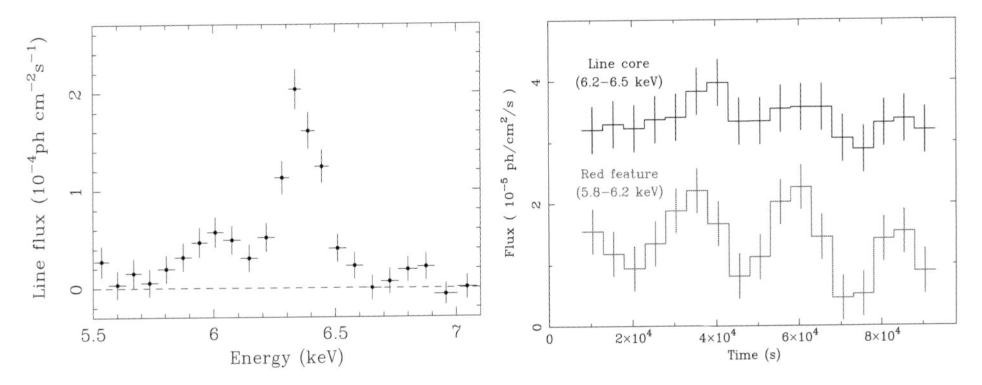


Fig. 9.23. *Left panel*: Time-averaged line(s) profile of NGC 3516. Two major emission features are seen, namely a line core at 6.4 keV and a red feature around 6 keV (rest-frame). *Right panel*: Light curves of the line core and red feature on a 5 ks timescale. The red feature seems to vary on a characteristic timescale of 25 ks while the line core is largely constant, apart from a possible increase delayed from the 'on' phase in the red feature by a few ks.

We first investigated the excess emission at resolutions of 5 ks in time and 100 eV in energy. A smoothed image of the excess emission in the time-energy plane is constructed from individual intervals of 5 ks. The detailed procedure of this method is described in Iwasawa, Miniutti & Fabian (2005). The light curves of the line core at 6.4 keV (6.2–6.5 keV) and of the red feature (5.8–6.2 keV) are obtained from the image and shown in the right panel of Fig. 9.23. The errors on the line fluxes are estimated from extensive simulations as discussed in Iwasawa, Miniutti & Fabian (2005). The red feature apparently shows a recurrent on-and-off behaviour which is suggestive of about four cycles with a timescale of 25 ks. In contrast, the 6.4 keV line core remains largely constant, apart from a possible increase delayed from the 'on' phase in the red feature by a few ks. Folding the light curve of the red feature confirms a characteristic interval of about 25 ks. The left panel of Fig. 9.24 shows the folded light curve of the red feature (grey), as well as the one obtained from the original unsmoothed data (black) by folding on a 25 ks interval.

Using the light curves as a guide (see right panel of Fig. 9.23), we constructed two spectra by selecting intervals in a periodic manner from the on and off phases to verify the implied variability in the red feature. The line profiles obtained from the two spectra are shown in the right panel of Fig. 9.24. The 6.4 keV core is resolved slightly ($\sim$5000 km s$^{-1}$ in FWHM) and found in both spectra with an equivalent width of 110 eV. While the 6.4 keV core remains similar between the two, there is a clear difference in the energy range of 5.7–6.2 keV due to the presence/absence of the red feature. The variability detected between the two spectra is significant at more than the $4\sigma$ level.

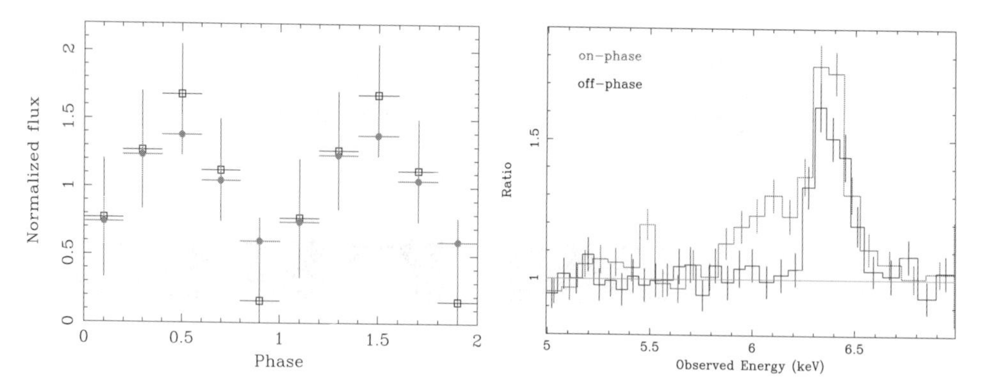


Fig. 9.24. *Left panel*: Red feature folded light curve on 25 ks from the smoothed (grey) and unsmoothed (black) images. See Iwasawa, Miniutti & Fabian (2005) for more details. *Right panel*: The line profile (shown as a ratio to the continuum) for the on- and off-phases. The red feature is present in the on-phase only, the difference with the off-phase spectrum being above the 4 $\sigma$ level.

We then investigated the variability of the red feature on shorter timescales to establish if not only flux, but also energy variation can be detected, with the ultimate goal of inferring the origin of the red feature. We constructed an image in the time–energy plane with 2 ks resolution in time and 100 eV in energy. The resulting image is seen in the left panel of Fig. 9.25 where the shade scale indicates the number of photons detected above the continuum (dark in the centre corresponding to about 15 counts). The image shows a relatively stable line core around 6.4 keV, while the red feature apparently moves with time during each on-phase: the feature emerges at around 5.7 keV, shifts its peak to higher energies with time, and joins the major line component at 6.4 keV, where there is marginal evidence for an increase of the 6.4 keV line flux. This evolution appears to be repeated for the on-phases.

The detection of only four cycles is not sufficient to establish any periodicity at high significance. The 25 ks is however a natural timescale of a black hole system with a black hole mass of a few times $10^7\ M_{\odot}$, as measured for NGC 3516 (e.g. Onken *et al.* 2003; Peterson *et al.* 2004). The finding could potentially be important and, especially considering the evolution of the line emission, warrant a theoretical study. We adopt a simple model in which a flare is located above an accretion disc, corotating with it at a fixed radius. The flare illuminates an underlying region on the disc (or spot) which produces a reflection spectrum, including an Fe K$\alpha$ line. The observed line flux and energy are both phase-dependent quantities and therefore, if the flare lasts for more than one orbital period, they will modulate periodically. We therefore interpret the characteristic timescale of 25 ks we measure as the orbital period. In the right panel of Fig. 9.25 we show a theoretical image constructed by considering a flare orbiting at 9 $r_g$ from the BH. The theoretical model matches the

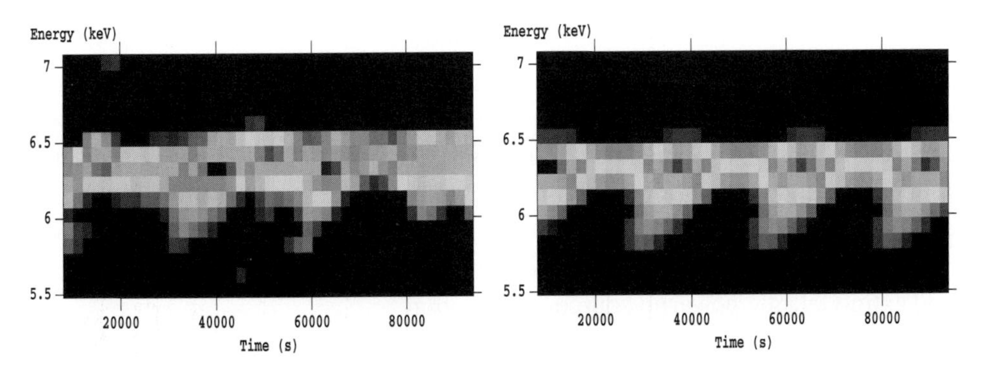


Fig. 9.25. *Left panel*: Excess emission image in the time–energy plane from the *XMM-Newton* observation. Each pixel is 2 ks in time and 100 eV in energy. *Right panel*: Theoretical image in the time–energy plane. The theoretical model assumes the presence of a flare orbiting the BH at 9 $r_g$ and at an height of 6 $r_g$ plus a constant line core at 6.4 keV. The match with the observed image (left) is surprisingly good.

data well enough. It is also possible that the emitting spot is not due to illumination from a flare but to some structure with enhanced emissivity (due to reflection of a more central continuum) on the disc itself (overdensity, spiral wave, etc.).

By constructing a large number of theoretical images and comparing them with the data we could constrain the location of the reflecting spot (or illuminating flare) in the range 7–16 $r_g$ from the BH. By combining this result with the timescale of 25 ks and by assuming this is the orbital period, the mass of the central BH in NGC 3516 can be estimated to be 1–5 $\times 10^7 M_\odot$, in excellent agreement with results obtained through reverberation mapping of optical emission lines. The estimates of $M_{\rm BH}$ in NGC 3516 from reverberation mapping lie in the range between $1 \times 10^7\ M_\odot$ and $4 \times 10^7\ M_\odot$. Onken *et al.* (2003) derive a value of $(1.68 \pm 0.33) \times 10^7\ M_\odot$, while Peterson *et al.* (2004) estimate $(4.3 \pm 1.5) \times 10^7\ M_\odot$. The previous analysis based on H$\beta$ alone gave $2.3 \times 10^7\ M_\odot$ (Ho 1999; no uncertainty is given). It is remarkable that our estimate of the black hole mass in NGC 3516 is in excellent agreement with the above results. Although the systematic flux and energy variability we report here is only tentative, the above agreement supports our interpretation.

Events like this happening at smaller radii where gravity is stronger are at present beyond the capabilities of X-ray observatories as is X-ray (Fe line) reverberation mapping (Reynolds *et al.* 1999; Young & Reynolds 2000). It is clear from the results presented above that the potential of future missions with much larger collecting area than *XMM-Newton* in the Fe K band, such as XEUS/*Constellation-X*, is outstanding. The prospects of probing the strong gravity regime of General Relativity via X-ray observations look stronger now than ever before.

## 9.9 Summary

A relativistically-broadened iron line is unambiguous in the spectra and behaviour of a few objects. The strength and breadth of reflection features is strong evidence for gravitational light bending and redshifts from a few $r_g$. They indicate that a dense disc extends close to the black hole, which must therefore be rapidly spinning ($a/m > 0.8$). Roy Kerr's solution to the Einstein equations in vacuum appears to be the astrophysically relevant one and the key ingredient in Galactic and extra-Galactic powerful sources of radiation.

The potential for understanding the accretion flow close to a black hole is enormous. Current observations are at the limit of *XMM-Newton* powers, which nevertheless has enabled a breakthrough in understanding the spectral behaviour of MCG–6-30-15 and similar objects. Similarities in the spectral and timing properties of AGN and BHCs is enabling further progress to be made. Studies in the near future with *Suzaku* followed by XEUS/*Constellation-X* in the next decade will continue to open up the immediate environment of accreting black holes, within just a few gravitational radii, to detailed study.

## Acknowledgments

Thanks to Kazushi Iwasawa, Jon Miller, Chris Reynolds and Simon Vaughan for collaboration and many discussions. ACF thanks the Royal Society and GM the PPARC for support.

## References

Agol, E. & Krolik, J. H. (2000), *Astroph. J.* **528**, 161.
Antonucci, R. R. (1993), *Ann. Rev. Astron. Astrophy.* **31**, 473.
Balestra, I., Bianchi, S. & Matt, G. (2004), *Astron. Astrophy.* **415**, 437.
Ballantyne, D. R., Ross, R. R. & Fabian, A. C. (2001), *Mon. Not. Roy. Astron. Soc.* **327**, 10.
Ballantyne, D. R., Fabian, A. C. & Iwasawa, K. (2004), *Mon. Not. Roy. Astron. Soc.* **354**, 839.
Bardeen, J. M., Press, W. H. & Teukolsky, S. A. (1972), *Astroph. J.* **178**, 347.
Beckwith, K. & Done, C. (2004), *Mon. Not. Roy. Astron. Soc.* **352**, 353.
Beloborodov, A. M. (1999), *Astroph. J.* **510**, L123.
Bianchi, S., Matt, G., Haardt, F., Maraschi, L., Nicastro, F., Perola, G. C., Petrucci, P. O. & Piro, L. (2001), *Astron. Astroph.* **376**, 77.
Bianchi, S., Miniutti, G., Fabian, A. C. & Iwasawa, K. (2005), *Mon. Not. Roy. Astron. Soc.* **360**, 380.
Blandford, R. D. & Znajek, R. L. (1977), *Mon. Not. Roy. Astron. Soc.* **179**, 433.
Boller, Th. *et al.* (2002), *Mon. Not. Roy. Astron. Soc.* **329**, L1.
Brusa, M., Gilli, R. & Comastri, A. (2005), *Astroph. J.* **621**, L5.
Carter, B. (1968), *Phys. Rev.* **174**, 1559.

Cunningham, C. T. (1975), *Astroph. J.* **202**, 788.
Cunningham, C. T. (1976), *Astroph. J.* **208**, 534.
Czerny, B., Nikolajuk, M., Rózańska, A. & Dumont, A.-M. (2003), *Astron. Astroph.* **412**, 317.
Dabrowski, Y., Fabian, A. C., Iwasawa, K., Lasenby, A. N. & Reynolds, C. S. (1997), *Mon. Not. Roy. Astron. Soc.* **288**, L11.
Dabrowski, Y. & Lasenby, A. N. (2001), *Mon. Not. Roy. Astron. Soc.* **321**, 605.
Dovčiak, M., Bianchi, S., Guainazzi, M., Karas, V. & Matt, G. (2004), *Mon. Not. Roy. Astron. Soc.* **350**, 745.
Dovčiak, M., Karas, V. & Yaqoob, T. (2004), *Astroph. J. Suppl.* **153**, 205.
Dumont, A.-M., Collin, S., Paletou, F., Coupé, S., Godet, O. & Pelat, D. (2003), *Astron. Astroph.* **407**, 13.
Fabian, A. C., Rees, M. J., Stella, L. & White, N. E. (1989), *Mon. Not. Roy. Astron. Soc.* **238**, 729.
Fabian, A. C., Nandra, K., Reynolds, C. S., Brandt, W. N., Otani, C., Tanaka, Y., Inoue, H. & Iwasawa, K. (1995), *Mon. Not. Roy. Astron. Soc.* **277**, L11.
Fabian, A. C., Iwasawa, K., Reynolds, C. S. & Young, A. J. (2000), *Publ. Astron. Soc. Pacific* **112**, 1145.
Fabian, A. C. *et al.* (2002a), *Mon. Not. Roy. Astron. Soc.* **335**, L1.
Fabian, A. C., Ballantyne, D. R., Merloni, A., Vaughan, S., Iwasawa, K. & Boller, Th. (2002b), *Mon. Not. Roy. Astron. Soc.* **331**, L35.
Fabian, A. C. & Vaughan, S. (2003), *Mon. Not. Roy. Astron. Soc.* **340**, L28.
Fabian, A. C., Miniutti, G., Gallo, L., Boller, Th., Tanaka, Y., Vaughan, S. & Ross, R. R. (2004), *Mon. Not. Roy. Astron. Soc.* **353**, 1071.
Fabian, A. C., Miniutti, G., Iwasawa, K. & Ross, R. R. (2005), *Mon. Not. Roy. Astron. Soc.* **361**, 795.
Fender, R. P., Belloni, T. M. & Gallo, E. (2004), *Mon. Not. Roy. Astron. Soc.* **355**, 1105.
Gallo, L., Tanaka, Y., Boller, Th., Fabian, A. C., Vaughan, S. & Brandt, W. N. (2004), *Mon. Not. Roy. Astron. Soc.* **353**, 1064.
Gammie, C. F. (1999), *Astroph. J.* **522**, L57.
Garofalo, D. & Reynolds, C. S. (2005), *Astroph. J.* **624**, 94.
George, I. M. & Fabian, A. C. (1991), *Mon. Not. Roy. Astron. Soc.* **249**, 352.
Ghisellini, G., Haardt, F. & Matt, G. (2004), *Astron. Astroph.* **413**, 535.
Gierliński, M. & Done, C. (2004), *Mon. Not. Roy. Astron. Soc.* **349**, L7.
Guainazzi, M. (2003), *Astron. Astroph.* **401**, 903.
Guilbert, P. W. & Rees, M. J. (1988), *Mon. Not. Roy. Astron. Soc.* **233**, 475.
Haardt, F. & Maraschi, L. (1991), *Astroph. J.* **380**, L51.
Haardt, F. & Maraschi, L. (1993), *Astroph. J.* **413**, 507.
Halpern, J. P. (1984), *Astroph. J.* **281**, 90.
Iwasawa, K., Fabian, A. C., Mushotzky, R. F., Brandt, W. N., Awaki, K. & Kunieda, H. (1996a), *Mon. Not. Roy. Astron. Soc.* **279**, 837.
Iwasawa, K. *et al.* (1996b), *Mon. Not. Roy. Astron. Soc.* **282**, 1038.
Iwasawa, K., Fabian, A. C., Young, A. J., Inoue, H. & Matsumoto, C. (1999), *Mon. Not. Roy. Astron. Soc.* **307**, 611.
Iwasawa, K., Lee, J. C., Young, A. J., Reynolds, C. S. & Fabian, A. C. (2004), *Mon. Not. Roy. Astron. Soc.* **347**, 411.
Iwasawa, K. & Miniutti, G. (2004), *Progr. Theor. Phys. Suppl.* **155**, 247.
Iwasawa, K., Miniutti, G. & Fabian, A. C. (2004), *Mon. Not. Roy. Astron. Soc.* **355**, 1073.
Jiménez-Bailón, E., Piconcelli, E., Guainazzi, M., Schartel, N., Rodríguez-Pascual, P. M. & Santos-Lleó, M. (2005), *Astron. Astroph.* **435**, 449.

Kaastra, J. S., Mewe, R., Liedhal, D. A., Komossa, S. & Brinkman, A. C. (2000), *Astron. Astroph.* **354**, L83.
Kallman, T. R. & Krolik, J. H. (1986), *NASA GSFC Lab. for High Energy Physics Spec. Rep.* [http://heasarc.gsfc.nasa.gov/docs/software/xstar/xstar.html].
Kaspi S. *et al.* (2001), *Astroph. J.* **554**, 216.
Kerr, R. P. (1963), *Phys. Rev. Lett.* **11**, 237.
Kinkhabwala, A. *et al.* (2002), *Astroph. J.* **575**, 732.
Krolik, J. H. (1999), *Astroph. J.* **515**, L73.
Krolik, J. H. & Hawley, J. F. (2002), *Astroph. J.* **573**, 754.
Lamer, G., McHardy, I. M., Uttley, P. & Jahoda, K. (2003), *Mon. Not. Roy. Astron. Soc.* **338**, 323.
Laor, A. (1991), *Astroph. J.* **376**, 90.
Li, L. (2003), *Phys. Rev.* **D 67**, 044007.
Lightman, A. P. & White, T. R. (1988), *Astroph. J.* **335**, 57.
Longinotti, A. L., Cappi, M., Nandra, K., Dadina, M. & Pellegrini, S. (2003), *Astron. Astroph.* **410**, 471.
Longinotti, A. L., Nandra, K., Petrucci, P. O. & O'Neill, P. M. (2004), *Mon. Not. Roy. Astron. Soc.* **355**, 929.
Markoff, S., Falcke, H. & Fender, R. P. (2001), *Astron. Astroph.* **372**, L25.
Martocchia, A., Karas, V. & Matt, G. (2000), *Mon. Not. Roy. Astron. Soc.* **312**, 817.
Martocchia, A., Matt, G. & Karas, V. (2002a), *Astron. Astroph.* **383**, L23.
Martocchia, A., Matt, G., Karas, V., Belloni, T. & Feroci, M. (2002b), *Astron. Astroph.* **387**, 215.
Matsumoto, C., Inoue, H., Fabian, A. C. & Iwasawa, K. (2003), *Publ. Astron. Soc. Japan* **55**, 615.
Matt, G., Perola, G. C. & Piro, L. (1991), *Astron. Astroph.* **247**, 25.
Matt, G. & Perola, G. C. (1992), *Mon. Not. Roy. Astron. Soc.* **259**, 433.
Matt, G., Fabian, A. C. & Ross, R. R. (1993), *Mon. Not. Roy. Astron. Soc.* **262**, 179.
Matt, G., Fabian, A. C. & Ross, R. R. (1996), *Mon. Not. Roy. Astron. Soc.* **278**, 1111.
McHardy, I. M., Gunn, K. F., Uttley, P. & Goad, M. R. (2005), *Mon. Not. Roy. Astron. Soc.* **359**, 1469.
Merloni, A. & Fabian, A. C. (2001a), *Mon. Not. Roy. Astron. Soc.* **321**, 549.
Merloni, A. & Fabian, A. C. (2001b), *Mon. Not. Roy. Astron. Soc.* **328**, 958.
Merloni, A. & Fabian, A. C. (2003), *Mon. Not. Roy. Astron. Soc.* **342**, 951.
Miller, J. M. *et al.* (2002a), *Astroph. J.* **570**, L69.
Miller, J. M. *et al.* (2002b), *Astroph. J.* **577**, L15.
Miller, J. M. *et al.* (2002c), *Astroph. J.* **578**, 348.
Miller, J. M. *et al.* (2003), *Mon. Not. Roy. Astron. Soc.* **338**, 7.
Miller, J. M. *et al.* (2004a), *Astroph. J.* **601**, 450.
Miller, J. M. *et al.* (2004b), *Astroph. J.* **606**, L131.
Miller, J. M., Fabian, A. C., Nowak, M. A. & Lewin, W. H. G. (2004), *Proc. 10th Marcel Grossman Meeting*, 20–26 July 2003, Rio de Janeiro, Brazil [arxiv:astro-ph/0402101].
Miniutti, G., Fabian, A. C., Goyder, R. & Lasenby, A. N. (2003), *Mon. Not. Roy. Astron. Soc.* **344**, L22.
Miniutti, G. & Fabian, A. C. (2004), *Mon. Not. Roy. Astron. Soc.* **349**, 1435.
Miniutti, G., Fabian, A. C. & Miller, J. M. (2004), *Mon. Not. Roy. Astron. Soc.* **351**, 466.
Miniutti, G. & Fabian, A. C. (2006), *Mon. Not. Roy. Astron. Soc.* **366**, 115.
Misner, C. W., Thorne, K. S. & Wheeler, J. A. (1973), *Gravitation* (W.H. Freeman, San Francisco).

Nandra, K., George, I. M., Mushotzky, R. F., Turner, T. J. & Yaqoob, T. (1997a), *Astroph. J.* **476**, 70.
Nandra, K., Mushotzky, R. F., Yaqoob, T., George, I. M. & Turner, T. J. (1997b), *Mon. Not. Roy. Astron. Soc.* **284**, L7.
Nandra, K., George, I. M., Mushotzky, R. F., Turner, T. J. & Yaqoob, T. (1999), *Astroph. J.* **524**, 707.
Nayakshin, S., Kazanas, D. & Kallman, T. R. (2000), *Astroph. J.* **537**, 833.
Nayakshin, S. & Kazanas, D. (2001), *Astroph. J.* **553**, 885.
Onken, C. A., Petereson, B. M., Dietrich, M., Robinson, A. & Salamanca, I. M. (2003), *Astroph. J.* **585**, 121.
Orosz, J. A., McClintock, J. E., Remillard, R. A. & Corbel, S. (2004), *Astroph. J.* **616**, 376.
Page, D. N. & Thorne, K. S. (1974), *Astroph. J.* **191**, 499.
Page, K. L., O'Brien, P. T., Reeves, J. N. & Turner, M. J. L. (2004), *Mon. Not. Roy. Astron. Soc.* **347**, 316.
Peterson B. M. *et al.* (2004), *Astroph. J.* **613**, 682.
Petrucci P. O. *et al.* (2002), *Astron. Astroph.* **338**, L5.
Ponti, G., Cappi, M., Dadina, M. & Malaguti, G. (2004), *Astron. Astroph.* **417**, 451.
Ponti, G., Miniutti, G., Cappi, M., Maraschi, L., Fabian, A. C. & Iwasawa, K. (2006), *Mon. Not. Roy. Astron. Soc.* **368**, 903.
Porquet, D., Reeves, J. N., Uttley, P. & Turner, T. J. (2004a), *Astron. Astroph.* **427**, 101.
Porquet, D., Reeves, J. N., O'Brien, P. & Brinkmann, W. (2004b), *Astron. Astroph.* **422**, 85.
Pounds, K. A., Reeves, J. N., Page, K. L. & O'Brien, P. T. (2004a), *Astroph. J.* **605**, 670.
Pounds, K. A., Reeves, J. N., King, A. R. & Page, K. L. (2004b), *Mon. Not. Roy. Astron. Soc.* **350**, 10.
Pringle, J. E. (1981), *Ann. Rev. Astron. Astroph.* **19**, 137.
Reeves, J. N., Nandra, K., George, I. M., Pounds, K. A., Turner, T. J. & Yaqoob, T. (2004), *Astroph. J.* **602**, 648.
Remillard, R. A. & McClintock, J. E. (2006), *Ann. Rev. Astron. Astroph.* **44**, 49.
Reynolds, C. S. (1996), PhD thesis, University of Cambridge.
Reynolds, C. S. (1997), *Mon. Not. Roy. Astron. Soc.* **286**, 513.
Reynolds, C. S. & Begelman, M. C. (1997), *Astroph. J.* **488**, 109.
Reynolds, C. S., Young, A. J., Begelman, M. C. & Fabian, A. C. (1999), *Astroph. J.* **514**, 164.
Reynolds, C. S. (2000), *Astroph. J.* **533**, 811.
Reynolds, C. S. & Nowak, M. A. (2003), *Phys. Rep.* **377**, 389.
Reynolds, C. S., Wilms, J., Begelman, M. C., Staubert, R. & Kendziorra, E. (2004), *Mon. Not. Roy. Astron. Soc.* **349**, 1153.
Ross, R. R. & Fabian, A. C. (1993), *Mon. Not. Roy. Astron. Soc.* **261**, 74.
Ross, R. R., Fabian, A. C. & Young, A. J. (1999), *Mon. Not. Roy. Astron. Soc.* **306**, 461.
Ross, R. R. & Fabian, A. C. (2005), *Mon. Not. Roy. Astron. Soc.* **358**, 211.
Rossi, S., Homan, J., Miller, J. M. & Belloni, T. (2005), *Mon. Not. Roy. Astron. Soc.* **360**, 763.
Rózańska, A., Dumont A.-M., Czerny, B. & Bollin, S. (2002), *Mon. Not. Roy. Astron. Soc.* **332**, 799.
Ruszkowski, M. (2000), *Mon. Not. Roy. Astron. Soc.* **315**, 1.
Schwarzschild, K. (1916), *Sitzungsber. Preuss. Akad. Wiss.*, 189.
Shakura, N. I. & Sunyaev, R. (1973), *Astron. Astroph.* **24**, 337.
Shemmer, O., Netzer, H., Maiolino, R., Oliva, E., Croom, S., Corbett, E. & di Fabrizio, L. (2004), *Astroph. J.* **614**, 547.

Shih, D. C., Iwasawa, K. & Fabian, A. C. (2001), *Mon. Not. Roy. Astron. Soc.* **333**, 687.
Stella, L. (1990), *Nature* **344**, 747.
Streblyanska, A., Hasinger, G., Finoguenov, A., Barcons, X., Mateos, S. & Fabian, A. C. (2005), *Astron. Astroph.* **432**, 395.
Tanaka, Y. *et al.* (1995), *Nature* **375**, 659.
Tanaka, Y., Gallo, L., Boller, Th., Keil, R. & Ueda, Y. (2004), *Publ. Astron. Soc. Japan* **56**, L9.
Taylor, R. D., Uttley, P. & McHardy, I. M. (2003), *Mon. Not. Roy. Astron. Soc.* **342**, L31.
Thorne, K. S. (1974), *Astroph. J.* **191**, 507.
Turner, A. K., Fabian, A. C., Vaughan, S. & Lee, J. C. (2003), *Mon. Not. Roy. Astron. Soc.* **346**, 833.
Turner, A. K., Fabian, A. C., Lee, J. C. & Vaughan, S. (2004), *Mon. Not. Roy. Astron. Soc.* **353**, 319.
Turner T. J. *et al.* (2002), *Astroph. J.* **574**, L123.
Turner, T. J., Kraemer, S. B. & Reeves, J. N. (2004), *Astroph. J.* **603**, 62.
Turner, T. J., Kraemer, S. B., George I. M., Reeves, J. N. & Bottorff, M. C. (2005), *Astroph. J.* **618**, 155.
Uttley, P. & McHardy, I. M. (2004), *Prog. Theor. Phys.* **S155**, 170.
Vaughan, S. & Edelson, R. (2001), *Astroph. J.* **548**, 694.
Vaughan, S., Fabian, A. C. & Nandra, K. (2003), *Mon. Not. Roy. Astron. Soc.* **339**, 1237.
Vaughan, S. & Fabian, A. C. (2004), *Mon. Not. Roy. Astron. Soc.* **348**, 1415.
Volonteri, M., Madau, P., Quataert, E. & Rees, M. J. (2005), *Astroph. J.* **620**, 69.
Walter, R. & Fink, H. H. (1993), *Astron. Astroph.* **274**, 105.
Wilms, J. *et al.* (2001), *Mon. Not. Roy. Astron. Soc.* **328**, L27.
Yaqoob, T., George, I. M., Kallman, T. R., Padmanabhan, U., Weaver, K. A. & Turner, T. J. (2003), *Astroph. J.* **596**, 85.
Yaqoob, T. & Padmanabhan, U. (2004), *Astroph. J.* **604**, 63.
Young, A. J., Ross, R. R. & Fabian, A. C. (1998), *Mon. Not. Roy. Astron. Soc.* **300**, L11.
Young, A. J. & Reynolds, C. S. (2000), *Astroph. J.* **529**, 101.
Young, A. J., Lee, J. C., Fabian, A. C., Reynolds, C. S., Gibson, R. R. & Canizares, C. R. (2005), *Astroph. J.* **631**, 733.
Zdziarski A. A. *et al.* (1994), *Mon. Not. Roy. Astron. Soc.* **269**, L55.
Zycki, P. T., Krolik, J. H., Zdziarski, A. A. & Kallman, T. R. (1994), *Astroph. J.* **437**, 597.

# 10

# Cosmological flashes from rotating black holes

Maurice H. P. M. van Putten

## 10.1 Introduction

Gamma-ray bursts (GRBs) were discovered serendipitously as "flashes in the sky" by the nuclear test-ban monitoring satellites Vela (US) and Konus (USSR). These were soon recognized to be a natural phenomenon. The first data were publicly released by Klebesadel *et al.* (1973) and Mazets *et al.* (1974). The mysterious origin of this new astrophysical transient, lasting up to tens of seconds in non-thermal gamma-ray radiation, gradually triggered a huge interest from the high-energy astrophysics community. The Burst and Transient Source Experiment (BATSE) on NASA's Compton Gamma-Ray Observatory, launched in 1991, identified an isotropic distribution in the sky and revealed the existence of short and long gamma-ray bursts, whose durations are broadly distributed around 0.3 and 30 s, respectively (Kouveliotou *et al.*, 1993; Paciesas, 1999). Moreover, the sensitivity of BATSE showed a deficit in faint bursts in the number-versus-intensity distribution distinct from a $-3/2$ power law. Meegan *et al.* (1992) hereby established that gamma-ray bursts are of cosmological origin. More recently, the Italian–Dutch satellite BeppoSax discovered X-ray flashes (Heise *et al.*, 2000), which may be closely related to long gamma-ray bursts. Their non-thermal emissions are best accounted for by internal shocks in ultrarelativistic outflows in the fireball model proposed by Rees & Mészáros (1992, 1994) – with an unknown inner engine producing these baryon-poor outflows. If gravitationally powered, their compact size, based on short-time scale variabilities in the gamma-ray lightcurves, may involve the formation of black holes surrounded by a high-mass torus (Paczyński, 1991; Woosley, 1993).

At cosmological distances, the isotropic equivalent luminosities are on the order of $10^{51}$ erg s$^{-1}$ – as bright as $10^{18}$ solar luminosities. These events, widely acclaimed

The Kerr Spacetime: Rotating Black Holes in General Relativity. Ed. David L. Wiltshire, Matt Visser and Susan M. Scott.
Published by Cambridge University Press. 

Table 10.1.

| Symbol | Expression | Comment |
|---|---|---|
| $\lambda$ | $\sin\lambda = a/M$ | |
| $\Omega_H$ | $\tan(\lambda/2)/2M$ | |
| $J_H$ | $M^2 \sin\lambda$ | |
| $E_{rot}$ | $2M \sin^2(\lambda/4)$ | $\leq 0.29M$ |
| $M_{irr}$ | $M \cos(\lambda/2)$ | $\geq 0.71M$ |

as the "Biggest Bang since the Big Bang", are now known to light up the Universe about once per minute. What, then, is their origin? The most powerful observational method to characterize the enigmatic energy source is through *calorimetry on all radiation channels*. This involves measuring the energy output in emissions in the electromagnetic spectrum, as well as in yet "unseen" emissions in neutrinos and gravitational waves. Perhaps we shall then be in a position to determine their constitution, and decide whether their inner engine is *baryonic*, such as a neutron star, or *non-baryonic*: a black hole energized by spin. The exact solution of Roy P. Kerr (1963) – see Table 10.1 for a parametrization – gives an energy per unit mass which is anomalously large at high spin rates, and exceeds that of a rapidly spinning neutron star by an order of magnitude. With this potentially large and baryon-free energy reservoir, a Kerr black hole is a leading candidate for the inner engine of GRBs, at least those of long durations, in view of the following observational results.

The Italian–Dutch BeppoSax satellite, launched in 1996, dramatically changed the landscape of long GRBs with the discovery by E. Costa (Costa *et al.*, 1997) of X-ray afterglows in GRB970228 (confirmed in observations by the ASCA and ROSAT satellites). These lower energy emissions permit accurate localizations, enabling follow-up by optical telescopes. Thus, J. van Paradijs (van Paradijs *et al.*, 1997) pointed the Isaac Newton Telescope and the William Herschel Telescope to further discover optical emissions associated with the same event. These lower energy X-ray and optical emissions agree remarkably well with the expected decay of shocks in ultrarelativistic, baryon-poor outflows in the previously developed fireball model. Even lower energy, radio afterglows have been observed in some cases, including GRB970228 (Frail *et al.*, 1997). The same optical observations have now led to tens of GRBs with individually measured redshifts, up to $z = 6.29$ in GRB050904 detected by the recently launched Swift satellite (Tagliaferri *et al.*, 2005; Haislip *et al.*, 2006).

Optical follow-up identified the association of long GRBs with supernovae. In particular, the unambiguous association of GRB030329 with SN2003dh ($z = 0.167$, $D = 800$ Mpc) (Stanek *et al.*, 2003) identified type Ib/c supernovae as the

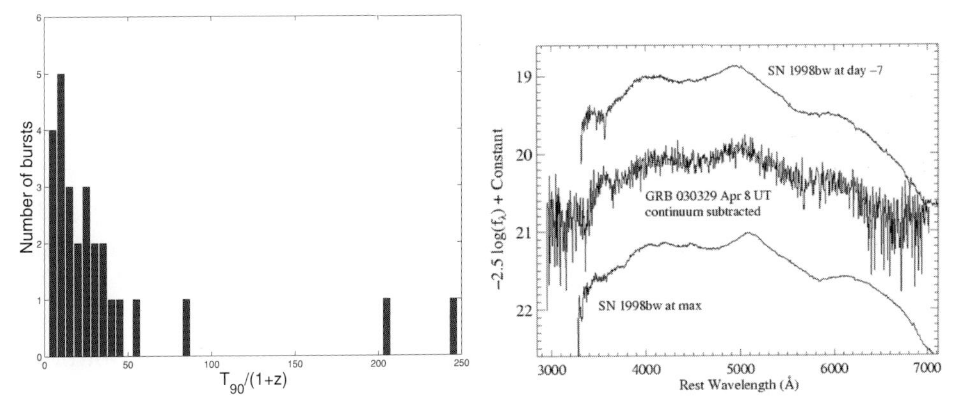


Fig. 10.1. (*Left*) A histogram of redshift-corrected distributions of 27 long bursts with individually determined redshifts from their afterglow emissions. It shows durations $T_{90}/(1+z)$ of tens of seconds at a mean redshift distance $<z>=1.25$, indicative of their cosmological distances. (*Right*) The optical spectrum of the type Ic SN2003dh associated with GRB030329 is remarkably similar to that of the type Ic SN1998bw of GRB980425 one week before maximum. GRB030329 displayed a gamma-ray luminosity of about $10^{-1}$ times typical at a distance of $z=0.167$ ($D\simeq 800$Mpc), whereas GRB980425 was observed at anomalously low gamma-ray luminosity ($10^{-4}$ times typical) in the local universe ($z=0.008$, $D\simeq 37$ Mpc). At the same time, their supernovae were very luminous with inferred $^{56}$Ni ejecta of about $0.5M_\odot$. (Reprinted from van Putten (2002); Stanek *et al.* (2003) ©The American Astronomical Society.)

parent population of long GRBs (Fig. 10.1). This observation confirms their association with core-collapse supernovae of massive stars, proposed by S. Woosley (Woosley, 1993; Katz, 1994; Paczyński, 1998). Massive stars have characteristically short lifetimes, with the result that GRB–supernovae track the cosmological event rate. This conclusion is quantitatively confirmed by consistent estimates of the true-to-observed GRB-SNe event rate of 450–500 deduced from two independent analyses, based on geometrical beaming factors (Frail *et al.*, 2001) and locking to the star formation rate (van Putten & Regimbau, 2003). These event rates correspond to a true cosmological event rate of about one per minute.

Core-collapse supernovae from massive stars are believed to produce neutron stars or black holes. The former are perhaps best known from the rapidly spinning neutron star (a pulsar) in the Crab nebula, a remnant of what was probably a type II supernova in 1054. The latter are more difficult to identify with certainty, such as in the example of the more recent type II event SN1987 in the Large Megellanic Cloud. At close proximity, its burst in MeV-neutrinos was detected by Kamiokanda and IMB (see Burrows & Lattimer (1987)) which provided direct evidence for the formation of matter at nuclear densities. The absence of a pulsar or a point-source of X-ray radiation in this case suggests continued collapse – probably into a stellar mass black hole.

Types II and Ib/c supernovae are both believed to represent the endpoint of massive stars (Filipenko, 1997; Turatto, 2003), and possibly so in binaries such as the type II/Ib event SN1993J (Maund *et al.*, 2004). This binary association suggests a hierarchy, wherein hydrogen-rich, envelope-retaining SNII are associated with wide binaries, while hydrogen-poor, envelope-stripped SNIb and SNIc are associated with increasingly compact binaries (Nomoto *et al.*, 1995; Turatto, 2003). By tidal coupling, the primary star in the latter will rotate at approximately the orbital period at the moment of core-collapse. With an evolved core (Bethe *et al.*, 2003), these type Ib/c events in particular are believed to produce a spinning black hole (Woosley, 1993; Paczyński, 1998; Brown *et al.*, 2000; Lee *et al.*, 2002). However, the branching ratio of type Ib/c into GRB–supernovae is small,

$$\mathcal{R}(\text{SNIb/c} \to \text{GRB}) = \frac{N(\text{GRB–SNe})}{N(\text{SNIb/c})} \simeq (2-4) \times 10^{-3} \qquad (10.1)$$

as calculated from the true GRB–supernova event rate relative to the observed event rates of supernovae of types II and Ib/c. This ratio is remarkably small, suggesting a higher-order down-selection process.

The identification of long GRBs with core-collapse supernovae brings into discussion their potential gravitational-wave emissions. These may arise from a variety of mechanisms associated with rapidly rotating fluids. In particular, emissions will be produced by bar-mode instabilities producing fragmentation of orbiting matter prior to the formation of a compact object (Nakamura & Fukugita, 1989; Bonnel & Pringle, 1995) or in the formation of multiple compact objects (Davies *et al.*, 2002), non-axisymmetries in accretion disks (Papadopoulos & Font, 2001; Mineshige *et al.*, 2002; Kobayashi & Mészáros, 2003a,b), as well as after the formation of a rapidly spinning black hole. The first mechanism may produce an initial "splash" of gravitational radiation (Rees *et al.*, 1974), the second a bi-modal spectrum containing high-frequency emissions produced by non-axisymmetric perturbations of the black hole and low-frequency emissions from the disk, whereas the latter creates a long-duration burst of low-frequency gravitational radiation. Calculations of this time-evolving spectrum of gravitational-radiation are only beginning to be addressed by computational hydro- and magnetohydrodynamics, during initial collapse (Rampp *et al.*, 1998; Fryer *et al.*, 1999; McFadyen & Woosley, 1999; Fryer *et al.*, 2002; Duez *et al.*, 2004) and in the formation of a non-axisymmetric torus (Bromberg *et al.*, 2006). Emissions from matter surrounding a stellar-mass black hole are in the frequency range of sensitivity of the gravitational-wave experiments LIGO (Abramovici *et al.*, 1992; Barish & Weiss, 1999) and Virgo (Bradaschia *et al.*, 1992; Acernese *et al.*, 2002; Spallici *et al.*, 2004). Understanding their wave-forms can serve strategic search-and-detection algorithms, triggered by gamma-ray observations (Finn *et al.*, 2004) or by their associated supernova. The latter is more common on account of the true-to-observed event ratio of GRB–supernovae mentioned above.

In this review, we shall discuss a theory of GRBs from rotating black holes. Current phenomenology on GRB–supernovae poses a challenge to model

- The durations of long GRBs of tens of seconds
- The formation of an associated aspherical supernova
- The launch of ultrarelativistic jets with beamed gamma-ray emissions
- A small branching ratio of type Ib/c supernovae into GRBs of less than 0.5%

This phenomenology will be directly linked to the spin energy of the black hole. Quite generally, we are left with the task of modeling all possible radiation channels produced by the putative Kerr black hole, including gravitational radiation, MeV-neutrino emissions, and magnetic winds.

Just as lower energy X-ray and optical afterglow emissions linked long GRBs to supernovae, we expect that the detection of a contemporaneous burst in gravitational-radiation will lift the veil on their enigmatic inner engine. A precise determination of this link will enable the identification of a Kerr black hole. The same might apply to types II and Ib/c supernovae.

Understanding the energy source of long GRBs might tell us also about the constitution of short GRBs. The first *faint* X-ray afterglows have been detected of the short event GRB050509B by Swift (Gehrels *et al.*, 2005) and GRB050709 by HETE-II (Fox *et al.*, 2005; Hjörth *et al.*, 2005; Villasenor *et al.*, 2005), providing further measurements of the low redshifts $z = 0.225$ and 0.16, respectively. The nature of host galaxies, preferably (older) elliptical galaxies for short bursts versus (young) star-forming galaxies for long bursts, supports the hypothesis that short events originate in the binary coalescence of black holes and neutron stars (Piro, 2005). After completion of this review, Kulkarni (2005) outlined a very interesting prospect for long-lasting supernova-like signatures to short bursts from the debris of a neutron star binary coalescence with a similar partner or black hole.

The dichotomy of short and long events has been explained in terms of hyper- and suspended accretion onto slowly and rapidly rotating black holes respectively (van Putten & Ostriker, 2001). Thus, we predicted similar X-ray afterglows with the property that those which relate to short events are relatively faint: *the short burst is identical to the final moment of a long burst of gamma-rays.* Hyperaccreting, slowly rotating black holes can be produced through various channels. Black-hole neutron star binaries are remnants of core-collapse events in binaries (Bethe *et al.* (2003); unless formed by capture in dense stellar clusters). Any rapidly spinning black hole will be spun-down in the core-collapse process, possibly representing a prior GRB–supernova event, as is discussed later in this review. The merger of a neutron star onto a black hole (Paczyński, 1991) could produce a state of hyperaccretion and produce a short gamma-ray burst. This scenario is consistent with the predominance of long over short events. Less clear is the potential for gamma-ray emissions from the merger of two neutron stars (e.g. Zhuge *et al.* (1994); Faber *et al.* (2004)) which, however, should also produce a slowly rotating black hole. Finally, slowly rotating black holes can also be produced in the core-collapse of a massive star in isolation or in a wide binary. Here as well, the detection of their gravitational-wave emissions may provide a probe to differentiate among these different types of short bursts.

In Section 10.2, we give an astrophysical overview of Kerr black holes and gravitational radiation. In Section 10.3, we review the formation and evolution (kick-velocities, growth and spin-down) of Kerr black holes in core-collapse supernovae. This includes the process of converting spin-energy into radiation and estimates of the lifetime of rapid spin. In Section 10.4, we calculate multipole mass-moments in a torus due to a hydrodynamical instability. By asymptotic analysis, the energies in the various radiation channels are given in Section 10.5. A mechanism for launching ultrarelativistic jets from rotating black holes by a small fraction of black hole spin is given in Section 10.6. Specific predictions for these "flashes" as sources for the LIGO and Virgo observatories are given in Section 10.7.

## 10.2 Theoretical background

Kerr black holes and gravitational radiation are two of the most dramatic predictions of general relativity (other than cosmology). While evidence for Kerr black holes, in particular in their role as powerful inner engines, remains elusive, the quadrupole formula for gravitational radiation has been confirmed observationally with great accuracy.

### *10.2.1 Energetics of Kerr black holes*

The Kerr metric describes an exact solution of a black hole spacetime with non-zero angular momentum. It demonstrates the remarkable property of *frame-dragging*: the angular velocity $\omega$ of otherwise zero-angular momentum observers in spacetime outside the black hole. Frame-dragging assumes a maximal and constant angular velocity $\Omega_H$ on the event horizon of the black hole, and decays with the cube of the distance to zero at large distances. This introduces an angular velocity $\Omega_H$ of the black hole, as well as *differential* frame-dragging by a non-zero gradient (whereby it is not gauge effect). The aforementioned energy reservoir in angular momentum satisfies,

$$E_{\rm rot} = 2M \sin^2(\lambda/4) = M_{irr}\left(\sqrt{1+(2M\Omega_H)^2}-1\right), \tag{10.2}$$

where $\sin\lambda$ denotes the specific angular momentum of the black hole per unit mass. Here, we use geometric units with Newton's constant $G = 1$ and velocity of light $c = 1$. These properties give Kerr black holes the potential to react energetically to their environment. They thereby have the potential of serving as universal sources of energy, distinct from any known baryonic object and with direct relevance to the phenomenology of GRBs.

Rotating objects have a general tendency to radiate away their energy and angular momentum in an effort to reach a lower energy state. In the dynamics of rotating fluids, this is described by the well-known Rayleigh stability criterion. In many ways, black hole radiation processes are governed by the same principle. The second law of thermodynamics $dS \geq 0$ for the entropy $S$ shows that the specific angular momentum of a Kerr black hole ($\Omega_H \leq 1/2M$) increases with radiation:

$$a_p \equiv \frac{-\delta J_H}{-\delta M} \geq \Omega_H^{-1} \geq 2M > M \geq a, \tag{10.3}$$

for a black hole of mass $M$ and angular momentum $J_H$ which emits a particle with specific angular momentum $a_p$. Generally, black holes in isolation are stable by exponential suppression of spontaneous emissions by canonical angular momentum barriers (Press & Teukolsky, 1972; Teukolsky, 1973, 1974; Unruh, 1974; Hawking, 1975). Fortunately, magnetic fields can modulate and suppress these angular momentum barriers.

The earliest studies of active black holes focus on energy-extraction processes in the ergosphere, i.e. scattering of positive energy waves onto rotating black holes – superradiant scattering of Zel'dovich (1971), Press & Teukolsky (1972), Starobinsky (1972) and Bardeen *et al.* (1972) – as a continuous-wave analogue to the process of Penrose (1969). A steady-state variation can be found in the process of spin-down through horizon Maxwell stresses, proposed by Ruffini & Wilson (1975) and extended to force-free magnetospheres by Blandford & Znajek (1977). Quite generally, therefore, magnetic fields are required as a mediating agent to stimulate luminosities which are of astrophysical interest.

Generally, stellar mass black holes produced in core-collapse of a massive star are parametrized by their mass, angular momentum and kick velocity $(M, J, K)$. The equilibrium electric charge of the black hole which arises in its lowest magnetic energy state, is relevant in sustaining adequate horizon magnetic flux, especially at high spin-rates, but is insignificant relative to the spin-energy of the black hole.

### 10.2.2 Linearized gravitational radiation

Gravitational-wave emissions are predominantly due to low multipole mass-moments. The quadrupole radiation formula according to the linearized theory of general relativity has been given by Peters & Mathews (1963)

$$L_{gw} = \frac{32}{5} (\omega\mathcal{M})^{10/3} F(e), \tag{10.4}$$

where $F(e)$ is a function of the ellipticity $e$ of a binary with orbital frequency $\omega$ and chirp mass $\mathcal{M}$.

This linearized result has been observationally confirmed to within 0.1% by decade-long observations of the Hulse-Taylor binary neutron star system PSR1913+16 (with ellipticity $e = 0.62$, Hulse & Taylor (1975); Taylor (1994)). Its gravitational-wave luminosity of about 0.15% of a solar luminosity $L_\odot = 4 \times 10^{33}$ erg s$^{-1}$ gives a binary lifetime of about 7.4 Gyr. Encouraged by this observational confirmation, we apply (10.4) to non-axisymmetric matter surrounding a black hole. With orbital period $\omega \simeq M_H^{1/2}/R^{3/2}$, $\mathcal{M} \simeq M_H(\delta M_T/M_H)^{3/5}$ and ellipticity $e \simeq 0$, this predicts

$$L_{gw} \simeq \frac{32}{5} \left(\frac{M_H}{R}\right)^5 \left(\frac{\delta M_T}{M_H}\right)^2. \tag{10.5}$$

The quadrupole moment is due to a mass-inhomogeneity $\delta M_T$ in the torus. A quadrupole moment appears spontaneously due to non-axisymmetric waves, such as a hydrodynamical or magnetohydrodynamical instability. A detailed description of radiation by all multipole mass-moments in a torus is given in Bromberg *et al.* (2006).

These existing approaches to energy extraction from rotating black holes do not shed light on the "loading problem": the diversity in radiation channels and their various energy outputs. This defines a novel challenge for Kerr black holes as models of the inner engines for gamma-ray bursts. These events, perhaps including core-collapse supernovae, could potentially provide a powerful link between Kerr black holes and gravitational radiation, as outlined below.

## 10.3 Formation and evolution of black holes in core-collapse SNe

While all SNIb/c might be producing black holes, only some are associated with GRB–supernovae in view of the observed small branching ratio (10.1). Black holes produced in *aspherical* core-collapse receive typical kick velocities of a few hundred km s$^{-1}$ (Bekenstein, 1973), measured relative to the center of mass of the progenitor star. Such objects inevitably escape from the central high-density region of the progenitor star, even before core-collapse is completed. A small sample of black holes receive low kick-velocities at birth. Remaining centered, these grow into rapidly spinning high-mass black holes by infall of a substantial

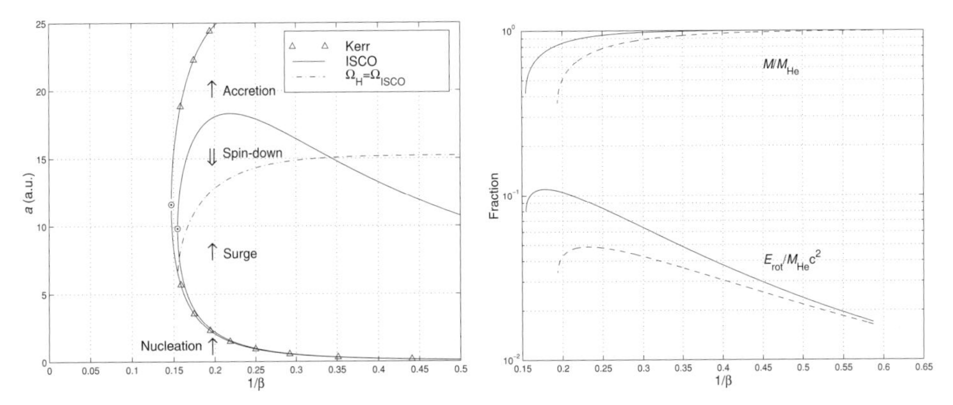


Fig. 10.2. (*Left*) Black holes with small kick velocities remain centered in core-collapse of a uniformly rotating massive star. We show the accumulated specific angular momentum of the central object (arbitrary units) versus dimensionless orbital period $1/\beta$. Arrows indicate the evolution as a function of time. Kerr black holes exist *inside* the outer curve (diamonds). A black hole forms in a first-order transition following the formation and collapse of a torus. This produces a short burst in gravitational radiation. When centered, the black hole surges to a high-mass object by direct infall of matter with relatively low specific angular momentum, up to the inner continuous curve (ISCO). At this point, the black hole either spins up by continuing accretion or spins down radiatively via gravitational radiation emitted by a surrounding non-axisymmetric torus. In this state, the black hole creates a baryon poor jet as input to GRB-afterglow emissions. This continues until the angular velocity of the black hole equals that of the torus (dot-dashed line). This scenario fails for black holes with typical kick velocities with inevitable escape from the high-density core, prohibiting the formation of a high mass black hole surrounded by a high-mass torus. The probability of small kick velocities defines the branching ratio of type Ib/c supernovae into long GRBs. (*Right*) The mass $M$ and rotational energy $E_{rot}$ of the black hole formed after the surge, in the case of small kick velocities, expressed relative to the mass $M_{He}$ of the progenitor He-star. The results are shown in cylindrical geometry (continuous) and spherical geometric (dashed). Note the broad distribution of high-mass black holes with large rotational energies of 5–10% (spherical to cylindrical) of $M_{He}c^2$. (Reprinted from van Putten (2004) ©The American Astronomical Society.)

fraction of the progenitor He-core mass (Fig. 10.2). This forms a starting point for the parametrization of rotating black holes as inner engines of GRB–supernovae.

The state of matter surrounding the newly formed black hole in core-collapse supernovae is unique, in attaining temperatures in excess of 1 MeV (Woosley, 1993) and masses of up to a few percent of the mass of the black hole (van Putten & Levinson, 2003). We further expect these accretion disks to be magnetized, representing a remnant magnetic field of the progenitor star and modified by a dynamo action in the disk.

The angular velocity $\Omega_H$ of the black hole easily exceeds that of surrounding matter, even at relatively low spin-energies. When a surrounding disk is largely

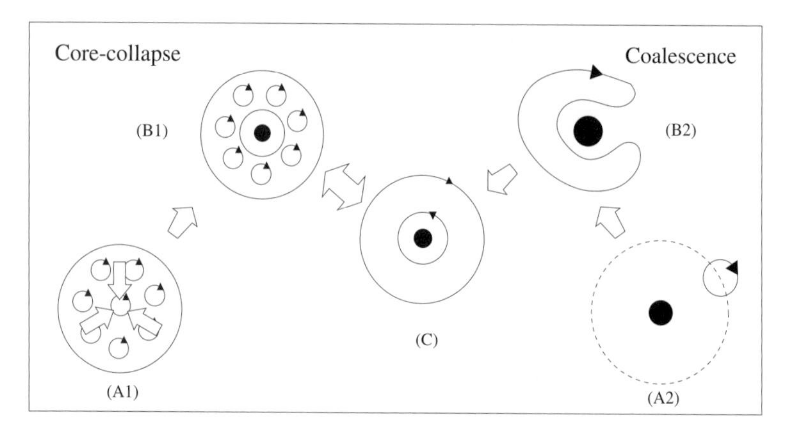


Fig. 10.3. A uniformly magnetized torus represented by two counter-oriented current rings around a black hole (C) forms out of both core-collapse (A1–B1) in massive stars and tidal break-up (A2–B2) in black hole–neutron star coalescence, followed by a single reconnection event (B2–C). (Reprinted from van Putten & Levinson (2003) ©The American Astronomical Society.)

unmagnetized, the black hole continues to accrete, enlarging it and spinning it up towards an extreme Kerr black hole along a *Bardeen trajectory* (Bardeen, 1970), shown in Fig. 10.2. Alternatively, the surrounding matter can be magnetized and, fairly generally, may contain an appreciable $m = 0$ component of poloidal flux on average or, when time-variable, with non-zero standard deviation. Topologically, this component represents a uniform magnetization, corresponding to two counter-oriented current rings (van Putten & Levinson, 2003). Under these conditions, a torus can form in a state of suspended accretion with angular velocity $\Omega_T < \Omega_H$. This introduces a channel for *catalytic* conversion of spin-energy into various radiation channels (van Putten & Levinson, 2003), forcing the black hole to spin down until the angular velocities become similar $\Omega_H \simeq \Omega_T$.

### *10.3.1 Catalyzing black-hole spin-energy*

A magnetized torus surrounding a black hole is expected to form in both black hole–neutron star coalescence and core-collapse of a massive star with very similar topological properties (Fig. 10.3). A magnetized star can be represented to leading order by a single current loop or equivalently a density of magnetic dipole moments. Stretching around the black hole, as in tidal break-up, or by excision of the center of the star, as in core-collapse into a new black hole, leaves a magnetized annulus consisting of two counter-oriented current rings. Thus, core-collapse supernovae and binary black hole–neutron-star coalescence both give the same outcome: a black hole surrounded by a magnetized torus. In practice, the resulting magnetic field will be modified by turbulence and, possibly, a dynamo.

We can look at the structure of the magnetic field of a torus by inspecting its shape in a poloidal cross-section, and by comparison with pulsar magnetospheres. In the case of a pulsar, we see magnetic flux-surfaces of closed magnetic field-lines that reach an outer light cylinder, while open magnetic field-lines extend outwards to infinity. The same structure is found with regards to the torus' outer face, when viewed in poloidal cross-section. Since the event horizon of the black hole is a null-surface, it has the same radiative boundary condition as asymptotic infinity (Blandford & Znajek, 1977; Thorne *et al.*, 1986; Okamoto, 1992), except for its finite surface area (and surface gravity). It follows that the structure of magnetic field-lines found in a magnetized neutron star also applies to the torus' inner face. For the same reason that a spinning neutron star transfers angular momentum to infinity, the inner face of the torus gains angular momentum from the black hole, whenever the latter spins more rapidly. This produces a *spin-connection by topological equivalence to pulsars.*

Angular momentum transport occurs by Alfvén waves emitted from the surface of the torus, carrying positive angular momentum waves off the outer face to infinity and negative angular momentum off the inner face into the black hole. In this process, the outer face becomes sub-Keplerian and the inner face becomes super-Keplerian. On balance of the competing torques between these two faces (van Putten & Levinson, 2003), this produces a state of suspended accretion (van Putten & Levinson, 2003). The angular velocity of a Kerr black hole can reach $\Omega_H = 1/2M$, which far surpasses that of an accretion disk or torus. The black-hole spin-energy (10.2) can be transferred to a torus with angular velocity $\Omega_T$ with efficiency

$$\eta = \frac{\Omega_H}{\Omega_T}. \tag{10.6}$$

The task is now to quantify the various radiation channels provided by the torus. We shall discuss these in the next sections.

The process of formation and spin-down of a Kerr black hole has been proposed to quantitatively model the supernovae by various groups (Brown *et al.*, 2000; Bethe *et al.*, 2003; van Putten & Levinson, 2003). This approach represents a special case of aspherical, magneto-rotationally driven core-collapse supernovae, notably discussed for type II supernovae by Bisnovatyi-Kogan (1970); LeBlanc & Wilson (1970); Bisnovatyi-Kogan *et al.* (1976); Kundt (1976); Wheeler *et al.* (2000); Akiyama *et al.* (2003).

### 10.3.2 Durations and lifetime of rapid spin

A given torus can support a finite poloidal magnetic field-energy. This is due to an instability, produced by magnetic moment-magnetic moment self-interactions in the fluid.

Instability criteria can be derived for both tilt and buckling (van Putten & Levinson, 2003). For a tilt instability between two counter-oriented current rings, representing a uniformly magnetized torus, these interactions are described by a potential

$$U_\mu(\theta) = -\mu B \cos\theta, \tag{10.7}$$

where $\mu$ denotes the magnetic moment dipole moment of the inner ring, $B$ denotes the magnetic field produced by the outer ring, and $\theta$ denotes the angle between $\mu$ and $B$. Note that $U_\mu(\theta)$ has period $2\pi$, is maximal (minimal) when $\mu$ and $B$ are parallel (antiparallel). For the buckling instability, the same interaction energies arise in the azimuthal partition of the torus into small current rings which, combined, are equivalent to the two counter-oriented current rings.

The central potential well of the black hole provides a stabilizing contribution. As the magnetic moment-magnetic moment interactions act primarily to introduce vertical displacements between two current rings (about their equilibrium configuration in the equatorial plane), we can focus on vertical displacements of fluid elements along surfaces of constant cylindrical radius $R$. This is distinct from motions of a rigid ring, whose fluid elements move on surfaces of constant spherical radius. In particular, the tilt of a current ring hereby changes the distance to the central black hole according to $\rho = \sqrt{R^2 + z^2} \simeq R(1 + z^2/2R^2)$. In the approximation of equal mass in the inner and outer face of the torus, simultaneous tilt of one ring upwards and the other ring downwards is associated with the potential energy (van Putten & Levinson (2003), corrected)

$$U_g(\theta) \simeq -\frac{M_T M_H}{R}\left(1 - \frac{1}{2}\tan^2(\theta/2)\right) \tag{10.8}$$

with $\tan(\theta/2) = z/R$ upon averaging over all segments of a ring. Note that $U_g(\theta)$ has period $\pi$ and is minimal at $\theta = 0$. Similar expressions hold for an azimuthal distribution of current rings. Stability is accomplished provided that the total potential energy $U(\theta) = U_\mu(\theta) + U_g(\theta)$ satisfies $U''(\theta) > 0$. We find the following poloidal magnetic field-to-kinetic energy ratios

$$\frac{\mathcal{E}_B}{\mathcal{E}_k} \simeq \begin{cases} \frac{1}{6} & \text{tilt instability} \\ \frac{1}{15} & \text{buckling instability} \end{cases} \tag{10.9}$$

in the approximation $\mathcal{E}_B = B^2 R^3/6$, representing the poloidal magnetic field-energy of the inner torus magnetosphere in a characteristic volume $4\pi R^3/3$ and $\mathcal{E}_k = M_T M_H/2R$. We next discuss the physical parameters at this point of critical stability, which we interpret as a practical limit on the magnetic field energy that the torus can support.

For a pair of rings of radii $R_\pm, (R_+ - R_-)/(R_+ + R_-) = O(1)$, we have $U_\mu(\theta) = (1/2)B^2R^3\cos\theta$, so that the point of critical stability, $U''(\theta) = 0$, gives for the critical magnetic field-strength

$$B_c^2 M_H^2 = (1/4)(M_H/R)^4(M_T/M_H), \tag{10.10}$$

or

$$B_c = 10^{16}\text{G}\left(\frac{7M_\odot}{M_H}\right)\left(\frac{6M_H}{R}\right)^2\left(\frac{M_T}{0.03M_H}\right)^{1/2}, \tag{10.11}$$

with critical poloidal magnetic field-energy (10.9).

Rotating black holes with $\Omega_H \gg \Omega_T$ dissipate most of their spin-energy "unseen" into the event horizon. The lifetime of rapid spin is thus

$$T_s \simeq \frac{E_{\rm rot}}{T\dot{S}_H}, \quad T\dot{S}_H \simeq \Omega_H^2 A_\phi^2, \tag{10.12}$$

where $2\pi A_\phi$ denotes the horizon flux, taking into account that most of the black-hole luminosity is incident onto the inner face of the torus. We then have

$$T_s \simeq 45\text{s}M_{H,7}\eta_{0.1}^{-8/3}\mu_{0.03}^{-1}E_{0.5}^{\rm rot}, \tag{10.13}$$

where the subscripts denote normalization constants, i.e.: $M_{H,7} = M_H/M_\odot, \eta_{0.1} = \eta/0.1$, $\mu_{0.03} = M_T/0.03M_H$ and $E_{0.5}^{\rm rot} = E_{\rm rot}/0.5E_{\rm rot,max}$ corresponding to $\sin\lambda = 0.8894$ in (10.2). This estimate agrees well with the observed durations of tens of seconds of long GRBs shown in Fig. (10.1) and statistics of the BATSE catalogue (Kouveliotou *et al.*, 1993).

## 10.4 Formation of multipole mass-moments

A torus tends to be unstable to a variety of symmetry-breaking instabilities. This provides a spontaneous mechanism for the creation of multipole mass-moments. Provided that the torus is not completely disrupted, such multipole mass-moments ensure that the torus is luminous in gravitational radiation, when its mass reaches a few percent of that of the black hole. These instabilities can take the form of hydrodynamic and magnetohydrodynamic instabilities.

Papaloizou & Pringle (1984) describe a hydrodynamic buckling instability in infinitesimally slender tori, due to a coupling of surface waves on a super-Keplerian inner face and a sub-Keplerian outer face. To be of practical relevance, this theory can be extended to tori of arbitrary slenderness (below). In a recent numerical study, the magnetic moment-magnetic moment self-interactions of a uniformly magnetized torus are also found to produce instabilities – buckling instabilities in the poloidal plane (Bromberg *et al.*, 2006). This fully non-linear numerical study, though one-dimensional, recovers our heuristic analytical bound on the maximal poloidal magnetic field energy-to-kinetic energy in the torus (10.9). It confirms that the dominant emission channel in gravitational radiation is through the quadrupole mass-moment. These are but two example calculations in the more general challenge of computing gravitational-wave spectra of magnetized tori in suspended accretion. Below, we discuss the hydrodynamic buckling instability for wide tori.

The dynamical stability of a torus with sub-Keplerian outer face and super-Keplerian inner face can be studied in the limit of an inviscid incompressible fluid with Newtonian angular velocity (Papaloizou & Pringle, 1984; Goldreich *et al.*, 1986)

$$\Omega(r) = \Omega_a \left(\frac{a}{r}\right)^q \quad (3/2 \le q \le 2). \tag{10.14}$$

Here, $q$ denotes the rotation index which is bounded between the Keplerian value $q = 3/2$ and Rayleigh's stability limit $q = 2$.

Irrotational perturbations to the underlying flow (vortical if $q \neq 2$) remain irrotational by Kelvin's theorem. In studying their stability we consider the harmonic velocity potential in cylindrical coordinates $(r, \theta, z)$

$$\phi = \Sigma_n a_n(r, \theta, z) z^n, \quad \Delta\phi = 0. \tag{10.15}$$

The equations of motion can be expressed in a local Cartesian frame $(x, y, z)$ with Newtonian angular velocity $\Omega_a = M^{1/2} a^{-3/2}$, equal to the angular velocity of the torus at its major radius $r = a$ about a central mass $M$. These Cartesian coordinates are related to cylindrical coordinates $(r, \theta)$ according to $x = r - a$, $\partial_x = \partial_r$ and $\partial_y = r^{-1}\partial_\theta$. Together with zero-enthalpy boundary conditions on the free inner and outer surface of the torus, we obtain a complete problem for linearized stability analysis. This problem can be solved semi-analytically, by searching numerically for points of change in stability (Keller, 1987) of harmonic perturbations $e^{im\theta - i\omega' t}$ of infinitesimal amplitude at frequency $\omega'$, as seen in a corotating frame at $r = a$. Fig. 10.4 shows the numerical stability diagram.

In general, we encounter for each $m$ a critical rotation index $q = q(b/a, m)$ which depends on the minor-to-major radius $b/a$ of the torus. These curves can be found using numerical continuation methods (Keller, 1987). The stability diagram for the rotation index is shown in Fig. 10.4. In particular, we mention the critical

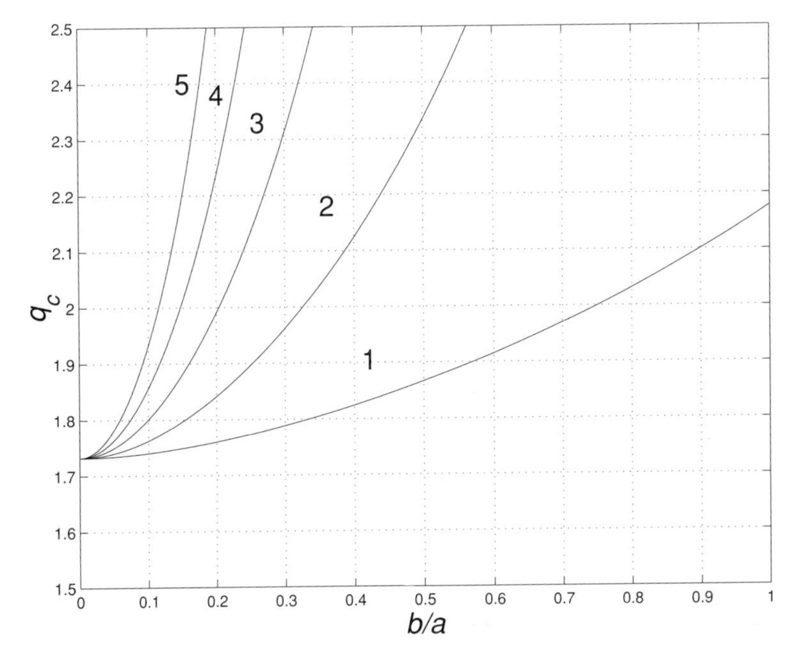


Fig. 10.4. The stability diagram showing the neutral stability curves for non-axisymmetric buckling modes in a torus of an inviscid incompressible fluid of arbitrary minor-to-major radius $b/a$. Curves of critical rotation index $q_c$ are labeled with azimuthal quantum number $m = 1, 2, \ldots,$ where instability sets in above and stability sets in below. (Reprinted from van Putten (2002) ©The American Astrophysical Society.)

values

$$b/a = 0.7506, 0.3260, 0.2037, 0.1473, 0.1152, \ldots, 0.56/m \tag{10.16}$$

for the various $m$-modes at the Rayleigh stability line $q = 2$ for the various $m$-modes at the Rayleigh stability line $q = 2$, where $0.56/m$ refers to the asymptotics for large $m$ observed in the numerical results.

## 10.5 Radiation-energies by a non-axisymmetric torus

Rotating black holes in suspended accretion dissipate most of their rotational energy (10.2) "unseen" in their event horizon, while a major fraction (10.6) is incident into the inner face of the surrounding torus. This lasts for the lifetime of rapid spin (10.13), whose tens of seconds represents a secular timescale relative to the millisecond period of the orbital motion of the torus. This reduces the problem of calculating the radiation output from the torus to algebraic equations of balance in energy and angular momentum flux, taking into account the channels of gravitational radiation, MeV-neutrino emissions and magnetic winds. The equations of suspended accretion are

$$\begin{aligned} \tau_+ &= \tau_- + \tau_{gw} \\ \Omega_+\tau_+ &= \Omega_-\tau_- + \Omega_T\tau_{gw} + P_\nu, \end{aligned} \tag{10.17}$$

where $(\tau_\pm, \Omega_\pm, \Omega_T = (\Omega_+ + \Omega_-)/2)$ denote the torques on and angular velocities of the inner and outer face of the torus due to surface Maxwell stresses as for those on a pulsar, $\tau_{gw}$ denotes the torque on the torus due to the emission of gravitational radiation with luminosity $L_{gw} = \Omega_T\tau_{gw}$, and $P_\nu$ the power in MeV-neutrino emissions due to dissipation. (The results show temperatures of a few MeV.) These equations are closed by a constitutive relation for the dissipation process. We set out by attributing dissipative heating to magnetohydrodynamical stresses. This introduces an overall scaling with the magnetic field-energy $E_B$. The resulting total energy emissions, $E_{gw}$, $E_w$ and $E_\nu$, produced over the lifetime of rapid spin of the black hole (10.13), thus become *independent* of $E_B$ and reduce to specific fractions of $E_{\text{rot}}$.

When the torus is sufficiently slender, $m = 2$ modes (two lumps swirling around the black hole) develop. These radiate at essentially twice the orbital frequency of the torus, when the minor-to-major radius is less than 0.3260 by (10.16).

By an asymptotic analysis of (10.17), the gravitational-wave emissions satisfy (van Putten & Levinson, 2003)

$$E_{gw} \simeq 2 \times 10^{53}\ \eta_{0.1} M_{H,7}\ E_{0.5}^{\text{rot}}\ \text{erg}, \quad f_{gw} \simeq 500\eta_{0.1} M_{H,7}^{-1}\ \text{Hz}. \tag{10.18}$$

Thus, an "unseen" energy output (10.18) surpasses the true energy $E_\gamma \simeq 3 \times 10^{50}$ erg in gamma-rays (Frail *et al.*, 2001) by several orders of magnitude.

The associated output in magnetic winds contains an additional factor $\eta$, i.e.

$$E_w \simeq 2 \times 10^{52}\ \eta_{0.1}^2 M_{H,7} E_{0.5}^{\text{rot}}\ \text{erg}. \tag{10.19}$$

This baryon-rich wind provides a powerful agent towards collimation of any outflows from the black hole (Levinson & Eichler, 2000), as well as a source of neutrinos for pick-up by the same (Levinson & Eichler, 2003). It is otherwise largely incident onto the remnant stellar envelope *from within*, which serves as energetic input to accompanying supernova ejecta. We estimate these kinetic energies to be

$$E_{SN} \simeq 1 \times 10^{51}\ \beta_{0.1} M_{H,7} \eta_{0.1}^2 E_{0.5}^{\text{rot}}\ \text{erg} \tag{10.20}$$

with $\beta = v_{ej}/c$ denoting the velocity $v_{ej}$ of supernova ejecta relative to the velocity of light $c$. In the expected aspherical geometry, $\beta = 0.1\beta_{0.1}$ refers to the mass-average taken over all angles. The canonical value $\beta = 0.1$ refers to the observed value in GRB011211 (Reeves *et al.*, 2002). Eventually, the expanding remnant envelope becomes optically thin, which permits the appearance of X-ray line-emissions excited by the underlying continuum emission $E_\gamma$ by dissipating $E_w$. The estimate (10.20) is in remarkable agreement with the kinetic energy $2 \times 10^{51}$ erg of the aspherical supernova SN1998bw associated with GRB980425 (Höflich *et al.*, 1999). The energy output in MeV-neutrinos is intermediate, in being smaller than $E_{gw}$ by an additional factor given by the slenderness parameter $\delta = qb/2R$, where $b/R$ denotes the ratio of minor-to-major radius of the torus and $q$ its rotation index (10.14). Thus, we have

$$E_\nu \simeq 1 \times 10^{53}\ \eta_{0.1}\delta_{0.30}M_{H,7}E_{0.5}^{\rm rot}\ \text{erg}, \tag{10.21}$$

where $\delta = 0.30\delta_{0.30}$. At the associated dissipation rate, the torus develops a temperature of a few MeV which stimulates the production of baryon-rich winds (van Putten & Levinson, 2003).

## 10.6 Launching an ultrarelativistic jet by differential frame-dragging

Frame-dragging induced by angular momentum extends through the environment of the black hole and includes the spin-axis. The Kerr metric provides an exact description of this gravitational induction process by the Riemann tensor (Chandrasekhar, 1983). In turn, the Riemann tensor couples to spin (Papapetrou, 1951a,b; Pirani, 1956; Misner *et al.*, 1973; Thorne *et al.*, 1986). Specific angular momentum (angular momentum per unit mass) represents a rate of change of surface area per unit of time, while the Riemann tensor is of dimension cm$^{-2}$ (in geometrical units). Therefore, curvature–spin coupling produces a force (dimensionless in geometrical units), whereby test particles with spin follow non-geodesic trajectories (Pirani, 1956). In practical terms, the latter holds promise as a mechanism for *linear acceleration*. By dimensional analysis once more, the gravitational potential for spin-aligned interactions satisfies

$$E = \omega J, \tag{10.22}$$

where $\omega$ denotes the local frame-dragging angular velocity produced by the black hole (or any other spinning object) and $J$ is the angular momentum of the particle at hand. Spinning bodies therefore couple to spinning bodies (O'Connel, 1972). Thus, (10.22) defines a mechanism for accelerating baryon-poor ejecta to ultrarelativistic velocities, provided $J$ is large.

The angular momentum $J$ of charged particles in strong magnetic fields, confined to individual magnetic flux-surfaces in radiative Landau states, is macroscopic in the form of orbital angular momentum

$$J = eA_\phi. \tag{10.23}$$

Here $e$ denotes the elementary charge and $2\pi A_\phi$ denotes the enclosed magnetic flux, where $A_a$ is the electromagnetic vector potential of an open magnetic flux-tube along the spin-axis of the black hole. Geometrically, the specific angular momentum represents the rate of change of surface area traced out by the orbital motion of the charged particle, by "helical motion" as seen in four-dimensional spacetime.

Combined, (10.22) and (10.23) describe a powerful mechanism for linear acceleration of baryon-poor matter. It radically differs from the common view, that black-hole energetic processes are limited exclusively to the action of frame-dragging in the ergosphere.

It is instructive to derive (10.22) by specializing to the Kerr metric on the spin-axis of the black hole. To this end, we may consider one-half the difference in potential energy of particles with angular momenta $\pm J$ suspended in a gravitational field about a rotating object. In the first case, we note that the Riemann tensor $R_{ijmn}$ expressed relative to tetrad elements associated with Boyer–Lindquist coordinates (Chandrasekhar, 1983) gives rise to a linear force

$$F_2 = JR_{3120} = -\partial_2 \omega J, \tag{10.24}$$

which can be integrated along the spin-axis to give

$$E = \int_r^\infty F_2 \, \mathrm{d}s = \omega J. \tag{10.25}$$

In the second case, we merely assume a metric $g_{ab}$ with time-like and azimuthal Killing vectors and consider two particles with velocity four-vectors $u^b$ according to angular momenta $J_\pm = g_{\phi\phi} u^t (\Omega_\pm - \omega)$,

$$J_\pm = \pm g_{\phi\phi} u^t \sqrt{\omega^2 - (g_{tt} + (u^t)^{-2}/g_{\phi\phi})} = \pm J, \tag{10.26}$$

which shows that $u^t$ is the same for both particles. The total energy of the particles $E_\pm = (u^t)^{-1} + \Omega_\pm$ gives rise to

$$E = \frac{1}{2}(E_+ - E_-) = \omega J. \tag{10.27}$$

To apply (10.22)–(10.23) to gamma-ray bursts from rotating black holes, consider a perfectly conducting blob of charged particles in a magnetic flux-tube subtended at a finite half-opening angle $\theta_H$ on the event horizon of the black hole and along its spin-axis. In the approximation of electrostatic equilibrium, it assumes a rigid rotation (Thorne *et al.*, 1986) described by an angular velocity $\Omega_b$. (By Faraday induction, differential rotation introduces potential differences along magnetic field-lines. Some differential rotation is inevitable over long length scales and, possibly, in the formation of gaps.) In the frame of zero-angular momentum observers, the equilibrium charge-density assumes the value of Goldreich & Julian (1969), as viewed by zero-angular momentum observers. Thus, given a number $N(s)$ of charged particles per unit scale height $s$ in the flux-tube,

$$N(s) = (\Omega_b - \omega)A_\phi/e. \tag{10.28}$$

A pair of blobs of scale height $h = h_M M_H$ in both directions along the spin-axis of the black hole each hereby receive a potential energy

$$E_b = \omega J N h = 10^{47} B_{15} h_M^3 H \text{ erg}, \tag{10.29}$$

where $B = B_{15} \times 10^{15}$G and $H = 4\hat{\omega}(\hat{\Omega}_b - \hat{\omega})$ is a quantity of order unity, expressed in terms of the normalized angular velocities $\hat{\omega} = \omega/\Omega_H$ and $\hat{\Omega}_b = \Omega_b/\Omega_H$.

Electrons and positrons in superstrong magnetic fields are essentially massless. The ejection of a pair of blobs with energy (10.29) thus takes place in a light-crossing time of about 0.3 ms of, e.g. a seven solar mass black hole of linear dimension $10^7$ cm. This corresponds to an instantaneous luminosity of about $3 \times 10^{50}$ erg s$^{-1}$. This produces a total kinetic energy output of up to $10^{52}$ erg in tens of seconds in long bursts, a fraction of which will be dissipated in gamma-rays and lower energy afterglow emissions in the internal-external shock model for GRBs (Rees & Mészáros, 1992, 1994; Mészáros & Rees, 1993; Piran, 2004).

Similar results obtain for the luminosity in a steady-state limit, by considering a horizon half-opening angle $\theta_H \simeq M_H/R$, set by the poloidal curvature $M_H/R$ of the magnetic field-lines. We note the paradoxical *small* energy output of GRB-afterglow emissions, when viewed relative to the total black-hole spin-energy (10.2). The luminosity in the jet scales with the square of the enclosed magnetic flux, while the latter scales with the enclosed surface area, and hence the square of the half-opening angle $\theta_H$. Thus, we encounter a geometrical scale-factor, which creates a jet luminosity

$$L_j \propto \theta_H^4. \tag{10.30}$$

Even when $\theta_H$ is not small, e.g. about 10 degrees, this scaling creates a small parameter. Integrated over time, the energy output in gamma-rays satisfies (van

Putten & Levinson, 2003)

$$E_\gamma \simeq 1 \times 10^{50} \epsilon_{0.3} \eta_{0.1}^{8/3} E_{0.5}^{\rm rot} \text{ erg}, \tag{10.31}$$

consistent with the observed true energy output $3 \times 10^{50}$ erg in gamma-rays of long bursts (Frail *et al.*, 2001), where $\eta = 0.1\eta_{0.1}$ and $\epsilon = 0.3\epsilon_{0.3}$ denotes the efficiency of converting kinetic energy to gamma-rays.

In our unification scheme, the durations of short and long bursts are attributed to different spin-down times of, respectively, a slowly and rapidly spinning black hole according to (10.13). The total energy output in gamma-rays of short GRBs is hereby significantly smaller than that of long GRBs, due to both shorter durations and smaller luminosities. In this light, the recent X-ray afterglow detections to short bursts, very similar but fainter than their counterparts to long bursts, are encouraging.

## 10.7 Long-duration bursts in gravitational waves

The output in gravitational radiation (10.18) is in the range of sensitivity of the broad band detectors LIGO and Virgo shown in Fig. 10.5, as well as GEO (Danzmann, 1995; Willke *et al.*, 2002) and TAMA (Ando *et al.*, 2002). Note that the match is better for higher-mass black holes, in view of their emissions at lower frequencies (towards the minimum in the detector noise curve) and their larger energy output.

Matched filtering gives a theoretical upper bound on the signal-to-noise ratio in the detection of long bursts in gravitational radiation from GRB–supernovae. In practice, the frequency will be unsteady at least on an intermittent timescale associated with the evolution of the torus, e.g. due to mass-loss in winds and possibly mass-gain by accretion from additional matter falling in. For this reason, a time-frequency trajectory method which correlates the coefficients of a Fourier transformation over subsequent windows of durations of seconds might apply. The ultimate signal-to-noise ratio will therefore be intermediate between that obtained through correlation – a second-order procedure – and matched filtering – a first-order procedure. The results shown in Fig. 10.5 show the maximal attainable signal-to-noise ratio (by matched filtering) for sources at a fiducial distance of 100 Mpc. This distance corresponds to an event rate of one per year.

GRB–supernovae are an astrophysical source population locked to the star-formation rate. Accordingly, we can calculate their contribution to the stochastic background in gravitational radiation given their band-limited signals, assuming $B = \Delta f / f_e$ of around 10%, where $f_e$ denotes the average gravitational-wave frequency in the comoving frame. In what follows, the following scaling relations are

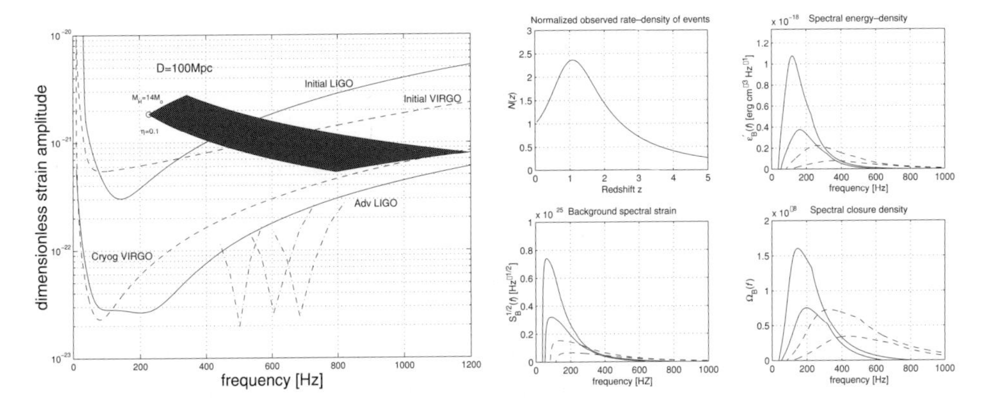


Fig. 10.5. (*Left*) GRB–supernovae from rotating black holes predict a contemporaneous long-duration burst in gravitational radiation within the range of sensitivity of upcoming gravitational-wave detectors LIGO and Virgo within a distance of about 100 Mpc. The corresponding event rate is about one per year. The "black bar" denotes the distribution of dimensionless strain amplitudes for a distribution of sources corresponding to a range of black hole masses and efficiency factors, assuming matched filtering. In practice, the sensitivity will depend less on using time frequency trajectory methods with correlations in the spectral domain. (*Right*) The cosmological distribution of GRB–supernovae is locked to the star-formation rate $N(z)$. This enables the calculation of the expected contribution to the stochastic background in gravitational radiation, here shown in terms of the spectral energy-density $\epsilon'_B(f)$, the strain amplitude $S_B^{1/2}(f)$ and the spectral closure density $\Omega_B(f)$. The results are calculated for uniform mass-distributions of the black hole, $M_H = (4 - 14) \times M_\odot$ (top curves) and $M_H = (5 - 8) \times M_\odot$ (lower curves) with $\eta = 0.1$ (solid curves) and $\eta = 0.2$ (dashed curves). The extremal value of $\Omega_B(f)$ is in the neighborhood of maximal sensitivity of LIGO and Virgo. (Reprinted from van Putten *et al.* (2004) ©2004 American Physical Society.)

applied,

$$E_{gw} = E_0 M_H / M_0, \quad f_e = f_0 M_0 / M_H \tag{10.32}$$

where $M_0 = 7M_\odot$, $E_0 = 0.203 M_\odot \eta_{0.1}$ and $f_0 = 455$ Hz $\eta_{0.1}$, assuming maximal spin-rates ($E_{\rm rot} = E_{\rm rot,max}$). For non-extremal black holes, a commensurate reduction factor in energy output can be inserted. This factor carries through proportionally to the final results, whence it is not taken into account explicitly.

Summation over a uniform distribution of black hole masses, e.g. $M_H = (4\text{–}14) \times M_\odot$, and assuming that the black hole mass and $\eta$ in (10.6), are uncorrelated, the expected spectral energy-density satisfies (van Putten *et al.*, 2004)

$$\langle \epsilon'_B(f) \rangle = 1.08 \times 10^{-18} \hat{f}_B(x) \text{ erg cm}^{-3} \text{ Hz}^{-1} \tag{10.33}$$

where $\hat{f}_B(x) = f_B(x)/\max f_B(\cdot)$ is a normalized frequency distribution. The associated dimensionless amplitude $\sqrt{S_B(f)} = \sqrt{2G/\pi c^3} f^{-1} \tilde{F}_B^{1/2}(f)$, where $\tilde{F}_B = c\epsilon'_B$ and $G$ denotes Newton's constant, satisfies

$$\sqrt{S_B(f)} = 7.41 \times 10^{-26} \eta_{0.1}^{-1} \hat{f}_S^{1/2}(x)\ \mathrm{Hz}^{-1/2} \tag{10.34}$$

where $\hat{f}_S(x) = f_S(x)/\max f_S(\cdot)$, and $f_S(x) = f_B(x)/x^2$. Likewise, we have for the spectral closure density $\Omega_B(f) = f\tilde{F}_B(f)/\rho_c c^3$ relative to the closure density $\rho_c = 3H_0^2/8\pi G$

$$\tilde{\Omega}_B(f) = 1.60 \times 10^{-8} \eta_{0.1} \hat{f}_\Omega(x), \tag{10.35}$$

where $\hat{f}_\Omega(x) = f_\Omega(x)/\max f_\Omega(\cdot)$, $f_\Omega(x) = x f_B(x)$ and $H_0$ denotes the Hubble constant.

These cosmological results show a simple scaling relation for the extremal value of the spectral closure density in its dependency on the model parameter $\eta$. The location of the maximum scales with $f_0$ in view of $x = f/f_0$. The spectral closure density thus becomes completely determined by the shape of the function representing the star-formation rate, the fractional GRB–supernova rate thereof, $\eta$, and the black-hole mass distribution. Fig. 10.5 shows the various distributions. The extremal value of $\Omega_B(f)$ is in the neighborhood of the location of maximal sensitivity of LIGO and Virgo. It would be of interest to search for this contribution to the stochastic background in gravitational waves by correlation in the spectral domain, following Fourier transformation over series of sub-windows on intermediate timescales of seconds.

## 10.8 Conclusions

The sixties saw two independent discoveries: the first GRB670702 by the Vela satellite and the exact solution of rotating black holes by Roy P. Kerr. Through observational campaigns with BATSE, BeppoSax, the Interplanetary Network (IPN), HETE-II and now Swift, we have come to understand the phenomenology of GRBs. Long bursts are associated with supernovae, representing a rare and extraordinary powerful cosmological transient phenomenon, taking place about once a minute and reaching the earliest epochs in the Universe. Independently, through the works of Penrose (1969); Ruffini & Wilson (1975); Blandford & Znajek (1977) and others, we have come to understand the potential significance of Kerr black holes as compact, baryon-free energy sources with certain universal properties. The applications to high-energy astrophysics should be enormous (e.g. Levinson (2004)). These include supermassive black holes in active galactic nuclei such as our own galaxy (Porquet *et al.*, 2004), stellar mass black holes in microquasars (Mirabel & Rodrǵuez, 1994) and, possibly, gamma-ray bursts.

While the formation-process of supermassive black holes remains inconclusive, the birthplace of stellar mass black holes is most probably core-collapse supernovae of massive stars. Evidence for Kerr black holes as the energy source for high-energy astrophysical processes remains elusive, however. Recent measurements of frame-dragging by X-ray spectroscopy, typically during inactive states of the black hole, are encouraging in this respect (Fabian & Miniutti, Chapter 9 of this volume).

In this review, we propose that GRB–supernovae are powered by rapidly rotating black holes, wherein (1) the durations of long GRBs of tens of seconds are identified with the lifetime of rapid spin of the black hole in a state of suspended accretion, (2) an accompanying supernova is radiation-driven by magnetic winds from a torus in suspended accretion, (3) ultrarelativistic outflows are launched by gravitational spin-orbit coupling with charged particles along open magnetic field-lines, and (4) a small branching ratio of type Ib/c supernovae into GRBs is attributed to the small probability of producing black holes with small kick velocities.

Modeling short GRBs from slowly rotating black holes, we predicted X-ray afterglows very similar but weaker (in total energies) to those of long bursts. The recent discovery of faint X-ray afterglows to GRB050509B and GRB050709 fit well within this scheme.

Gamma-ray bursts present a potentially powerful link between rotating black holes and gravitational radiation. Strategic searches for their chirps in binary coalescence of neutron stars and black holes, or long-duration bursts in gravitational radiation during radiative spin-down of a high-mass black hole, can be pursued by the advanced detectors LIGO and Virgo. Strategic searches are preferrably pursued in combination with upcoming optical-radio supernova surveys, e.g. Pan-Starrs in Hawaii (Kudritzki, 2003) in combination with the Low Frequency Array (LOFAR, 2005), see also Gal-Yam (2005). In light of the proposed supernova-like signatures from the debris of a neutron star (Kulkarni, 2005), these strategies might apply to both short and long bursts.

With a serendipitous discovery and the tools of general relativity, LIGO and Virgo promise to provide a unique method for identifying Kerr black holes as the most luminous objects in the Universe (van Putten & Levinson, 2002).

## *Note added in proof*

Since the Kerr Fest was held in 2004, two notable discoveries put Kerr black holes into the spotlight as potential inner engines to ultrarelativistic phenomena in the universe: the Swift event GRB060614 (Della Valle *et al.*, 2006; Fynbo *et al.*, 2006; Gal-Yam *et al.*, 2006; Gehrels *et al.*, 2006) and a detection by the Pierre

Auger Collaboration (2007) of a correlation between ultra-high-energy cosmic rays (UHECRs) with active galactic nuclei (AGN) in the catalogue of Véron-Cetty & Véron (2006). The AGN are sufficiently nearby to circumvent the GZK cut-off in the observable energy spectrum from cosmological sources (Greisen, 1966; Zatsepin & Kuzmin, 1966).

With a duration of over 100 seconds, GRB060614 defines a remarkable new class of long-duration gamma-ray burst *sans* supernova. Its gamma-ray emissions and intermittency are consistent with the existing class of long-duration GRB–supernovae. It may, therefore, represent the binary coalescence of a stellar mass black hole and a neutron star or, possibly, the merger of two neutron stars. Either way, the result is expected to produce a black hole surrounded by high-density remnant neutron star matter. However, mergers of compact objects pose a challenge: how to account for the long-duration timescale on the order of one minute, which is far in excess of the dynamical timescale of milliseconds. If the black hole spins rapidly, and more rapidly so than the surrounding matter, then long-durations can be found in the dissipative timescale of black hole spin-energy in a suspended accretion state. A detailed description of the lifetime of rapid spin and intermittency in the suspended merger of a spinning black hole and a neutron star (van Putten, 1999) is in good agreement with the observational properties of GRB060614. An association of long-duration GRBs with rapidly spinning black holes that interact with high-density matter, regardless of the presence of an accompanying supernova, predicts a luminous output in gravitational radiation. At present detection sensitivity, the ground-based detectors LIGO and Virgo, promise detections with signal-to-noise ratios of about 1 at an event rate of about 1 per year (within a distance of 100 Mpc; for a review, see van Putten (2005)). Detection prospects are expected to improve significantly in transition to Advanced LIGO with a planned sensitivity enhancement of about one order of magnitude in the high-frequency, shot-noise dominated region (above approximately 100 Hz). A characteristic feature of a long-duration burst of gravitational waves powered by black hole spin-energy is expected to be a *negative chirp*, whose frequency shows exponential decay to a finite value which is representative for the mass of the black hole (van Putten, 2008a).

A correlation of UHECRs with AGN in the local universe supports the widely held notion, that particles of energies in excess of 1 exa eV (1 EeV=$10^{18}$ eV) are somehow produced by the activity of supermassive black holes. (All galaxies appear to harbor supermassive black holes. Most of these were probably active at some point in the evolution of their host galaxy, and some nearby are active at the present time.) UHECRs, probably protons and possibly including some heavier nuclei, may hereby represent the most exceptional non-thermal emissions from (supermassive) black holes. A non-thermal spectrum can be derived in a modified Hawking process in the presence of an external magnetic field, representing a Fermi-level on the event horizon of a black hole due to Faraday-induction (van Putten, 2000). Here, the Fermi-level represents an unprocessed, "raw" potential of the black hole inner engine which assumes UHECR-type energies when surrounded by a magnetized accretion disk. Earlier works proposed that any such Faraday-induction leads to pair-cascade processes sufficiently pervasive to produce overall charge-separation (Blandford & Znajek, 1977), thus inhibiting the formation of a suitably clean site for linear acceleration. However, global analysis on a semi-infinite open flux-tube extending from the event horizon to infinity shows a completely different picture. The Fermi-level on the event horizon is now communicated to an outgoing Alfvén front. Upstream, the magnetic field remains sub-critical (where canonical pair-cascade processes cease to operate) and is well-described by the vacuum solution of Wald (1974). Here, (10.22) describes a "relativistic capillary effect" acting along open magnetic flux-tubes, thus providing a mechanism for accelerating baryonic contaminants to UHECR energies (van Putten, 2008b). For rapidly spinning supermassive black holes in active galactic nuclei (AGN), the UHECR energies produced by a flux tube with horizon half-opening angle $\theta_H$, set by the poloidal curvature $M/R$ of magnetic field lines in the inner disk of radius $R$, can be expressed in terms of the mass $M = M_9 \times 10^9 M_\odot$ and lifetime of spin $T = T_8 \times 10^8$ yr. An estimate gives

$$\varepsilon = 3 \times 10^{19} \sqrt{\frac{M_9}{T_8}} \left(\frac{\theta_H}{0.30}\right)^2 \text{ eV}. \tag{10.36}$$

The result identifies a potential link between the Greisen–Zatsepin–Kuzmin (GZK) cutoff (Zatsepin & Kuzmin, 1966; Greisen, 1966) of $6 \times 10^{19}$ and the lifetime $T_{\rm AGN}$ of an AGN, here identified with the spin of supermassive black holes in galactic nuclei. Clearly, it would be of interest to pursue a detailed observational study on $T_{\rm AGN}$, in combination with the somewhat better known distribution of black hole masses in galactic centres.

**Acknowledgment**

The author thanks A. Levinson, R. P. Kerr, R. Preece, and D. L. Wiltshire for constructive comments. This research is supported by the LIGO Observatories, constructed by Caltech and MIT with funding from NSF under cooperative agreement PHY 9210038. The LIGO Laboratory operates under cooperative agreement PHY-0107417. This paper has been assigned LIGO document number LIGO-P040013-00-R.

## References

Abramovici, A., *et al.* (1992), *Science* **292**, 325.
Acernese, F., *et al.* (2002), *Class. Quantum Grav.* **19**, 1421.
Akiyama, S., Wheeler, J. C., Meier, D. L. & Lichtenstadt, I. (2003), *Astroph. J.* **584**, 954.
Ando, M., and the TAMA Collaboration (2002), *Class. Quantum Grav.* **19**, 1409.
Bardeen, J. M. (1970), *Nature* **226**, 64.
Bardeen, J. M., Press, W. H. & Teukolsky, S. A. (1972), *Astroph. J.* **178**, 347.
Barish, B. & Weiss, R. (1999), *Phys. Today* **52**, 44.
Bekenstein, J. D. (1973), *Astroph. J.* **183**, 657.
Bethe, H. A., Brown, G. E. & Lee, C.-H. (2003), *Selected Papers: Formation and Evolution of Black Holes in the Galaxy* (World Scientific, Singapore), p. 262.
Bisnovatyi-Kogan, G. S. (1970), *Astron. Zh.* **47**, 813.
Bisnovatyi-Kogan, G. S., Popov, Yu. P. & Samochin, A. A. (1976), *Astroph. Space Sci.* **41**, 287.
Blandford, R. D. & Znajek, R. L. (1977), *Mon. Not. Roy. Astron. Soc.* **179**, 433.
Bonnell, I. A. & Pringle, J. E. (1995), *Mon. Not. Roy. Astron. Soc.* **273**, L12.
Bradaschia, C., Del Fabbro, R., di Virgilio, A., *et al.* (1992), *Phys. Lett.* **A 163**, 15.
Bromberg, O., Levinson, A. & van Putten, M. H. P. M. (2006), *New Astron.* **11**, 619.
Brown, G. E., Lee, C.-H., Wijers, R. A. M. J., Lee, H. K., Israelian, G. & Bethe, H. A. (2000), *New Astron.* **5**, 191.
Burrows, A. & Lattimer, J. M. (1987), *Astroph. J.* **318**, L63.
Chandrasekhar, S. (1983), *The Mathematical Theory of Black Holes* (Oxford University Press, Oxford).
Costa, E., *et al.* (1997), *Nature* **387**, 878.
Danzmann, K. (1995), *First Edoardo Amaldi Conference on Gravitational Wave Experiments*, ed. E. Coccia, G. Pizella & F. Ronga (World Scientific, Singapore), p. 100.
Davies, M. B., King, A., Rosswog, S. & Wynn, G. (2002), *Astroph. J.* **579**, L63.
Della Valle, M., *et al.* (2006), *Nature* **444**, 1050.
Duez, M. D., Shapiro, S. L. & Yo, H.-J. (2004), *Phys. Rev.* **D 69**, 104016.
Faber, J. A., Grandclément, P. & Rasio, F. A. (2004), *Phys. Rev.* **D 69**, 124036.
Filipenko, A. V. (1997), *Ann. Rev. Astron. Astroph.* **35**, 309.
Finn, L. S., Krishnan, B. & Sutton, P. J. (2004), *Astroph. J.* **607**, 384.
Fox, D. B., *et al.* (2005), *Nature* **437**, 845.
Frail, D. A., *et al.* (1997), *Nature* **389**, 261.
Frail, D. A., *et al.* (2001), *Astroph. J.* **562**, L55.
Fryer, C. L., Woosley, S. E., Herant, M. & Davies, M. B. (1999), *Astroph. J.* **520**, 650.
Fryer, C. L., Holz, D. E. & Hughes, S. A. (2002), *Astroph. J.* **565**, 430.
Fynbo, J. P. U., *et al.* (2006), *Nature*, **444**, 1047.
Gal-Yam, A., *et al.* (2005), *Astroph. J.* **639**, 331.
Gal-Yam, A., *et al.* (2006), *Nature* **444**, 1053.
Gehrels, N., *et al.* (2005), *Nature* **437**, 851.
Gehrels, N., *et al.* (2006), *Nature* **444**, 1044.
Goldreich, P. & Julian, W. H. (1969), *Astroph. J.* **157**, 869.
Goldreich, P., Goodman, J. & Narayan, R. (1986), *Mon. Not. Roy. Astron. Soc.* **221**, 339.
Greisen, K. (1966), *Phys. Rev. Lett.* **16**, 748.
Haislip, J. B., *et al.* (2006), *Nature* **440**, 181.
Hawking S. W. (1975), *Commun. Math. Phys.* **43**, 199.

Heise, J., 't Zand, J., Kippen, R. M. & Woods, P. M. (2001), *Gamma-ray Bursts in the Afterglow Era, CNR (2000)*, ed. E. Costa, F. Frontera & J. Hjörth (Springer, Berlin), p. 16.
Hjörth, J., *et al.* (2005), *Nature* **437**, 859.
Höflich, P. J., Wheeler, J. C., Wang, L. (1999), *Astroph. J.* **521**, 179.
Hulse, R. A. & Taylor, J. H. (1975), *Astroph. J.* **195**, L51.
Katz, J. I. (1994), *Astroph. J.* **432**, L27.
Keller, H. B. (1987), *Numerical Methods in Bifurcation Problems* (Springer/Tata Inst. Fundam. Res., Berlin).
Kerr, R. P. (1963), *Phys. Rev. Lett.* **11**, 237.
Klebesadel, R., Strong, I. and Olson, R. (1973), *Astroph. J.* **182**, L85.
Kobayashi, S. & Mészáros, P. (2003), *Astroph. J.* **582**, L89.
Kobayashi, S. & Mészáros, P. (2003), *Astroph. J.* **582**, L89.
Kouveliotou, C., *et al.* (1993), *Astroph. J.* **413**, L101.
Kudritzki, R. (2003), Private communication (see http://www.ifa.hawaii.edu/pan-starrs).
Kulkarni, S. R. (2005), astro-ph/0510256.
Kundt, W. (1976), *Nature* **261**, 673.
LOFAR (see http://www.lofar.org).
LeBlanc, J. M. & Wilson, J. R. (1970), *Astroph. J.* **161**, 541.
Lee, C.-H., Brown, G. E. & Wijers, R. A. M. J. (2002), *Astroph. J.* **575**, 996.
Levinson, A. & van Putten, M. H. P. M. (1997), *Astroph. J.* **488**, 69.
Levinson, A. & Eichler, D. (2000), *Phys. Rev. Lett.* **85**, 236.
Levinson, A. & Eichler, D. (2003), *Astroph. J.* **594**, L19.
Levinson, A. (2004), *Astroph. J.* **608**, 411.
McFadyen, A. I. & Woosley, S. E. (1999), *Astroph. J.* **524**, 262.
Maund, J. R., Smartt, S. J., Kudritzki, R. P., Podsiadiowski, P. & Gilmore, G. F. (2004), *Nature* **427**, 129.
Mazets, E. P., Golenetskii, S. V. & Ilinskii, V. N. (1974), *JETP Lett.* **19**, 77.
Meegan, C. A., *et al.* (1992), *Nature* **355**, 143.
Mészáros, P. & Rees, M. J. (1993), *Astroph. J.* **405**, 278.
Mineshige, S., Takashi, H., Mami, M. & Matsumoto, R. (2002), *Publ. Astron. Soc. Japan* **54**, 655.
Mirabel, I. F. & Rodríguez, L. F. (1994), *Nature* **371**, 46.
Misner, C. W., Thorne, K. S. & Wheeler, J. A. (1973), *Gravitation* (W. H. Freeman, San Francisco).
Nakamura, T. & Fukugita, M. (1989), *Astroph. J.* **337**, 466.
Nomoto, K., Iwamoto, K. & Suzuki, T. (1995), *Phys. Rep.* **256**, 173.
O'Connel, R. F. (1972), *Phys. Rev.* **D 10**, 3035.
Okamoto, I. (1992), *Mon. Not. Roy. Astron. Soc.* **253**, 192.
Paciesas, W. S. (1999), *Astroph. J. Suppl.* **122**, 465.
Paczyński, B. P. (1991), *Acta. Astron.* **41**, 257.
Paczyński, B. P. (1998), *Astroph. J.* **494**, L45.
Papadopoulos, P. & Font, J. A. (2001), *Phys. Rev.* **D 63**, 044016.
Papaloizou, J. C. B. & Pringle, J. E. (1984), *Mon. Not. Roy. Astron. Soc.* **208**, 721.
Papapetrou, A. (1951a), *Proc. Roy. Soc. Lond.* **209**, 248.
Papapetrou, A. (1951b), *Proc. Roy. Soc. Lond.* **209**, 259.
Penrose, R. (1969), *Rev. Nuovo Cimento* **1**, 252.
Peters, P. C. & Mathews, J. (1963), *Phys. Rev.* **131**, 435.
Pierre Auger Collaboration (2007), *Science* **318**, 938.
Piran, T. (2004), *Rev. Mod. Phys.* **76**, 1143.

Pirani, F. A. E. (1956), *Acta Phys. Pol.* **15**, 389.
Piro, L. (2005), *Nature* **437**, 822.
Porquet, D., *et al.* (2004), *Astron. Astroph.* **407**, L17.
Press, W. H. & Teukolsky, S. A. (1972), *Astroph. J.* **178**, 347.
Rampp, M., Müller, E. & Ruffert, M. (1998), *Astron. Astroph.* **332**, 969.
Rees, M. J., Ruffini, R. & Wheeler, J. A. (1974), *Black Holes, Gravitational Waves and Cosmology: an Introduction to Current Research* (Gordon and Breach, New York), Section 7.
Rees, M. J. & Mészáros, P. (1992), *Mon. Not. Roy. Astron. Soc.* **258**, P41.
Rees, M. J. & Mészáros, P. (1994), *Astroph. J.* **430**, L93.
Reeves, J. N., Watson, D. & Osborne, J. P. (2002), *Nature* **416**, 512.
Ruffini, R. & Wilson, J. R. (1975), *Phys. Rev.* **D 12**, 2959.
Spallici, A. D. A. M., Aoudia, S., de Freitas Pacheco, J. A., *et al.* (2004), *Class. Quantum Grav.* **22**, S461.
Stanek, K. Z., *et al.* (2003), *Astroph. J.* **591**, L17.
Starobinsky, A. A. (1972), *Zh. Eksp. Teor. Fiz.* **64**, 48 [*Sov. Phys. JETP* **37**, 28 (1973)].
Tagliaferri, G., *et al.* (2005), *Astron. Astroph.* **443**, L1.
Taylor, J. H. (1994), *Rev. Mod. Phys.* **66**, 711.
Teukolsky, S. A. (1973), *Astroph. J.* **185**, 635.
Teukolsky, S. A. & Press, W. H. (1974), *Astroph. J.* **193**, 443.
Thorne, K. S., Price, R. H. & MacDonald, D. A. (1986), *Black Holes: the Membrane Paradigm* (Yale University Press, New Haven).
Turatto, M. (2003), *Supernovae and Gamma-ray Bursters*, ed. K. W. Weiler (Springer, Heidelberg), p. 21.
Unruh, W. G. (1974), *Phys. Rev.* **D 10**, 3194.
van Paradijs, J., *et al.* (1997), *Nature* **386**, 686.
van Putten, M. H. P. M. (1999), *Science* **284**, 115.
van Putten, M. H. P. M. (2000), *Phys. Rev. Lett.* **84**, 3752.
van Putten, M. H. P. M. & Ostriker, E. C. (2001), *Astroph. J.* **552**, L31.
van Putten, M. H. P. M. (2002), *Astroph. J.* **575**, L71.
van Putten, M. H. P. M. & Levinson, A. (2002), *Science* **294**, 1837.
van Putten, M. H. P. M. & Levinson, A. (2003), *Astroph. J.* **584**, 953.
van Putten, M. H. M. P. & Regimbau, T. (2003), *Astroph. J.* **593**, L15.
van Putten, M. H. P. M. (2004), *Astroph. J.* **611**, L81.
van Putten, M. H. P. M., Levinson, A., Lee, H.-K., Regimbau, T. & Harry, G. (2004), *Phys. Rev.* **D 69**, 044007.
van Putten, M. H. P. M. (2005), *Gravitational Radiation, Luminous Black Holes and Gamma-Ray Burst Supernovae* (Cambridge University Press, Cambridge).
van Putten, M. H. P. M. (2008a), *Astroph. J.* **684**, L91.
van Putten, M. H. P. M. (2008b), *Astroph. J. Lett.* **685**, to appear.
Véron-Cetty, M.-P. & Véron, P. (2006), *Astron. Astroph.* **455**, 773.
Villasenor, J. S., *et al.* (2005), *Nature* **437**, 855.
Wald, R. M. (1974), *Phys. Rev.* **D 10**, 1680.
Wheeler, J. C., Yi, I., Höflich, P. & Wang, L. (2000), *Astroph. J.* **537**, 810.
Willke, B., *et al.* (2002), *Class. Quantum Grav.* **19**, 1377.
Woosley, S. E. (1993), *Astroph. J.* **405**, 273.
Zatsepin, G. T. & Kuzmin, V. A. (1966), *Sov. Phys. JETP Lett.* **4**, 78.
Zel'dovich, Ya. B. (1971), *Zh. Eksp. Teor. Fiz.* **14**, 270 [*JETP Lett.* **14**, 180 (1971)].
Zhuge, X., Centrella, J. M. & McMillam, S. L. W. (1994), *Phys. Rev.* **D 50**, 6247.

# Part III

## Quantum gravity: Rotating black holes at the theoretical frontiers

# 11

# Horizon constraints and black hole entropy

Steve Carlip

## 11.1 Introduction

It has been more than thirty years since Bekenstein [1] and Hawking [2] first taught us that black holes are thermodynamic objects, with characteristic temperatures

$$T_H = \frac{\hbar \kappa}{2\pi c}, \tag{11.1}$$

and entropies

$$S_{BH} = \frac{A}{4\hbar G}, \tag{11.2}$$

where $\kappa$ is the surface gravity and $A$ is the area of the event horizon. Extensive experience with thermodynamics in less exotic settings encourages us to believe that these quantities should reflect some kind of underlying statistical mechanics. The Bekenstein–Hawking entropy (11.2), for example, should count the number of microscopic states of the black hole. But by Wheeler's famous dictum, "a black hole has no hair": a classical, equilibrium black hole is determined completely by its mass, charge, and angular momentum, with no room for additional microscopic states to account for thermal behavior.

If black hole thermodynamics has a statistical mechanical origin, the relevant states must therefore be nonclassical. Indeed, they should be quantum gravitational – the Hawking temperature and Bekenstein–Hawking entropy depend upon both Planck's constant $\hbar$ and Newton's constant $G$. Thus the problem of black hole statistical mechanics is not just a technical question about some particular configurations of matter and gravitational fields; if we are lucky, it may give us new insight into the profound mysteries of quantum gravity.

The Kerr Spacetime: Rotating Black Holes in General Relativity. Ed. David L. Wiltshire, Matt Visser and Susan M. Scott. Published by Cambridge University Press. 

This chapter will focus on an attempt to find a "universal" description of black hole statistical mechanics, one that involves quantum gravity but does not depend on fine details of any particular model of quantum gravity. This work is incomplete and tentative, and might ultimately prove to be wrong. But even if it is wrong, my hope is that we will learn something of value along the way.

## 11.2 Black hole entropy and the problem of universality

Ten years ago, the question of what microstates were responsible for black hole thermodynamics would have met with an almost unanimous answer: "We don't know." There were interesting ideas afloat, involving entanglement entropy [3] (the entropy coming from correlations between states inside and outside the horizon) and entropy of an "atmosphere" of external fields near the horizon [4], but we had nothing close to a complete description.

Today, by contrast, we suffer an embarrassment of riches. We have many candidates for the microscopic states of a black hole, all different, but all apparently giving the same result [5]. In particular, black hole entropy may count:

- Weakly coupled string and D-brane states [6, 7]: black holes can be constructed in semiclassical string theory as bound states of strings and higher dimensional D-branes. For supersymmetric (BPS) configurations, thermodynamic properties computed at weak couplings are protected by symmetries as one dials couplings up to realistic values, so weakly coupled states can be counted to determine the entropy.
- Nonsingular geometries [8]: The weakly coupled D-brane excitations that account for black hole entropy in string theory may correspond to certain nonsingular, horizonless geometries; a typical black hole state would then be a "fuzzball" superposition of such states.
- States in a dual conformal field theory "at infinity" [9]: for black holes whose near-horizon geometry looks like anti-de Sitter space, AdS/CFT duality can be used to translate questions about thermodynamics to questions in a lower dimensional dual, nongravitational conformal field theory.
- Spin network states crossing the horizon [10]: in loop quantum gravity, one can isolate a boundary field theory at the horizon and relate its states to the states of spin networks that "puncture" the horizon. The entropy apparently depends on one undetermined parameter, but once that parameter is fixed for one type of black hole, the approach yields the correct entropy for a wide range of other black holes.
- "Heavy" degrees of freedom in induced gravity [11]: as Sakharov first suggested [12], the Einstein–Hilbert action can be induced in a theory with no gravitational action by integrating out heavy matter fields. The Bekenstein–Hawking entropy can then be computed by counting these underlying massive degrees of freedom.
- No local states [13]: in the Euclidean path integral approach, black hole thermodynamics is determined by global topological features of spacetime rather than any local properties

of the horizon. Perhaps no localized degrees of freedom are required to account for black hole entropy.

- Nongravitational states: Hawking's original calculation of black hole radiation was based on quantum field theory in a fixed, classical black hole background. Perhaps the true degrees of freedom are not gravitational at all, but represent "entanglement entropy" [3] or the states of matter near the horizon [4].

None of these pictures has yet given us a complete model for black hole thermodynamics. But each can be used to count states for at least one class of black holes, and within its realm of applicability, each seems to give the correct Bekenstein–Hawking entropy (11.2). In an open field of investigation, the existence of competing explanations may be seen as a sign of health. But the existence of competing explanations that all *agree* is also, presumably, a sign that we are missing some deeper underlying structure.

This problem of universality occurs even within particular approaches to black hole statistical mechanics. In string theory, for example, one does not typically relate black hole entropy directly to horizon area. Rather, one constructs a particular semiclassical black hole as an assemblage of strings and D-branes; computes the entropy at weak coupling as a function of various charges; and then, separately, computes the horizon area as a function of those same charges. The results agree with the Bekenstein–Hawking formula (11.2), but the agreement has to be checked case by case. We may "know" that the next case will agree as well; but in a deep sense, we do not know why.

A simple approach to this problem is to note that black hole temperature and entropy can be computed semiclassically, so any quantum theory of gravity that has the right classical limit will have to give the "right" answer. But while this may be true, it does not really address the fundamental question: how is it that the classical theory enforces these restrictions on a quantum theory? In ordinary thermodynamics, we can determine entropy classically, by computing the volume of the relevant region of phase space; the correspondence principle then ensures that the quantum mechanical answer will be the same to lowest order. For black holes, no such calculation seems possible: once the mass, charge, and spin are fixed, there is no classical phase space left. Once again, we are missing a vital piece of the puzzle.

## 11.3 Symmetries and state-counting: conformal field theory and the Cardy formula

One possibility is that the missing piece is a classical symmetry of black hole spacetimes. Such a symmetry would be inherited by any quantum theory, regardless of the details of the quantization. At first sight, this explanation seems unlikely:

we are not used to the idea that a symmetry can be strong enough to determine such detailed properties of a quantum theory as the density of states. In at least one instance, though, this is known to happen.

Consider a two-dimensional conformal field theory, defined initially on the complex plane with coordinates $(z, \bar{z})$. The holomorphic diffeomorphisms $z \to f(z)$, $\bar{z} \to \bar{f}(\bar{z})$ are symmetries of such a theory. Denote by $L_n^{cl}$ and $\bar{L}_n^{cl}$ the generators of the transformations

$$z \to z + \epsilon z^{n+1}, \qquad \bar{z} \to \bar{z} + \epsilon \bar{z}^{n+1} \tag{11.3}$$

(the superscript $cl$ means "classical"). The Poisson brackets of these generators are almost uniquely determined by the symmetry: they form a Virasoro algebra,

$$\begin{aligned}
\left\{L_m^{cl}, L_n^{cl}\right\} &= i(n-m)L_{m+n}^{cl} + \frac{ic^{cl}}{12}n(n^2-1)\delta_{m+n,0} \\
\left\{\bar{L}_m^{cl}, \bar{L}_n^{cl}\right\} &= i(n-m)\bar{L}_{m+n}^{cl} + \frac{ic^{cl}}{12}n(n^2-1)\delta_{m+n,0} \\
\left\{L_m^{cl}, \bar{L}_n^{cl}\right\} &= 0,
\end{aligned} \tag{11.4}$$

where $c^{cl}$ is a constant, the central charge [14].

When $c^{cl} = 0$, equation (11.4) is just a representation of the ordinary algebra of holomorphic vector fields,

$$\left[z^{m+1}\frac{d}{dz}, z^{n+1}\frac{d}{dz}\right] = (n-m)z^{m+n+1}\frac{d}{dz}. \tag{11.5}$$

Up to field redefinitions, the Virasoro algebra is the only central extension of this algebra. The central charge, $c$, commonly appears quantum mechanically, arising from operator re-orderings. But it can occur classically as well. There, it appears because the canonical generators are unique only up to the addition of constants; a central charge represents a nontrivial cocycle [15], that is, a set of constants that cannot be removed by field redefinitions.

In 1986, Cardy found a remarkable property of such conformal field theories [16, 17]. Let $\Delta_0$ be the smallest eigenvalue of $L_0$ in the spectrum, and define an "effective central charge"

$$c_{\mathit{eff}} = c - 24\Delta_0. \tag{11.6}$$

Then for large $\Delta$, the density of states with eigenvalue $\Delta$ of $L_0$ has the asymptotic form

$$\rho(\Delta) \sim \exp\left\{2\pi\sqrt{\frac{c_{\mathit{eff}}\Delta}{6}}\right\}\rho(\Delta_0), \tag{11.7}$$

independent of any other details of the theory. The asymptotic behavior of the density of states is thus determined by a few features of the symmetry, the central charge $c$ and the ground state conformal weight $\Delta_0$. In particular, theories with very different field contents can have exactly the same asymptotic density of states.

A careful proof of this result, using the method of steepest descents, is given in [18]. One can derive the logarithmic corrections to the entropy by the same methods [19]; indeed, by exploiting results from the theory of modular forms, one can obtain even higher order corrections [20, 21]. But although the mathematical derivation of the Cardy formula is relatively straightforward, I do not know of any good, intuitive *physical* explanation for (11.7). Standard derivations rely on a duality between high and low temperatures, which arises because of modular invariance: by interchanging cycles on a torus, one can trade a system on a circle of circumference $L$ with inverse temperature $\beta$ for a system on a circle of circumference $\beta$ with inverse temperature $L$. Such a transformation relates states at high "energies" $\Delta$ to the ground state, leading ultimately to Cardy's result. But it would be valuable to have a more direct understanding of why the density of states should be so strongly constrained by symmetry.

Note that upon quantization, after making the usual substitutions

$$\{\,,\,\} \to [\,,\,]/i\hbar \quad \text{and} \quad L^{cl}_m \to L_m/\hbar, \tag{11.8}$$

the Virasoro algebra (11.4) becomes

$$\begin{aligned}
&[L_m, L_n] = (m-n)L_{m+n} + \frac{c^{cl}}{12\hbar}m(m^2-1)\delta_{m+n,0} \\
&[\bar{L}_m, \bar{L}_n] = (m-n)\bar{L}_{m+n} + \frac{c^{cl}}{12\hbar}m(m^2-1)\delta_{m+n,0} \\
&[L_m, \bar{L}_n] = 0.
\end{aligned} \tag{11.9}$$

A nonvanishing classical central charge $c^{cl}$ thus contributes $c^{cl}/\hbar$ to the quantum central charge, and a classical conformal "charge" $\Delta^{cl}$ gives a quantum conformal weight $\Delta^{cl}/\hbar$. Hence the classical contribution to the entropy $\log\rho(\Delta)$ coming from (11.7) goes as $1/\hbar$, matching the behavior of the Bekenstein–Hawking entropy (11.2) and giving us a first hint that an approach of this sort might be productive.

The black holes we are interested in are not two dimensional, of course, and despite some interesting speculation [22], there is no proven higher dimensional analog to the Cardy formula. But there is reason to hope that the two-dimensional result (11.7) might be relevant to the near-horizon region of an arbitrary black hole. For instance, it is known that near a horizon, matter can be described by a two-dimensional conformal field theory [23, 24]. Indeed, in "tortoise coordinates," the near-horizon metric in any dimension becomes

$$ds^2 = N^2(dt^2 - dr_*{}^2) + ds_\perp{}^2, \tag{11.10}$$

where the lapse function $N$ goes to zero at the horizon. The Klein–Gordon equation then reduces to

$$(\Box - m^2)\varphi = \frac{1}{N^2}(\partial_t^2 - \partial_{r_*}^2)\varphi + \mathcal{O}(1) = 0. \tag{11.11}$$

The mass and transverse excitations become negligible near the horizon: they are essentially redshifted away relative to excitations in the $r_*$-$t$ plane, leaving an effective two-dimensional conformal field theory. A similar reduction occurs for spinor and vector fields. Moreover, Jacobson and Kang have observed that the surface gravity and temperature of a stationary black hole are conformally invariant [25], and Medved *et al.* have recently shown that a generic stationary black hole metric has an approximate conformal symmetry near the horizon [26, 27].

## 11.4 The BTZ black hole

The first concrete evidence that conformal symmetry can determine black hole thermodynamics came from studying the (2+1)-dimensional black hole of Bañados, Teitelboim, and Zanelli [28–30]. A solution of the vacuum Einstein equations with a negative cosmological constant $\Lambda = -1/\ell^2$, the BTZ black hole is the (2+1)-dimensional analog of the Kerr–AdS metric. In Boyer–Lindquist-like coordinates, the BTZ metric takes the form

$$\begin{aligned} ds^2 &= N^2\, dt^2 - N^{-2}\, dr^2 - r^2(d\phi + N^\phi dt)^2 \\ \text{with } N &= \left(-8GM + \frac{r^2}{\ell^2} + \frac{16G^2J^2}{r^2}\right)^{1/2}, \quad N^\phi = -\frac{4GJ}{r^2}, \end{aligned} \tag{11.12}$$

where $M$ and $J$ are the anti-de Sitter analogs of the ADM mass and angular momentum. As in 3+1 dimensions, the apparent singularities at $N = 0$ are merely coordinate singularities, and an analog of Kruskal–Szekeres coordinates can be found.

Like all vacuum spacetimes in 2+1 dimensions, the BTZ geometry has the peculiar feature of having constant curvature, and can be in fact expressed as a quotient of anti-de Sitter space by a discrete group of isometries. Nevertheless, it is a genuine black hole:

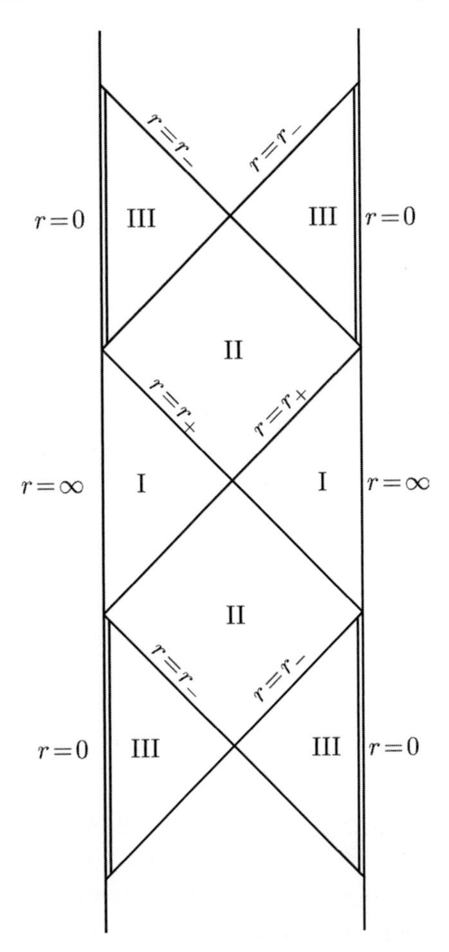


Fig. 11.1. The Carter–Penrose diagram for a nonextremal BTZ black hole.

- It has a true event horizon at $r = r_+$ and, if $J \neq 0$, an inner Cauchy horizon at $r = r_-$, where

$$r_\pm^2 = 4GM\ell^2 \left\{ 1 \pm \left[ 1 - \left( \frac{J}{M\ell} \right)^2 \right]^{1/2} \right\}; \tag{11.13}$$

- it occurs as the end point of gravitational collapse of matter;
- its Carter–Penrose diagram, Fig. 11.1, is essentially the same as that of an ordinary Kerr–AdS black hole;
- and, most important for our purposes, it exhibits standard black hole thermodynamics, with a temperature and entropy given by (11.1) and (11.2), where the horizon "area" is the circumference $A = 2\pi r_+$.

The thermodynamic character of the BTZ black hole may be verified in much the same way that it is for the Kerr black hole: by looking at quantum field theory in a BTZ background [31, 32]; by examining the Euclidean path integral [33] and the Brown–York microcanonical path integral [34]; by appealing to Wald's Noether charge approach [30, 35]; and by investigating tunneling through the horizon [36, 37]. In 2+1 dimensions, a powerful new method is also available [38]: one can consider quantum gravitational perturbations induced by a classical scalar source, and then use detailed balance arguments to obtain thermodynamics. The quantitative agreement of all of these approaches gives us confidence that the thermal properties are real.

We now come to a deep mystery. Vacuum general relativity in 2+1 dimensions has no dynamical degrees of freedom. This is most easily seen by a simple counting argument – in the canonical formalism, the field is described by a spatial metric (three degrees of freedom per point) and its canonical momentum (three more degrees of freedom per point), but we also have three constraints that restrict initial values and three arbitrary coordinate choices, leaving $6 - 6 = 0$ dynamical degrees of freedom. The same conclusion may be reached by noting that the (2+1)-dimensional curvature tensor is algebraically determined by the Einstein tensor,

$$G^{\mu}{}_{\nu} = -\frac{1}{4}\epsilon^{\mu\pi\rho}\,\epsilon_{\nu\sigma\tau}\,R^{\sigma\tau}{}_{\pi\rho}, \tag{11.14}$$

so a vacuum spacetime necessarily has constant curvature. While the theory admits a few global "topological" excitations [39], there is no local dynamics. Where, then, can the degrees of freedom responsible for thermal behavior come from?

One piece of the answer, discovered independently by Strominger [40] and Birmingham, Sachs, and Sen [41], can be found by looking at boundary conditions at infinity. The conformal boundary of a (2+1)-dimensional asymptotically anti-de Sitter spacetime is a cylinder, so it is not surprising that the asymptotic symmetries of the BTZ black hole are described by a Virasoro algebra (11.4). It is a bit more surprising that this algebra has a central extension, but as Brown and Henneaux showed in 1986 [42], the classical central charge, computed from the standard ADM constraint algebra, is nonzero:

$$c = \frac{3\ell}{2G}. \tag{11.15}$$

Confirmation of this result has come from a path integral analysis [43], from investigating the constraint algebra in the Chern–Simons formalism [44], and from examining the conformal anomaly of the boundary stress-energy tensor [45, 46].

Moreover, the classical Virasoro "charges" $L_0$ and $\bar{L}_0$ can be computed within ordinary canonical general relativity, employing the same techniques that are used to determine the ADM mass [42]. Indeed, in this context the zero-modes of the diffeomorphisms (11.3) are simply linear combinations of time translations and rotations, and the corresponding conserved quantities are linear combinations of the ordinary ADM mass and angular momentum. For the BTZ black hole, in particular, one finds

$$\Delta = \frac{1}{16G\ell}(r_+ + r_-)^2, \quad \bar{\Delta} = \frac{1}{16G\ell}(r_+ - r_-)^2. \tag{11.16}$$

By the Cardy formula (11.7), with the added assumption that $\Delta_0$ and $\bar{\Delta}_0$ are small, one then obtains an entropy

$$S = \log \rho \sim \frac{2\pi}{8G}(r_+ + r_-) + \frac{2\pi}{8G}(r_+ - r_-) = \frac{2\pi r_+}{4G}, \tag{11.17}$$

in precise agreement with the Bekenstein–Hawking entropy.

This argument is incomplete, of course: it tells us that the entropy is related to symmetries and boundary conditions at infinity, but does not explain the underlying quantum degrees of freedom. I will argue later that this is a *good* feature, since it allows us to explain the "universality" described in Section 11.2. For the BTZ black hole, though, we can go a bit further.

As first observed by Achucarro and Townsend [47] and subsequently extensively developed by Witten [48, 49], vacuum Einstein gravity in 2+1 dimensions with a negative cosmological constant is equivalent to a Chern–Simons gauge theory, with a gauge group $SL(2, \mathbb{R}) \times SL(2, \mathbb{R})$. On a compact manifold, a Chern–Simons theory is a "topological field theory", described by a finite number of global degrees of freedom. On a manifold with boundary, however, boundary conditions can partially break the gauge invariance. As a consequence, field configurations that would ordinarily be considered gauge-equivalent become distinct at the boundary. In a manner reminiscent of the Goldstone mechanism [50], new "would-be pure gauge" degrees of freedom appear, providing new dynamical degrees of freedom.

For a Chern–Simons theory, the resulting induced boundary dynamics can be described by a Wess–Zumino–Witten model [51, 52]. For (2+1)-dimensional gravity, slightly stronger boundary conditions can further reduce the boundary theory to Liouville theory [53, 54], a result that may also be obtained directly in the metric formalism [55–58]. Whether the resulting degrees of freedom reproduce the Bekenstein–Hawking entropy remains an open question – for a recent review, see [59] – but Chen has found strong hints that a better understanding of Liouville theory might allow an explicit microscopic description of BTZ black hole thermodynamics [60].

## 11.5 Horizon constraints

The BTZ black hole offers a test case for the hypothesis that black hole entropy might be controlled by an underlying classical symmetry. But it is clearly not good enough. To start with, the computations described in the preceding section relied heavily on a very particular feature of (2+1)-dimensional asymptotically anti-de Sitter spacetime, the fact that the "boundary" at which asymptotic diffeomorphisms are defined is a two-dimensional timelike surface. A few other black holes have a similar character: the near-extremal black holes considered in string theory often have a near-horizon geometry that looks like that of a BTZ black hole, and two-dimensional methods can be used to obtain their entropy [40, 61]. But for the generic case, there is no reason to expect such a nice structure.

Moreover, the standard computations of BTZ black hole entropy use conformal symmetries at infinity. For a single, isolated black hole in 2+1 dimensions, this choice is probably harmless: there are no propagating degrees of freedom between the horizon and infinity, so it may not matter where we count the states. But even in 2+1 dimensions, the interpretation becomes unclear when there is more than one black hole present, or when the black hole is replaced by a "star" for which the BTZ solution is only the exterior geometry. If we wish to isolate the microscopic states of a particular black hole, it will be very difficult to do so using only symmetries at conformal infinity; we should presumably be looking near the horizon instead.

We are thus left with several hints. We should

- look for "broken gauge invariance" to provide new degrees of freedom;
- hope for an effective two-dimensional picture, which would allow us to use the Cardy formula;
- but look near horizon.

Before proceeding, though, we need to take a step back and ask a more general question. We want to investigate the statistical mechanics of a black hole. But how, exactly, do we tell that a black hole is present in the context of a fully quantum mechanical theory of gravity?

This question is frequently overlooked, because in the usual semiclassical approaches to black hole thermodynamics the answer is obvious: we fix a definite black hole background and then ask about quantum fields and gravitational fluctuations in that background. But in a full quantum theory of gravity, we cannot do that: there is no fixed background, the geometry is quantized, and the uncertainty principle prevents us from exactly specifying the metric. In general, it becomes difficult to tell whether a black hole is present or not. At best, we can ask a *conditional* question: "If a black hole with characteristic X is present, what is the probability of phenomenon Y?"

There are two obvious ways to impose a suitable condition to give such a conditional probability. One, discussed in [62], is to treat the horizon as a "boundary" and impose appropriate black hole boundary conditions. The horizon is not, of course, a true boundary: a falling observer can cross a horizon without seeing anything special happen, and certainly doesn't drop off the edge of the Universe. Nonetheless, we can ensure the presence of a black hole by specifying "boundary conditions" at a horizon. In a path integral formulation, for example, we can divide the manifold into two pieces along a hypersurface $\Sigma$ and perform separate path integrals over the fields on each piece, with fields restricted at the "boundary" by the requirement that $\Sigma$ be a suitable black hole horizon. This kind of split path integral has been studied in detail in 2+1 dimensions [63], where it yields the correct counting for the boundary degrees of freedom.

Alternatively, we can impose "horizon constraints" directly, either classically or in the quantum theory. We might, for example, construct an operator $\vartheta$ representing the expansion of a particular null surface, and restrict ourselves to states annihilated by $\vartheta$. As we shall see below, such a restriction can alter the algebra of diffeomorphisms, allowing us to exploit the Cardy formula.

The "horizon as boundary" approach has been explored by a number of authors; see, for instance, [64–72]. A conformal symmetry in the $r$–$t$ plane appears naturally, and one can obtain a Virasoro algebra with a central charge that leads to the correct Bekenstein–Hawking entropy. But the diffeomorphisms whose algebra yields that central charge – essentially, the diffeomorphisms that leave the lapse function invariant – are generated by vector fields that blow up at the horizon [73–75], and it is not clear whether they should be permitted in the theory. A closely related approach looks for an approximate conformal symmetry in the neighborhood of the horizon [76–79]. Again, one finds a Virasoro algebra with a central charge that apparently leads to the correct Bekenstein–Hawking entropy.

The "horizon constraint" approach is much newer, and is not yet fully developed. Let us now examine it further. Suppose we wish to constrain our theory of gravity by requiring that some surface $\Sigma$ be the horizon of a black hole. We must first decide exactly what we mean by a "horizon." Demanding that $\Sigma$ be a true event horizon seems impractical: an event horizon is determined globally, and requires that we know the entire future development of the spacetime. The most promising alternative is probably offered by the "isolated horizon" program [80]. An isolated horizon is essentially a null surface with vanishing expansion, with a few added technical conditions. Such a horizon shares many of the fundamental features of an event horizon [81] – in particular, it leads to standard black hole thermodynamics – and seems to do a good job of capturing the idea of a "local" horizon.

Isolated horizon constraints are constraints on the allowed data on a hypersurface, and in principle we should be able to use the well-developed apparatus

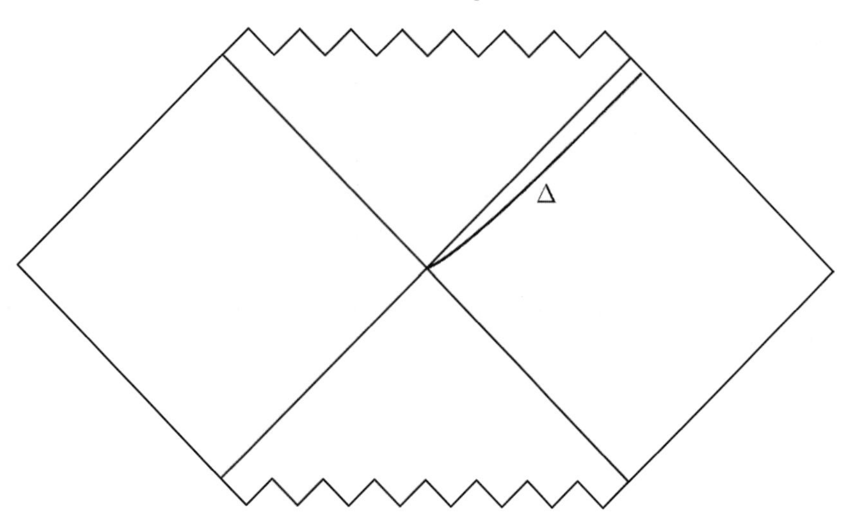


Fig. 11.2. A spacelike "stretched horizon" $\Delta$.

of constrained Hamiltonian dynamics [82–84] to study such conditions. Unfortunately, though, an isolated horizon is by definition a null surface, and we would require an approach akin to light cone quantization. Light cone quantization of gravity is difficult, and at this writing, such an analysis of horizon constraints has not been carried out.

We can much more easily impose constraints requiring the presence of a spacelike "stretched horizon" that becomes nearly null, as illustrated in Fig. 11.2. On such a hypersurface, it is possible to employ standard methods of constrained dynamics. As we shall see below, at least for the relatively simple model of two-dimensional dilaton gravity, the horizon constraints lead to a Virasoro algebra with a calculable central charge, allowing us to use the Cardy formula to obtain the correct Bekenstein–Hawking entropy.

## 11.6 Dilaton black holes and the Bekenstein–Hawking entropy

As noted in Section 11.3, it is plausible that the thermodynamic properties of a black hole are determined by dynamics in the "$r$–$t$ plane." As a first step, we can therefore look at a dimensionally reduced theory. The full Einstein–Hilbert action in any dimension may be written exactly as an action for a two-dimensional theory: standard Kaluza–Klein techniques allow us to express the effect of "extra dimensions" in terms of scalar and gauge fields, albeit with an enormous gauge group [85]. Near a black hole horizon, though, most of these fields become negligible, and one obtains a simple two-dimensional theory, dilaton gravity [86, 87], with an action of the form

$$I = \frac{1}{16\pi G}\int d^2x\sqrt{-g}\,[AR + V(A)]\,. \tag{11.18}$$

Here, $R$ is the two-dimensional scalar curvature, while $A$ is a scalar field, the dilaton (often denoted $\varphi$). $V(A)$ is a potential whose detailed form depends on the higher dimensional theory we started with; we will not need an exact expression. As the notation suggests, $A$ is the transverse area in the higher dimensional theory, and the expansion – the fractional rate of change of the transverse area along a null curve with null normal $l^a$ – becomes

$$\vartheta = l^a\nabla_a A/A. \tag{11.19}$$

It is useful to reexpress the action (11.18) in terms of a null dyad $(l^a, n^a)$ with $l^2 = n^2 = 0$, $l \cdot n = -1$. These determine "surface gravities" $\kappa$ and $\bar{\kappa}$, defined by the conditions

$$\begin{aligned} \nabla_a l_b &= -\kappa n_a l_b - \bar{\kappa} l_a l_b \\ \nabla_a n_b &= \kappa n_a n_b + \bar{\kappa} l_a n_b. \end{aligned} \tag{11.20}$$

By an easy computation, the action (11.18) becomes

$$I = \int d^2x\left[\hat{\epsilon}^{ab}\left(2\kappa n_b\partial_a A - 2\bar{\kappa} l_b\partial_a A\right) + \sqrt{-g}V\right], \tag{11.21}$$

where $\hat{\epsilon}^{ab}$ is the two-dimension Levi–Civita density and I have adopted units such that $16\pi G = 1$. If we now define components of our dyad with respect to coordinates $(u, v)$,

$$l = \sigma\ du + \alpha\ dv, \qquad n = \beta\ du + \tau\ dv, \tag{11.22}$$

it is straightforward to find the Hamiltonian form of the action. Details may be found in [88]. The key feature is that the system contains three first-class constraints,

$$\begin{aligned} C_\perp &= \pi_\alpha{}' - \frac{1}{2}\pi_\alpha\pi_A - \tau V(A) \\ C_\parallel &= \pi_A A' - \alpha\pi_\alpha{}' - \tau\pi_\tau{}' \\ C_\pi &= \tau\pi_\tau - \alpha\pi_\alpha + 2A', \end{aligned} \tag{11.23}$$

where a prime a derivative with respect to $v$ and $\pi_X$ is the momentum conjugate to the field $X$. $C_\perp$ and $C_\parallel$ are ordinary Hamiltonian and momentum constraints of general relativity, that is, the canonical versions of the generators of diffeomorphisms; $C_\pi$ is a disguised version of the generator of local Lorentz invariance, appearing because the pair $\{l, n\}$ is invariant under the boost $l \to fl$, $n \to f^{-1}n$.

We can now impose our "stretched horizon" constraints at the surface $\Sigma$ defined by the condition $u = 0$. We first demand that $\Sigma$ be "almost null," i.e. that its normal be nearly equal to the null vector $l^a$. By (11.22), this requires that $\alpha = \epsilon_1 \ll 1$.

We next demand that $\Sigma$ be "almost nonexpanding," that is, that the expansion $\vartheta$ be "almost zero" on $\Sigma$. This condition is slightly more subtle, since the absolute scale of $l^a$ is not fixed. While the restriction $\vartheta = 0$ is independent of this scale, a restriction of the form $\vartheta \ll 1$ clearly is not. Fortunately, though, the surface gravity $\kappa$ scales identically under constant rescalings of $l^a$, so we can consistently require that $l^v \nabla_v A / \kappa A = \epsilon_2 \ll 1$. Rewriting these conditions in terms of canonical variables, we obtain two constraints:

$$\begin{aligned} K_1 &= \alpha - \epsilon_1 = 0 \\ K_2 &= A' - \frac{1}{2}\epsilon_2 A_+ \pi_A + \frac{a}{2} C_\pi = 0, \end{aligned} \tag{11.24}$$

where $A_+$ is the value of the dilaton at the horizon. The term proportional to $C_\pi$ in $K_2$ is not necessary, but has been added for later convenience. By looking at a generic exact solution, one can verify that for $\epsilon_2 < 0$, these constraints do, in fact, define a spacelike stretched horizon of the type illustrated in Fig. 11.2.

$K_1$ and $K_2$ are not quite "constraints" in the ordinary sense of constrained Hamiltonian dynamics. In particular, they restrict allowable data only on the stretched horizon $\Sigma$, and cannot be imposed elsewhere. Nevertheless, they are similar enough to conventional constraints that many existing techniques can be used. In particular, note that the $K_i$ have nontrivial brackets with the momentum and boost generators $C_\parallel$ and $C_\pi$, so these can no longer be considered generators of invariances of the constrained theory.[1] But we can fix this in a manner suggested by Bergmann and Komar many years ago [84]: we define new generators

$$\begin{aligned} C_\parallel &\to C_\parallel^* = C_\parallel + a_1 K_1 + a_2 K_2 \\ C_\pi &\to C_\pi^* = C_\pi + b_1 K_1 + b_2 K_2 \end{aligned} \tag{11.25}$$

with coefficients $a_i$ and $b_i$ chosen so that $\{C^*, K_i\} = 0$. Since the $K_i$ vanish on admissible geometries – those for which our initial surface is a suitable stretched horizon – the constraints $C^*$ are physically equivalent to the original $C$; but they now preserve the horizon constraints as well.

[1] The $K_i$ have nontrivial brackets with $C_\perp$ as well, but that is not a problem: $C_\perp$ generates diffeomorphisms that move us off the initial surface $\Sigma$, and we should not expect the horizon constraints to be preserved.

We now make the crucial observation that the redefinitions (11.25) affect the Poisson brackets of the constraints. With the choice $a = -2$ in (11.24), it is not hard to check that

$$\begin{aligned}
\{C^*_\parallel[\xi], C^*_\parallel[\eta]\} &= -C^*_\parallel[\xi\eta' - \eta\xi'] + \frac{1}{2}\epsilon_2 A_+ \int dv(\xi'\eta'' - \eta'\xi'') \\
\{C^*_\parallel[\xi], C^*_\pi[\eta]\} &= -C^*_\pi[\xi\eta'] \\
\{C^*_\pi[\xi], C^*_\pi[\eta]\} &= -\frac{1}{2}\epsilon_2 A_+ \int dv(\xi\eta' - \eta\xi')
\end{aligned} \tag{11.26}$$

where $C[\xi]$ means $\int dv\, \xi C$. The algebra (11.26) has a simple conformal field theoretical interpretation [14]: the $C^*_\parallel$ generate a Virasoro algebra with central charge

$$\frac{c}{48\pi} = -\frac{1}{2}\epsilon_2 A_+, \tag{11.27}$$

while $C^*_\pi$ is an ordinary primary field of weight one.

To take advantage of the Cardy formula (11.7), the central charge (11.27) is not enough; we also need the classical Virasoro "charge" $\Delta$. As in conventional approaches to black hole mechanics, this charge comes from the contribution of a boundary term that must be added to make the Virasoro generator $C^*_\parallel$ "differentiable" [89]. Under a variation of the fields, the momentum constraint (11.23) picks up a boundary term from integration by parts,

$$\delta C_\parallel[\xi] = \cdots + \xi\pi_A \delta A|_{v=v_+}, \tag{11.28}$$

at the boundary $v = v_+$. For Poisson brackets with $C_\parallel$ to be well-defined, this term must be canceled. We therefore add a boundary term to $C_\parallel$,

$$C^*_{\parallel\, bdry}[\xi] = -\, \xi\pi_A A|_{v=v_+}, \tag{11.29}$$

which will give a nonvanishing classical contribution to $\Delta$.

We also need a "mode expansion" to define the Fourier component $L_0$, or, equivalently, a normalization for the "constant translation" $\xi_0$. For a conformal field theory defined on a circle, or on a full complex plane with a natural complex coordinate, the mode expansion (11.3) is essentially unique. Here, though, it is not so obvious how to choose a coordinate $z$. As argued in [77], however, there is one particularly natural choice,

$$z = e^{2\pi i A/A_+}, \qquad \xi_n = \frac{A_+}{2\pi A'} z^n, \tag{11.30}$$

where the normalization is chosen so that $[\xi_m, \xi_n] = i(n-m)\xi_{m+n}$.

Equation (11.29) then implies that

$$\Delta = C^*_{\parallel\, bdry}[\xi_0] = -\frac{A_+}{2\pi A'}\pi_A A_+ = -\frac{A_+}{\pi\epsilon_2}, \tag{11.31}$$

where I have used the constraint $K_2 = 0$ to eliminate $\pi_A$ in the last equality. Inserting (11.27) and (11.31) into the Cardy formula, assuming that $\Delta_0$ is small, and restoring the factors of $16\pi G$ and $\hbar$, we obtain an entropy

$$S = \frac{2\pi}{16\pi G}\sqrt{\left(-\frac{24\pi\epsilon_2 A_+}{6\hbar}\right)\left(-\frac{A_+}{\pi\epsilon_2\hbar}\right)} = \frac{A_+}{4\hbar G}, \tag{11.32}$$

exactly reproducing the Bekenstein–Hawking entropy (11.2).

## 11.7 What are the states?

I argued in Section 11.2 that one of the main strengths of a symmetry-based derivation is that it is "universal," that is, that it does not depend on the details of a quantum theory of gravity. Nevertheless, such a derivation does allow us to say *something* about the relevant states.

Standard approaches to canonical gravity require that physical states satisfy the condition

$$C_\parallel|\text{phys}\rangle = C_\pi|\text{phys}\rangle = 0, \tag{11.33}$$

that is, that they be annihilated by the constraints. But the condition (11.33) is not consistent with a Virasoro algebra with nonvanishing central charge: schematically,

$$[C^*, C^*]|\text{phys}\rangle \sim C^*|\text{phys}\rangle + \text{const.}|\text{phys}\rangle \neq 0. \tag{11.34}$$

We must therefore weaken the physical state condition, for example by requiring only that

$$C^{*(+)}_\parallel|\text{phys}\rangle = 0 \tag{11.35}$$

where $C^{*(+)}_\parallel$ is the positive-frequency component of the momentum constraint.

It is well known in conformal field theory that such a loosening of the constraints leads to a collection of new "descendant" states $L_{-n}|\text{phys}\rangle$, which would be excluded by the stronger constraint (11.33). This phenomenon is closely analogous to the appearance of "boundary states" for the BTZ black hole, as discussed in Section 11.4. By relaxing the physical state constraints, we have allowed states that would formerly have been considered to be gauge-equivalent to differ physically, thus introducing a new set of "would-be pure gauge" states into our state-counting. It is an interesting open question whether the "Goldstone-like" description of Section 11.4 can be extended to this setting.

## 11.8 Where do we go from here?

The "horizon constraint" program described in Section 11.6 seems promising. But it is clear that important pieces are still missing. In particular:

- Neither the constraints (11.24) nor the mode choices (11.30) are unique. It is important to understand how sensitive the final results are to these choices. Some flexibility certainly exists – for example, it can be shown that the final expression for the black hole entropy does not depend on the parameter $a$ in (11.24) – but a good deal of unexplored freedom remains.
- At this writing, the analysis has been completed only for the two-dimensional dilaton black hole. While it may be argued that this case captures the essential features of an arbitrary black hole, an explicit check of this claim is clearly needed.
- While the final expression (11.32) for the entropy is well-behaved, and exists in the limit that the "stretched horizon" approaches the true horizon, several intermediate quantities – including the central charge $c$, the conformal weight $\Delta$, and the vector fields $\xi_n$ – behave badly at the horizon. Of course, the canonical approach used in Section 11.6 itself only makes sense on a spacelike stretched horizon, so this breakdown is not necessarily a sign of an underlying problem. But it would clearly be desirable to perform a similar analysis in light cone quantization on a true horizon.

This last problem is reminiscent of the "horizon as boundary" approach of, for instance, [65], in which the relevant diffeomorphisms are generated by vector fields that blow up at the horizon. If this proves to be a general feature of conformal symmetry methods, it could be telling us something about "black hole complementarity" [90]: perhaps the Bekenstein–Hawking entropy is only well-defined for an observer who remains outside the horizon.

Beyond these particular issues, several more general questions must be answered before the conformal symmetry program can be taken too seriously. First, we need to understand much more about the coupling of horizon degrees of freedom to matter. Black hole thermodynamics is, after all, more than the Bekenstein–Hawking entropy; one must also demonstrate that any putative horizon degrees of freedom couple in a way that explains Hawking radiation. In this matter, the string theory computations clearly have the lead, reproducing not only the Hawking temperature but the correct gray body factors as well [7, 91]. But there are a few hints that similar results can be obtained from more general conformal symmetries [38, 92].

Second, if the contention that near-horizon symmetry is "universal" is correct, then such a symmetry must be present – albeit, perhaps, hidden – in other derivations of black hole thermodynamics. There are a few cases in which this is known to be true: for example, certain string theoretical derivations of near-extremal black hole thermodynamics exploit the conformal symmetry of the near-horizon BTZ geometry [40, 61], and the induced gravity approach can be related to a two-dimensional conformal symmetry [93]. But huge gaps still remain. One interesting avenue would be to explore the Euclidean version of the horizon constraint program, perhaps allowing us to relate the symmetry-derived count of states more directly to the Euclidean path integral. For the Euclidean black hole, the "stretched horizon" constraints (11.24) have a nice geometric interpretation, defining a circle in imaginary time around the horizon with a proper radius proportional to $\epsilon_2$. It may be that some of the choices we have made in defining the constraints become more natural in such a setting.

Third, we will eventually have to move away from "isolated horizons" to consider dynamical black holes, with horizons that grow as matter falls in and shrink as Hawking radiation carries away energy. It is only in such a setting that the horizon constraint program will be able to analyze such crucial problems as the information loss paradox and the final fate of an evaporating black hole. Whether this will eventually be possible remains to be seen.

## Acknowledgments

This work was supported in part by the U.S. Department of Energy under grant DE-FG02-99ER40674.

## References

[1] Bekenstein, J. D. (1973), *Phys. Rev.* **D 7**, 2333.
[2] Hawking, S. W. (1974), *Nature* **248**, 30.
[3] Bombelli, L., Koul, R. K., Lee, J. & Sorkin, R. (1986), *Phys. Rev.* **D 34**, 373.
[4] 't Hooft, G. (1985), *Nucl. Phys.* **B 256**, 727.
[5] Wald, R. M. (2001), *Living Rev. Rel.* **4**, 6.
[6] Strominger, A. & Vafa, C. (1996), *Phys. Lett.* **B 379**, 99 [arXiv:hep-th/9601029].
[7] Peet, A. W. (1998), *Class. Quantum Grav.* **15**, 3291 [arXiv:hep-th/9712253].
[8] Mathur, S. D. (2005), *Fortsch. Phys.* **53**, 793 [arXiv:hep-th/0502050].
[9] Aharony, O., Gubser, S. S., Maldacena, J. M., Ooguri, H. & Oz, Y. (2000), *Phys. Rep.* **323**, 183 [arXiv:hep-th/9905111].
[10] Ashtekar, A., Baez, J., Corichi, A. & Krasnov, K. (1998), *Phys. Rev. Lett.* **80**, 904 [arXiv:gr-qc/9710007].
[11] Frolov, V. P. & Fursaev, D. V. (1997), *Phys. Rev.* **D 56**, 2212 [arXiv:hep-th/9703178].
[12] Sakharov, A. D. (1968), *Sov. Phys. Dokl.* **12**, 1040; reprinted in *Gen. Rel. Grav.* **32**, 365 (2000).
[13] Hawking, S. W. & Hunter, C. J. (1999), *Phys. Rev.* **D 59**, 044025 [arXiv:hep-th/9808085].

[14] Di Francesco, P., Mathieu, P. & Sénéchal, D. (1997), *Conformal Field Theory* (Springer, Berlin).
[15] Arnold, V. (1978), *Mathematical Methods of Classical Mechanics* (Springer, Berlin), appendix 5.
[16] Cardy, J. A. (1986), *Nucl. Phys.* **B 270**, 186.
[17] Blöte, H. W. J., Cardy, J. A. & Nightingale, M. P. (1986), *Phys. Rev. Lett.* **56**, 742.
[18] Carlip, S. (1998), *Class. Quantum Grav.* **15**, 3609 [arXiv:hep-th/9806026].
[19] Carlip, S. (2000), *Class. Quantum Grav.* **17**, 4175 [arXiv:gr-qc/0005017].
[20] Birmingham, D., Sachs, I. & Sen, S. (2001), *Int. Mod, J.. Phys.* **D 10**, 833 [arXiv:hep-th/0102155].
[21] Dijkgraaf, R., Maldacena, J. M., Moore, G. W. & Verlinde, E. [arXiv:hep-th/0005003].
[22] Verlinde, E., arXiv:hep-th/0008140.
[23] Birmingham, D., Gupta, K. S. & Sen, S. (2001), *Phys. Lett.* **B 505**, 191 [arXiv:hep-th/0102051].
[24] Gupta, K. S. & Sen, S. (2002), *Phys. Lett.* **B 526**, 121 [arXiv:hep-th/0112041].
[25] Jacobson, T. & Kang, G. (1993), *Class. Quantum Grav.* **10**, L201 [arXiv:gr-qc/9307002].
[26] Medved, A. J. M., Martin, D. & Visser, M. (2004), *Class. Quantum Grav.* **21**, 3111 [arXiv:gr-qc/0402069].
[27] Medved, A. J. M., Martin, D. & Visser, M. (2004), *Phys. Rev.* **D 70**, 024009 [arXiv:gr-qc/0403026].
[28] Bañados, M., Teitelboim, C. & Zanelli, J. (1992), *Phys. Rev. Lett.* **69**, 1849 [arXiv:hep-th/9204099.
[29] Bañados, M., Henneaux, M., Teitelboim, C. & Zanelli, J. (1993), *Phys. Rev.* **D 48**, 1506 [arXiv:gr-qc/9302012].
[30] Carlip, S. (1995), *Class. Quantum Grav.* **12**, 2853 [arXiv:gr-qc/9506079].
[31] Lifschytz, G. & Ortiz, M. (1994), *Phys. Rev.* **D 49**, 1929 [arXiv:gr-qc/9310008].
[32] Hyun, S., Lee, G. H. & Yee, J. H. (1994), *Phys. Lett.* **B 322**, 182.
[33] Carlip, S. & Teitelboim, C. (1995), *Phys. Rev.* **D 51**, 622 [arXiv:gr-qc/9405070].
[34] Brown, J. D., Creighton, J. & Mann, R. B. (1994), *Phys. Rev.* **D 50**, 6394 [arXiv:gr-qc/9405007].
[35] Carlip, S., Gegenberg, J. & Mann, R. B. (1995), *Phys. Rev.* **D 51**, 6854 [arXiv:gr-qc/9410021].
[36] Englert, F. & Reznik, B. (1994), *Phys. Rev.* **D 50**, 2692 [arXiv:gr-qc/9401010].
[37] Medved, A. J. M. (2002), *Class. Quantum Grav.* **19**, 589 [arXiv:hep-th/0110289].
[38] Emparan, R. & Sachs, I. (1998), *Phys. Rev. Lett.* **81**, 2408 [arXiv:hep-th/9806122].
[39] Carlip, S. (2005), *Living Rev. Rel.* **8**, 1 [arXiv:gr-qc/0409039].
[40] Strominger, A. (1998), *JHEP* **02**, 009 [arXiv:hep-th/9712251].
[41] Birmingham, D., Sachs, I. & Sen, S. (1998), *Phys. Lett.* **B 424**, 275 [arXiv:hep-th/9801019].
[42] Brown, J. D. & Henneaux, M. (1986), *Commun. Math. Phys.* **104**, 207.
[43] Terashima, H. (2001), *Phys. Lett.* **B 499**, 229 [arXiv:hep-th/0011010].
[44] Bañados, M. (1996), *Phys. Rev.* **D 52**, 5816 [arXiv:hep-th/9405171].
[45] Henningson, M. & Skenderis, K. (1998), *JHEP* **07**, 023 [arXiv:hep-th/9806087].
[46] Balasubramanian, V. & Kraus, P. (1999), *Commun. Math. Phys.* **208**, 413 [arXiv:hep-th/9902121].
[47] Achúcarro, A. & Townsend, P. K. (1986), *Phys. Lett.* **B 180**, 89.
[48] Witten, E. (1988), *Nucl. Phys.* **B 311**, 46.
[49] Witten, E. (1989), *Commun. Math. Phys.* **121**, 351.

[50] Kaloper, N. & Terning, J., personal communication.
[51] Witten, E. (1984), *Commun. Math. Phys.* **92**, 455.
[52] Elitzur, S., Moore, G. W., Schwimmer, A. & Seiberg, N. (1989), *Nucl. Phys.* **B 326**, 108.
[53] Coussaert, O., Henneaux, M. & van Driel, P. (1995), *Class. Quantum Grav.* **12**, 2961 [arXiv:gr-qc/9506019].
[54] Rooman, M. & Spindel, Ph. (2001), *Nucl. Phys.* **B 594**, 329 [arXiv:hep-th/0008147.
[55] Skenderis, K. & Solodukhin, S. N. (2000), *Phys. Lett.* **B 472**, 316 [arXiv:hep-th/9910023].
[56] Bautier, K., Englert, F., Rooman, M. & Spindel, Ph. (2000), *Phys. Lett.* **B 479**, 291 [arXiv:hep-th/0002156].
[57] Rooman, M. & Spindel, Ph. (2001), *Class. Quantum Grav.* **18**, 2117 [arXiv:gr-qc/0011005].
[58] Carlip, S. (2005), *Class. Quantum Grav.* **22**, 3055 [arXiv:gr-qc/0501033].
[59] Carlip, S. (2005), *Class. Quantum Grav.* **22**, R85 [arXiv:gr-qc/0503022].
[60] Chen, Y.-J. (2004), *Class. Quantum Grav.* **21**, 1153 [arXiv:hep-th/0310234].
[61] Skenderis, K. (2000), *Lect. Notes Phys.* **541**, 325 [arXiv:hep-th/9901050].
[62] Carlip, S. (1997), *Constrained Dynamics and Quantum Gravity 1996*, ed. V. de Alfaro *et al.*, *Nucl. Phys. Proc. Suppl.* **57**, 8 [arXiv:gr-qc/9702017].
[63] Witten, E. (1992), *Commun. Math. Phys.* **144**, 189.
[64] Carlip, S. (1999), *Phys. Rev. Lett.* **82**, 2828 [arXiv:hep-th/9812013].
[65] Carlip, S. (1999), *Class. Quantum Grav.* **16**, 3327 [arXiv:gr-qc/9906126].
[66] Navarro, D. J., Navarro-Salas, J. & Navarro, P. (2000), *Nucl. Phys.* **B 580**, 311 [arXiv:hep-th/9911091].
[67] Jing, J. & Yan, M.-L. (2001), *Phys. Rev.* **D 63**, 024003 [arXiv:gr-qc/0005105].
[68] Izquierdo, J. M., Navarro-Salas, J. & Navarro, P. (2002), *Class. Quantum Grav.* **19**, 563 [arXiv:hep-th/0107132].
[69] Park, M.-I. (2002), *Nucl. Phys.* **B 634**, 339 [arXiv:hep-th/0111224].
[70] Silva, S. (2002), *Class. Quantum Grav* **19**, 3947 [arXiv:hep-th/0204179].
[71] Cvitan, M., Pallua, S. & Prester, P. (2003), *Phys. Lett.* **B 555**, 248 [arXiv:hep-th/0212029].
[72] Cvitan, M., Pallua, S. & Prester, P. (2004), *Phys. Rev.* **D 70**, 084043 [arXiv:hep-th/0406186].
[73] Dreyer, O., Ghosh, A. & Wisniewski, J. (2001), *Class. Quantum Grav.* **18**, 1929 [arXiv:hep-th/0101117].
[74] Koga, J. (2001), *Phys. Rev.* **D 64**, 124012 [arXiv:gr-qc/0107096].
[75] Pinamonti, N. & Vanzo, L. (2004), *Phys. Rev.* **D 69**, 084012 [arXiv:hep-th/0312065].
[76] Solodukhin, S. N. (1999), *Phys. Lett.* **B 454**, 213 [arXiv:hep-th/9812056].
[77] Carlip, S. (2002), *Phys. Rev. Lett.* **88**, 241301 [arXiv:gr-qc/0203001].
[78] Giacomini, A. & Pinamonti, N. (2003), *JHEP* **02**, 014 [arXiv:gr-qc/0301038].
[79] Solodukhin, S. N. (2004), *Phys. Rev. Lett.* **92**, 061302 [arXiv:hep-th/0310012].
[80] Ashtekar, A., Beetle, C. & Fairhurst, S. (1999), *Class. Quantum Grav.* **16**, L1 [arXiv:gr-qc/9812065].
[81] Ashtekar, A. & Krishnan, B. (2004), *Living Rev. Rel.* **7**, 10 [arXiv:gr-qc/0407042].
[82] Dirac, P. A. M. (1950), *Can. J. Math.* **2**, 129.
[83] Dirac, P. A. M. (1951), *Can. J. Math.* **3**, 1.
[84] Bergmann, P. G. & Komar, A. B. (1960), *Phys. Rev. Lett.* **4**, 432.
[85] Yoon, J. H. (1999), *Phys. Lett.* **B 451**, 296 [arXiv:gr-qc/0003059].
[86] Louis-Martinez, D. & Kunstatter, G. (1995), *Phys. Rev.* **D 52**, 3494 [arXiv:gr-qc/9503016].

[87] Grumiller, D., Kummer, W. & Vassilevich, D. V. (2002), *Phys. Rep.* **369**, 327 [arXiv:hep-th/0204253].
[88] Carlip, S. (2005), *Class. Quantum Grav.* **22**, 1303 [arXiv:hep-th/0408123].
[89] Regge, T. & Teitelboim, C. (1974), *Ann. Phys. (N.Y.)* **88**, 286.
[90] Susskind, L., Thorlacius, L. & Uglum, J. (1993), *Phys. Rev.* **D 48**, 3743 [arXiv:hep-th/9306069].
[91] Maldacena, J. & Strominger, A. (1997), *Phys. Rev.* **D 55**, 861 [arXiv:hep-th/9609026].
[92] Larsen, F. (1997), *Phys. Rev.* **D 56**, 1005 [arXiv:hep-th/9702153].
[93] Frolov, V. P., Fursaev, D. & Zelnikov, A. (2003), *JHEP* **03**, 038 [arXiv:hep-th/0302207].

# 12

# Higher dimensional generalizations of the Kerr black hole

Gary T. Horowitz

## 12.1 Introduction

When I was a graduate student at the University of Chicago in the late 1970s, I often heard Chandrasekhar raving about the Kerr solution [1]. He was amazed by all of its remarkable properties and even its mere existence. As he said at the time: "In my entire scientific life . . . the most shattering experience has been the realization that an exact solution of general relativity, discovered by the New Zealand mathematician Roy Kerr, provides the absolutely exact representation of untold numbers of massive black holes that populate the Universe" [2].

It took me a while to understand Chandra's fascination, but I have come to agree. One can plausibly argue that the black hole solution discovered by Roy Kerr is the most important vacuum solution ever found to Einstein's equation. To honor Kerr's 70th birthday, I would like to describe some recent generalizations of the Kerr solution to higher spatial dimensions.

Before I begin, let me say a word about the motivation for this work. There are two main reasons for studying these generalizations. The first comes from string theory, which is a promising approach to quantum gravity. String theory predicts that spacetime has more than four dimensions. For a while it was thought that the extra spatial dimensions would be of order the Planck scale, making a geometric description unreliable, but it has recently been realized that there is a way to make the extra dimensions relatively large and still be unobservable. This is if we live on a three-dimensional surface (a "brane") in a higher dimensional space. String theory contains such higher dimensional extended objects, and it turns out that nongravitational forces are confined to the brane, but gravity is not. In such a scenario, all gravitational objects such as black holes are higher dimensional. The second reason for studying these solutions has nothing to do with string theory.

The Kerr Spacetime: Rotating Black Holes in General Relativity. Ed. David L. Wiltshire, Matt Visser and Susan M. Scott. Published by Cambridge University Press. 

Four-dimensional black holes have a number of remarkable properties. It is natural to ask whether these properties are general features of black holes or whether they crucially depend on the world being four dimensional. We will see that many of them are indeed special properties of four dimensions and do not hold in general.

## 12.2 Nonrotating black holes in $D > 4$

To become familiar with black holes in higher dimensions, I will start by discussing nonrotating black holes. (For more extensive reviews of the material in this section, see [3, 4].) For simplicity, we will focus on $D = 5$. There are two possible boundary conditions to consider: asymptotically flat in five dimensions, or the Kaluza–Klein choice – asymptotically $M_4 \times S^1$. In the asymptotically flat case, the only static black hole is the five-dimensional Schwarzschild–Tangherlini solution [5]

$$\mathrm{d}s^2 = -\left(1 - \frac{r_0^2}{r^2}\right)\mathrm{d}t^2 + \left(1 - \frac{r_0^2}{r^2}\right)^{-1}\mathrm{d}r^2 + r^2\,\mathrm{d}\Omega_3. \qquad (12.1)$$

In the Kaluza–Klein case, there are more possibilities. Let $L$ be the length of the circle at infinity. The simplest solution with an event horizon is just the product of four-dimensional Schwarzschild with radius $r_0$ and $S^1$:

$$\mathrm{d}s^2 = -\left(1 - \frac{r_0}{r}\right)\mathrm{d}t^2 + \left(1 - \frac{r_0}{r}\right)^{-1}\mathrm{d}r^2 + r^2\,\mathrm{d}\Omega + \mathrm{d}z^2. \qquad (12.2)$$

This has horizon topology $S^2 \times S^1$ and is sometimes called a black string, since it looks like a one-dimensional extended object surrounded by an event horizon. Gregory and Laflamme (GL) showed that this spacetime is unstable to linearized perturbations with a long wavelength along the circle [6]. More precisely, there is a critical size for the circle, $L_0$, of order $r_0$ such that black strings with $L \leq L_0$ are stable and those with $L > L_0$ are unstable. The unstable mode is spherically symmetric, but causes the horizon to oscillate in the $z$ direction. Gregory and Laflamme also compared the total entropy of the black string with that of a five-dimensional spherical black hole with the same total mass, and found that when $L > L_0$, the black hole had greater entropy. They thus suggested that the full nonlinear evolution of the instability would result in the black string breaking up into separate black holes which would then coalesce into a single black hole. Classically, horizons cannot bifurcate, but the idea was that under classical evolution, the event horizon would pinch off and become singular. When the curvature became large enough, it was plausible that quantum effects would smooth out the transition between the black string and spherical black holes.

However, it turns out that an event horizon cannot pinch off in finite time [7]. In particular, if one perturbs (12.2), an $S^2$ on the horizon cannot shrink to zero size in finite affine parameter. The reason is the following. Hawking's famous area theorem [8] is based on a local result that the divergence $\theta$ of the null geodesic generators of the horizon cannot become negative, i.e. the null geodesics cannot start to converge. If an $S^2$ on the horizon tries to shrink to zero size, the null geodesics on that $S^2$ must be converging. The total $\theta$ can stay positive only if the horizon is expanding rapidly in the circle direction, but this produces a large shear. If the $S^2$ were to shrink to zero size in finite time, one can show this shear would drive $\theta$ negative. When it was realized that the black string cannot pinch off in finite time, it was suggested that the solution should settle down to a static nonuniform black string.

A natural place to start looking for these new solutions is with the static perturbation of the uniform black string that exists with wavelength $L_0$. Gubser [9] did a perturbative calculation and found evidence that the nonuniform solutions with small inhomogeneity could not be the endpoint of the GL instability. Recent numerical work has found vacuum solutions describing static black strings with large inhomogeneity [10]. Surprisingly, all of these solutions have a mass which is larger than the unstable uniform black strings. So they cannot be the endpoint of the GL instability.[1] Solutions describing topologically spherical black holes in Kaluza–Klein theory have also been found numerically [12, 13]. When the black hole radius is much less than $L$, it looks just like (12.1). As you increase the radius one finds that the size of the fifth dimension near the black hole grows and the black hole remains approximately spherical. It then reaches a maximum mass. Remarkably, one can continue past this point and find another branch of black hole solutions with lower mass and squashed horizons. It was conjectured by Kol [14] that the nonuniform black strings should meet the squashed black holes at a point corresponding to a static solution with a singular horizon, and this appears to be the case [15]. This yields a nice consistent picture of static Kaluza–Klein solutions with horizons, but it doesn't answer the question of what is the endpoint of the GL instability. An attempt to numerically evolve a perturbed black string is underway. An earlier attempt could not be followed far enough to reach the final state [16].

It was suggested by Wald [17] that the black string horizon might pinch off in infinite affine parameter (avoiding the above no-go theorem), but still occur at finite advanced time as seen from the outside. This is possible since the spacetime is singular when the horizon pinches off, and some evidence for this has been found [18]. If this were the case, then the original suggestion of Gregory and Laflamme that the black string will break up into spherical black holes might still be correct.

[1] In dimensions greater than 13 this changes and the nonuniform black string can be the endpoint of the instability [11].

## 12.3 Rotating black holes in $D > 4$

With Kaluza–Klein boundary conditions, the only known solution is the rotating black string obtained by taking the product of the Kerr metric and a circle. Most of the recent work on higher dimensional rotating black holes has been in the context of asymptotically flat spacetimes, so from now on we will focus on this case.

The direct generalization of the Kerr metric to higher dimensions was found by Myers and Perry in 1986 [19]. In more than three spatial dimensions, black holes can rotate in different orthogonal planes, so the general solution has several angular momentum parameters. The general solution, with all possible angular momenta nonzero is known explicitly. Like the Kerr metric, these solutions are all of the Kerr–Schild form [20]

$$g_{\mu\nu} = \eta_{\mu\nu} + h\, k_\mu\, k_\nu, \tag{12.3}$$

where $k_\mu$ is null.

If we set all but one of the angular momentum parameters to zero, we can write the metric in Boyer–Lindquist like coordinates. In $D$ spacetime dimensions, the solution is

$$\begin{aligned} ds^2 = &-dt^2 + \sin^2\theta(r^2 + a^2)\, d\varphi^2 + \frac{\mu}{r^{D-5}\rho^2}(dt - a\sin^2\theta\, d\varphi)^2 \\ &+ \Psi^{-1} dr^2 + \rho^2\, d\theta^2 + r^2\cos^2\theta\, d\Omega_{D-4}, \end{aligned} \tag{12.4}$$

where $\rho^2 = r^2 + a^2\cos^2\theta$, and

$$\Psi = \frac{r^2 + a^2}{\rho^2} - \frac{\mu}{r^{D-5}\rho^2}. \tag{12.5}$$

Like the Kerr metric, this solution has two free parameters $\mu$, $a$ which determine the mass and angular momentum:

$$M = \frac{(D-2)\Omega_{D-2}}{16\pi}\mu; \qquad J = \frac{2}{D-2} M a; \tag{12.6}$$

where $\Omega_{D-2}$ is the area of a unit $S^{D-2}$. One of the most surprising properties of these solutions is that for $D > 5$, there is a regular horizon for all $M, J$! There is no extremal limit. This follows from the fact that the horizon exists where $\Psi = 0$, but for $D > 5$ this equation always has a solution. However, when the angular momentum is much bigger than the mass (using quantities of the same dimension, this is $J^{D-3} \gg M^{D-2}$) the horizon is like a flat pancake: it is spread out in the plane of rotation, but very thin in the orthogonal directions. Locally, it looks like the product of $(D-2)$-dimensional Schwarzschild and $R^2$. It is probably subject to a GL instability [21]. If so, there would be an effective extremal limit for stable black holes in all dimensions.

In five dimensions, there is an extremal limit, but the horizon area goes to zero in this limit. The extremal limit corresponds to $\mu = a^2$. Setting $D = 5$ in (12.6), the mass and angular momentum are

$$M = \frac{3\pi}{8}\mu; \qquad J = \frac{2}{3}Ma; \tag{12.7}$$

so in the extremal limit

$$J^2 = \frac{32}{27\pi}M^3. \tag{12.8}$$

These solutions have recently been generalized to include a cosmological constant [22, 23].

## 12.4 Black rings

The above solutions all have a horizon which is topologically spherical $S^{D-2}$. There is a qualitatively new type of rotating black hole that arises in $D = 5$ (and possibly higher dimensions). One can take a black string, wrap it into a circle, and spin it along the circle direction just enough so that the gravitational attraction is balanced by the centrifugal force. Remarkably, an exact solution has been found describing this [24]. It was not found by actually carrying out the above proceedure, but by a trick involving analytically continuing a Kaluza–Klein C-metric. The solution is independent of time, $t$, and two orthogonal rotations parameterized by $\varphi$ and $\psi$, so the isometry group is $R \times U(1)^2$. Introducing two other spatial coordinates, $-1 \leq x \leq 1$ and $y \leq -1$, the metric is:

$$\begin{aligned} \mathrm{d}s^2 = &-\frac{F(y)}{F(x)}\left[\mathrm{d}t + C(\nu)R\frac{1+y}{F(y)}\,\mathrm{d}\psi\right]^2 \\ &+ \frac{R^2}{(x-y)^2}F(x)\left[-\frac{G(y)}{F(y)}\,\mathrm{d}\psi^2 - \frac{\mathrm{d}y^2}{G(y)} + \frac{\mathrm{d}x^2}{G(x)} + \frac{G(x)}{F(x)}\,\mathrm{d}\varphi^2\right], \end{aligned} \tag{12.9}$$

where

$$\begin{aligned} F(\xi) = 1 + \frac{2\nu}{1+\nu^2}\xi; \qquad G(\xi) = (1-\xi^2)(1+\nu\xi); \\ C(\nu) = \frac{\nu}{1+\nu^2}\left[\frac{2(1+\nu)^3}{1-\nu}\right]^{1/2}; \end{aligned} \tag{12.10}$$

and the angular variables are periodically identified with period

$$\Delta\varphi = \Delta\psi = \frac{2\pi}{\sqrt{1+\nu^2}}. \tag{12.11}$$

Although it is not obvious in these coordinates, this solution is asymptotically flat. The asymptotic region corresponds to $x \to -1,\ y \to -1$. The solution depends on two parameters $R > 0$ and $0 < \nu < 1$ which determine the mass and angular momentum

$$M = \frac{3\pi R^2}{2}\frac{\nu}{(1-\nu)(1+\nu^2)}; \qquad J = \frac{\pi R^3}{2}\frac{C(\nu)}{(1-\nu)\sqrt{1+\nu^2}}. \tag{12.12}$$

The coordinates $(x, \varphi)$ parameterize two-spheres, so on a constant $t$ slice, surfaces of constant $y < -1$ are topologically $S^2 \times S^1$. The limiting value $y = -1$ corresponds to the axis for the $\psi$ rotations. The solution has an event horizon at $y = -1/\nu$ (where $G(y) = 0$) with topology $S^2 \times S^1$ and a curvature singularity at $y = -\infty$. It has no inner horizon, but it does have an ergoregion given by $-1/\nu < y < -(1+\nu^2)/2\nu$. This is just like the solution one would obtain by boosting the black string (12.2) in the $z$ direction. This solution is called a *black ring* rather than a black string since it is wrapped around a topologically trivial circle in space.

Not surprisingly, the angular momentum now has a lower bound, but no upper bound. Solutions exist only when $J^2 \geq M^3/\pi \equiv J^2_{min}$. Something interesting happens when $J$ is near its minimum value. If $J^2_{min} < J^2 < \frac{32}{27}J^2_{min}$ there are two stationary black ring solutions with different horizon area. In addition, there is a Myers–Perry black hole! So for $M,\ J$ in this range, there are three black holes clearly showing the black hole uniqueness theorems do not extend to higher dimensions. Near $J_{min}$, the Myers–Perry black hole has the largest horizon area. Near $(32/27)^{1/2}J_{min}$, one of the black rings has the largest area.

It is not yet known if these solutions are stable, although the solutions with large $J$ are likely to be unstable. This is because in this limit, the ring becomes very long and thin, so one expects an analog of the GL instability. It cannot settle down to an inhomogeneous ring, since the rotation would cause the system to lose energy due to gravitational waves. The outcome of this instability (assuming it exists) is unknown.

## 12.5 Einstein–Maxwell solutions

So far, we have considered only vacuum solutions. The higher dimensional generalization of the Kerr–Newman solution is still not known analytically (but special cases have been found numerically [25]). However exact solutions describing generalizations of the neutral black ring have been found. In fact, the discrete nonuniqueness discussed above becomes a continuously infinite nonuniqueness when we add a Maxwell field. Suppose we consider Einstein–Maxwell theory in five dimensions

$$S = \int \mathrm{d}^5x \sqrt{-g} \left( R - \frac{1}{4} F_{\mu\nu} F^{\mu\nu} \right). \tag{12.13}$$

In this theory, there is a global electric charge

$$Q = \frac{1}{4\pi} \int_{S^3} {}^*F, \tag{12.14}$$

but no global magnetic charge since one cannot integrate the two-form $F$ over the three-sphere at infinity. However, there is a local magnetic charge carried by a string. Given a point on a string, one can surround it by a two-sphere and define

$$q = \frac{1}{4\pi} \int_{S^2} F. \tag{12.15}$$

If the string is wrapped into a circle, there is no net magnetic charge at infinity. The asymptotic field resembles a magnetic dipole.

It was suggested in [26] that there should exist a generalization of (12.10) which has a third independent parameter labeling the magnetic dipole moment $q$. The explicit solution was found about a year later [27]. Note that unlike four-dimensional black holes, the dipole moment is an independent adjustable parameter. Since the only global charges are $M$ and $J$, there is continuous nonuniqueness. Solutions with $q \neq 0$ have a smooth inner horizon as well as an event horizon. There is now an upper bound on the angular momentum as well as a lower bound $J^2_{min} \leq J^2 \leq J^2_{max}$. Solutions with $J^2 = J^2_{max}$ are extremal in that the inner and outer horizons coincide. The extremal solution has a smooth degenerate horizon, with zero surface gravity.

This dipole charge enters the first law in much the same way as an ordinary global charge. There is a corresponding potential $\phi$ and one can show [28] that for any perturbation satisfying the linearized constraints

$$\delta M = \frac{\kappa}{8\pi}\, \delta A_H + \Omega_i\, \delta J^i + \Phi\, \delta Q + \phi\, \delta q, \tag{12.16}$$

where, as usual, $\kappa$ is the surface gravity, $A_H$ is the horizon area, $\Omega_i$ $(i = 1, 2)$ are the angular velocities in the two orthogonal planes, and $\Phi$ is the electrostatic potential. We have included the possibility of a global electric charge and rotation in both orthogonal planes even though the dipole ring found in [27] has zero electric charge and only one nonzero angular momentum.

## 12.6 Supersymmetry

Let us first recall the situation with supersymmetric black holes in four dimensions. In $D = 4$, supersymmetry requires $M = Q$, so the only supersymmetric black holes in Einstein–Maxwell theory are extremal Reissner–Nordström and superpositions of them. The Kerr–Newman solution with $M = Q$, and $J \neq 0$ is supersymmetric but describes a naked singularity. (If one adds additional matter fields, there exist supersymmetric multi black hole solutions with angular momentum [29].)

The situation is different in $D = 5$. To begin, the bosonic sector of the minimal supergravity theory is not just Einstein–Maxwell, but also contains a Chern–Simons term

$$S = \int \mathrm{d}^5x\,\sqrt{-g}\Big(R - \frac{1}{4}F_{\mu\nu}F^{\mu\nu} - \frac{1}{12\sqrt{3}}\epsilon^{\mu\nu\rho\sigma\eta}F_{\mu\nu}F_{\rho\sigma}A_\eta\Big). \tag{12.17}$$

Surprisingly, the addition of the Chern–Simons term makes it easier to solve the field equations,[2] and analytic solutions describing charged, rotating black holes in this theory are known [30]. Their extremal limit provides a two parameter family of rotating supersymmetric black holes [31]. One parameter determines the angular momenta: the angular momenta in both orthogonal planes are non-zero and equal, $J_1 = J_2$. The second determines the electric charge. The mass is determined in terms of the charge $M = (\sqrt{3}/2)Q$. As expected, the surface gravity vanishes for these extremal solutions, but surprisingly, the angular velocity also vanishes. The fact that there is no ergoregion follows from the fact that supersymmetric, asymptotically flat solutions must have a Killing field that remains timelike (or null) everywhere outside the horizon [32]. These horizons are topologically $S^3$, and Reall has shown that these are the only supersymmetric solutions with spherical horizons [26].

However, there are also supersymmetric black rings with $S^2 \times S^1$ horizons [33]. The dipole rings discussed above are also solutions to (12.17) since the Chern–Simons term does not contribute in this case, but they are not supersymmetric. To obtain supersymmetric solutions, one must add electric charge and take a limit so that $M = (\sqrt{3}/2)Q$. It turns out that the supersymmetric solutions can have two independent angular momenta, so there are three free parameters $Q, J_1, J_2$. There is also a magnetic dipole charge but it is not independent.

The near horizon limit of a $D = 4$ extreme Reissner–Nordström black hole is $AdS_2 \times S^2$ (where AdS denotes anti-de Sitter spacetime). Similarly, the near

[2] This is because one can employ various solution generating techniques known for supergravity.

horizon limit of the supersymmetric black ring is $AdS_3 \times S^2$. Interestingly enough, the near horizon limit of the extreme Kerr metric and some Myers–Perry solutions also resemble (warped) products of AdS and a sphere [34].

One of the advantages of considering supersymmetric solutions is that they minimize the mass for given charges. Since there is no lower mass solution to decay to, these supersymmetric solutions are expected to be stable. Another advantage is that it is easier to count their microstates in string theory and compare with the Hawking–Bekenstein entropy. This was done successfully for the topologically spherical black hole [31] and has also been discussed for the black rings [35, 36].

In this example of five-dimensional minimal supergravity, the supersymmetric solution is uniquely determined by the conserved charges at infinity. However, in ten-dimensional supergravity, supersymmetric solutions have been found with three types of electric charges and three local magnetic charges but only one constraint [37–39]. So together with the two angular momenta there are a total of seven parameters, but only five global charges. So even among supersymmetric solutions, there is continuous nonuniqueness.

## 12.7 Discussion

In four dimensions, black holes enjoy a number of quite special properties, including:

(1) they are uniquely specified by $M$, $J$, $Q$;
(2) they are topologically spherical;
(3) they are stable.

We have seen that none of these properties are preserved in higher dimensions. The solutions can be labelled by local charges rather than global charges. In five dimensions, black hole horizons can have topology $S^2 \times S^1$ as well as $S^3$. Finally, we have seen that the straight black string is unstable, and highly rotating black rings and black holes in more than five dimensions are likely to be also. With regard to the nonuniqueness, it should be noted that there is still a finite number of parameters characterizing the solution. When this is compared with the vast number of ways of forming these objects, one can take the view that the spirit of the no hair theorem is preserved.[3]

A surprisingly large number of higher dimensional generalizations of the Kerr solution have been found. But the list is far from complete, and a great deal remains to be understood. For example, we do not yet have good restrictions on the topology of stationary black holes in $D > 4$. Clearly $S^n$ and $S^2 \times S^1$ are possible. Are there

[3] I thank B. Carter for stressing this point.

others? There should also be more general black ring solutions in five dimensions. The vacuum solution given in (12.10) has angular momentum only along the circle. One expects that it should also be possible to have angular momentum on the two-spheres. This would be analogous to first taking the product of the $D = 4$ Kerr metric and a line to obtain a rotating black string, and then wrapping the string into a circle and spinning it to obtain another stationary black ring. Finally, all of the known higher dimensional black holes have a great deal of symmetry: in addition to time translation invariance, they are all invariant under rotations in mutually orthogonal planes. It was pointed out in [26] that on general grounds, one only expects rotating black holes to have one additional Killing field. So there may well exist much more general black holes in $D > 4$ with only two Killing fields.

In short, the space of black holes is much richer in more than four spacetime dimensions. It remains to be seen whether nature takes advantage of this richness.

## Acknowledgments

I would like to thank the organizers of the Kerr Fest (Christchurch, New Zealand, Aug. 26–28, 2004) for a stimulating meeting. I also want to thank H. Elvang and H. Reall for their explanations of some of the results discussed here. This work was supported in part by NSF grant PHY-0244764.

## References

[1] Kerr, R. P. (1963), "Gravitational field of a spinning mass as an example of algebraically special metrics", *Phys. Rev. Lett.* **11**, **237**.
[2] Chandrasekhar, S. (1987), Lecture reprinted in *Truth and Beauty* (University of Chicago Press, Chicago), p. 54.
[3] Kol, B. (2006), "The phase transition between caged black holes and black strings: a review", *Phys. Rep.* **422**, 119 [arXiv:hep-th/0411240].
[4] Harmark, T. & Obers, N. A., "Phases of Kaluza–Klein black holes: a brief review", arXiv:hep-th/0503020.
[5] Gibbons, G. W., Ida, D. & Shiromizu, T. (2002), "Uniqueness and non-uniqueness of static black holes in higher dimensions", *Phys. Rev. Lett.* **89**, 041101 [arXiv:hep-th/0206049].
[6] Gregory, R. & Laflamme, R. (1993), "Black strings and $p$-branes are unstable", *Phys. Rev. Lett.* **70**, 2837 [arXiv:hep-th/9301052].
[7] Horowitz, G. T. & Maeda, K. (2001), "Fate of the black string instability", *Phys. Rev. Lett.* **87**, 131301 [arXiv:hep-th/0105111].
[8] Hawking, S. W. (1971), "Gravitational radiation from colliding black holes", *Phys. Rev. Lett.* **26**, 1344.
[9] Gubser, S. S. (2002), "On non-uniform black branes", *Class. Quantum Grav.* **19**, 4825 [arXiv:hep-th/0110193].
[10] Wiseman, T. (2003), "Static axisymmetric vacuum solutions and non-uniform black strings", *Class. Quantum Grav.* **20**, 1137 [arXiv:hep-th/0209051].

[11] Sorkin, E. (2004), "A critical dimension in the black-string phase transition", *Phys. Rev. Lett.* **93**, 031601 [arXiv:hep-th/0402216].
[12] Kudoh, H. & Wiseman, T. (2004), "Properties of Kaluza–Klein black holes", *Prog. Theor. Phys.* **111**, 475 [arXiv:hep-th/0310104].
[13] Sorkin, E., Kol, B. & Piran, T. (2004), "Caged black holes: Black holes in compactified spacetimes. II: 5d numerical implementation", *Phys. Rev.* **D 69**, 064032 [arXiv:hep-th/0310096].
[14] Kol, B. (2005), "Topology change in general relativity and the black-hole black-string transition", *JHEP* **10**, 049 [arXiv:hep-th/0206220].
[15] Kudoh, H. & Wiseman, T. (2005), "Connecting black holes and black strings", *Phys. Rev. Lett.* **94**, 161102 [arXiv:hep-th/0409111].
[16] Choptuik, M. W., Lehner, L., Olabarrieta, I., Petryk, R., Pretorius, F. & Villegas, H. (2003), "Towards the final fate of an unstable black string", *Phys. Rev.* **D 68**, 044001 [arXiv:gr-qc/0304085].
[17] Wald, R., private communication.
[18] Marolf, D. (2005), "On the fate of black string instabilities: an observation", *Phys. Rev.* **D 71**, 127504 [arXiv:hep-th/0504045].
[19] Myers, R. C. & Perry, M. J. (1986), "Black holes in higher dimensional space-times", *Ann. Phys. (N.Y.)* **172**, 304.
[20] Kerr, R. P. & Schild, A. (1965), "Some algebraically degenerate solutions of Einstein's gravitational field equations", *Proc. Symp. Appl. Math.* **17**, 199.
[21] Emparan, R. & Myers, R. C. (2003), "Instability of ultra-spinning black holes", *JHEP* **09**, 025 [arXiv:hep-th/0308056].
[22] Hawking, S. W., Hunter, C. J. & Taylor-Robinson, M. M. (1999), "Rotation and the AdS/CFT correspondence", *Phys. Rev.* **D 59**, 064005 [arXiv:hep-th/9811056].
[23] Gibbons, G. W., Lu, H., Page, D. N. & Pope, C. N. (2004), "Rotating black holes in higher dimensions with a cosmological constant", *Phys. Rev. Lett.* **93**, 171102 [arXiv:hep-th/0409155].
[24] Emparan, R. & Reall, H. S. (2002), "A rotating black ring in five dimensions", *Phys. Rev. Lett.* **88**, 101101 [arXiv:hep-th/0110260].
[25] Kunz, J., Navarro-Lerida, F. & Petersen, A. K. (2005), "Five-dimensional charged rotating black holes", *Phys. Lett.* **B 614**, 104 [arXiv:gr-qc/0503010].
[26] Reall, H. S. (2003), "Higher dimensional black holes and supersymmetry", *Phys. Rev.* **D 68**, 024024 [Erratum *Phys. Rev.* **D 70**, 089902 (2004)] [arXiv:hep-th/0211290].
[27] Emparan, R. (2004), "Rotating circular strings, and infinite non-uniqueness of black rings", *JHEP* **03**, 064 [arXiv:hep-th/0402149].
[28] Copsey, K. & Horowitz, G. T. (2006), "The role of dipole charges in black hole thermodynamics", *Phys. Rev.* **D 73**, 024015 [arXiv:hep-th/0505278].
[29] Bates, B. & Denef, F., "Exact solutions for supersymmetric stationary black hole composites", arXiv:hep-th/0304094.
[30] Breckenridge, J. C., Lowe, D. A., Myers, R. C., Peet, A. W., Strominger, A. & Vafa, C. (1996), "Macroscopic and microscopic entropy of near-extremal spinning black holes", *Phys. Lett.* **B 381**, 423 [arXiv:hep-th/9603078].
[31] Breckenridge, J. C., Myers, R. C., Peet, A. W. & Vafa, C. (1997), "D-branes and spinning black holes", *Phys. Lett.* **B 391**, **93** [arXiv:hep-th/9602065].
[32] Gauntlett, J. P., Myers, R. C. & Townsend, P. K. (1999), "Black holes of $D = 5$ supergravity", *Class. Quantum Grav.* **16**, 1 [arXiv:hep-th/9810204].
[33] Elvang, H., Emparan, R., Mateos, D. & Reall, H. S. (2004), "A supersymmetric black ring", *Phys. Rev. Lett.* **93**, 211302 [arXiv:hep-th/0407065].

[34] Bardeen, J. M. & Horowitz, G. T. (1999), "The extreme Kerr throat geometry: a vacuum analog of AdS(2) $\times$ S(2)", *Phys. Rev.* **D 60**, 104030 [arXiv:hep-th/9905099].
[35] Cyrier, M., Guica, M., Mateos, D. & Strominger, A. (2005), "Microscopic entropy of the black ring", *Phys. Rev. Lett.* **94**, 191601 [arXiv:hep-th/0411187].
[36] Bena, I. & Kraus, P. (2004), "Microscopic description of black rings in AdS/CFT", *JHEP* **12**, 070 [arXiv:hep-th/0408186].
[37] Elvang, H., Emparan, R., Mateos, D. & Reall, H. S. (2005), "Supersymmetric black rings and three-charge supertubes", *Phys. Rev.* **D 71**, 024033 [arXiv:hep-th/0408120].
[38] Gauntlett, J. P. & Gutowski, J. B. (2005), "Concentric black rings", *Phys. Rev.* **D 71** (2005) 025013 [arXiv:hep-th/0408010]; "General concentric black rings", *Phys. Rev.* **D 71**, 045002 [arXiv:hep-th/0408122].
[39] Bena, I. & Warner, N. P. (2005), "One ring to rule them all . . . and in the darkness bind them?", *Adv. Theor. Math. Phys.* **9**, 667 [arXiv:hep-th/0408106].

# Part IV

Appendices

# 13

# Gravitational field of a spinning mass as an example of algebraically special metrics

Roy P. Kerr

VOLUME 11, NUMBER 5 PHYSICAL REVIEW LETTERS 1 SEPTEMBER 1963

## GRAVITATIONAL FIELD OF A SPINNING MASS AS AN EXAMPLE OF ALGEBRAICALLY SPECIAL METRICS

Roy P. Kerr*
University of Texas, Austin, Texas and Aerospace Research Laboratories, Wright-Patterson Air Force Base, Ohio
(Received 26 July 1963)

Goldberg and Sachs[1] have proved that the algebraically special solutions of Einstein's empty-space field equations are characterized by the existence of a geodesic and shear-free ray congruence, $k_\mu$. Among these spaces are the plane-fronted waves and the Robinson-Trautman metrics[2] for which the congruence has nonvanishing divergence, but is hypersurface orthogonal.

In this note we shall present the class of solutions for which the congruence is diverging, and is not necessarily hypersurface orthogonal. The only previously known example of the general case is the Newman, Unti, and Tamburino metrics,[3] which is of Petrov Type D, and possesses a four-dimensional group of isometries.

If we introduce a complex null tetrad ($t^*$ is the complex conjugate of $t$), with

$$ds^2 = 2tt^* + 2mk,$$

then the coordinate system may be chosen so that

$$t = P(r + i\Delta)d\zeta,$$

$$k = du + 2\,\mathrm{Re}(\Omega d\zeta),$$

$$m = dr - 2\,\mathrm{Re}\{[(r - i\Delta)\dot{\Omega} + iD\Delta]d\zeta\} + \Big\{r\dot{P}/P + \mathrm{Re}[P^{-2}D(D^*\ln P + \dot{\Omega}^*)] + \frac{m_1 r - m_2\Delta}{r^2 + \Delta^2}\Big\}k; \qquad (1)$$

where $\zeta$ is a complex coordinate, a dot denotes differentiation with respect to $u$, and the operator $D$ is defined by

$$D = \partial/\partial\zeta - \Omega\partial/\partial u.$$

$P$ is real, whereas $\Omega$ and $m$ (which is defined to be $m_1 + im_2$) are complex. They are all independent of the coordinate $r$. $\Delta$ is defined by

$$\Delta = \mathrm{Im}(P^{-2}D^*\Omega).$$

There are two natural choices that can be made for the coordinate system. Either (A) $P$ can be chosen to be unity, in which case $\Omega$ is complex, or (B) $\Omega$ can be taken pure imaginary, with $P$ different from unity. In case (A), the field equations are

$$(m - D^*D^*D\Omega) = |\partial_u D\Omega|^2, \qquad (2)$$

$$\mathrm{Im}(m - D^*D^*D\Omega) = 0, \qquad (3)$$

$$D^*m = 3m\dot{\Omega}. \qquad (4)$$

The second coordinate system is probably better, but it gives more complicated field equations.

It will be observed that if $m$ is zero then the field equations are integrable. These spaces correspond to the Type-III and null spaces with

VOLUME 11, NUMBER 5 PHYSICAL REVIEW LETTERS 1 SEPTEMBER 1963

nonzero divergence. If $m \neq 0$, then there are certain integrability conditions which must be satisfied by Eqs. (2)-(4). These may be solved for $m$ as a function of $\Omega$ and its derivatives provided that either $\dot{\Delta}$ or $\ddot{\Omega}$ is nonzero. This expression for $m$ may then be substituted back into the field equations giving conditions on $\Omega$ and its derivatives, from which further integrability conditions are extracted.

If both $\dot{\Delta}$ and $\ddot{\Omega}$ are zero, then we may transform the metric to a coordinate system in which $\Omega$ is pure imaginary and $P \neq 1$, with $\dot{\Omega} = \dot{P} = 0$. The field equations then become

$$m = cu + A + iB,$$

where $c$ is a real constant, and

$$P^{-2}\nabla[P^{-2}\nabla(\ln P)] = 2c, \quad \nabla = \partial^2/\partial\zeta\partial\zeta^*.$$

$A$, $B$, and $\Omega$, which are all independent of $u$ and $r$, are determined by

$$iB = \tfrac{1}{2}P^{-2}\nabla(P^{-2}\partial\Omega/\partial\zeta) - P^{-4}(\partial\Omega/\partial\zeta)\nabla(\ln P),$$

$$\nabla B = ic\partial\Omega/\partial\zeta,$$

$$(\partial/\partial\zeta)(A - iB) = c\Omega,$$

where $\zeta = \xi + i\eta$. If $c$ is zero, then $\partial/\partial u$ is a Killing vector.

Among the solutions of these equations, there is one which is stationary ($c = 0$) and also is axially symmetric. Like the Schwarzschild metric, which it contains, it is Type D. Also, $B$ is zero, and $m$ is a real constant, the Schwarzschild mass. The metric is

$$ds^2 = (r^2 + a^2\cos^2\theta)(d\theta^2 + \sin^2\theta d\phi^2) + 2(du + a\sin^2\theta d\phi) \times (dr + a\sin^2\theta d\phi) - \left(1 - \frac{2mr}{r^2 + a^2\cos^2\theta}\right) \times (du + a\sin^2\theta d\phi)^2,$$

where $a$ is a real constant. This may be transformed to an asymptotically flat coordinate system by the transformation

$$(r - ia)e^{i\phi}\sin\theta = x + iy, \quad r\cos\theta = z, \quad u = t + r,$$

the metric becoming

$$ds^2 = dx^2 + dy^2 + dz^2 - dt^2 + \frac{2mr^3}{r^4 + a^2z^2}(k)^2,$$

$$(r^2 + a^2)rk = r^2(xdx + ydy) + ar(xdy - ydx) + (r^2 + a^2)(zdz + rdt). \quad (5)$$

This function $r$ is defined by

$$r^4 - (R^2 - a^2)r^2 - a^2z^2 = 0, \quad R^2 = x^2 + y^2 + z^2,$$

so that asymptotically $r = R + O(R^{-1})$. In this coordinate system the solution is analytic everywhere, except at $R = a$, $z = 0$.

If we expand the metric in Eq. (5) as a power series in $m$ and $a$, assuming $m$ to be of order two and $a$ of order one, and compare it with the third-order Einstein-Infeld-Hoffmann approximation for a spinning particle, we find that $m$ is the Schwarzschild mass and $ma$ the angular momentum about the $z$ axis. It has no higher order multipole moments in this approximation. Since there is no invariant definition of the moments in the exact theory, one cannot say what they are, except that they are small. It would be desirable to calculate an interior solution to get more insight into this.

---

*Supported in part by U. S. Air Force through the Aerospace Research Laboratory of the Office of Aerospace Research.

[1]J. N. Goldberg and R. K. Sachs, Acta Phys. Polon. 22, 13 (1962).

[2]I. Robinson and A. Trautman, Proc. Roy. Soc. (London) A265, 463 (1961).

[3]E. Newman, L. Tamburino, and T. Unti (to be published).



## *Errata*

(i) Equation (2) should read

$$\partial_u \left(m - D^* D^* D\Omega\right) = |\partial_u D\Omega|^2 .$$

(ii) Equation (4) should read

$$D^* m = 3m\dot{\Omega}^* .$$

(iii) In the displayed equations on lines 19–20 of column 1, p. 238, the factors of $\zeta$ in the partial derivatives should be $\xi$; i.e. these equations should read:

$$iB = \tfrac{1}{2}P^{-2}\nabla\left(P^{-2}\partial\Omega/\partial\xi\right) - P^{-4}\left(\partial\Omega/\partial\xi\right)\nabla(\ln P);$$
$$\nabla B = ic\,\partial\Omega/\partial\xi.$$

# 14

# Gravitational collapse and rotation

Roy P. Kerr

Reprinted with permission from: *Quasi-stellar sources and gravitational collapse: Including the proceedings of the First Texas Symposium on Relativistic Astrophysics*, edited by Ivor Robinson, Alfred Schild, and E. L. Schücking (University of Chicago Press, Chicago, 1965), pages 99–102.
The conference was held in Austin, Texas, on 16–18 December 1963.

# GRAVITATIONAL COLLAPSE AND ROTATION

R. P. Kerr

In the past all exact solutions of collapsing gravitational systems have been spherically symmetric and have been based on the exterior Schwarzschild solution. This solution may be written in a form first given by Eddington (1924):

$$ds^2 = dx^2 + dy^2 + dz^2 - dt^2 + \frac{2m}{r}(dr + dt)^2, \tag{1}$$

where $r^2 = x^2 + y^2 + z^2$, and units have been chosen so that the velocity of light $c = 1$ and the gravitational constant $G = 1$.

This metric has a true singularity at the origin $r = 0$. However, it has peculiar physical properties inside and on the Schwarzschild sphere, $S$,

$$r = 2m\,. \tag{2}$$

$S$ is a null surface, outside ($r > 2m$) of which the metric is static, the $t$-axis being timelike. Inside $S(r < 2m)$ the metric is *not* static since the $t$-axis is spacelike.

For matter collapsing all the way to and beyond the Schwarzschild sphere we have the following behavior: matter and energy can pass from the exterior to the interior of $S$, but can never move out again, and so a spherically symmetric system collapsing beyond the Schwarzschild sphere can no longer radiate energy to the outside. It cannot be seen by an outside observer; only its gravitational field can be felt.

A collapsing particle will reach the Schwarzschild sphere and pass into its interior in a finite proper time, i.e., in a finite time as measured by a comoving clock. However, for a distant observer the time of collapse, measured by his clocks, is infinite. He will never observe the stage where the collapsing matter reaches the Schwarzschild sphere and passes to the inside.

This behavior causes difficulties in theories which attempt to explain the large energies emitted by quasi-stellar sources in terms of the gravitational collapse of large masses into the Schwarzschild sphere $S$.

In this paper we wish to show that the topological and physical properties of $S$ may

This research has been sponsored by the Aerospace Research Laboratory, Office of Aerospace Research, and the Office of Scientific Research, U.S. Air Force.

R. P. Kerr, The University of Texas.

change radically when rotation is taken into account. This suggests that it would be worthwhile to re-examine the problem of gravitational collapse for a mass whose external gravitational field is the stationary field of a rotating body.

An exact solution of Einstein's gravitational field equation for empty space is given by the metric (Kerr 1963):[1]

$$ds^2 = dx^2 + dy^2 + dz^2 - dt^2 + \frac{2m\rho^3}{\rho^4 + a^2 z^2}(k_\mu dx^\mu)^2, \tag{3}$$

where $k_\mu$ is a null vector field given by

$$k_\mu dx^\mu = dt + \frac{z}{\rho}dz + \frac{\rho}{\rho^2 + a^2}(xdx + ydy) + \frac{a}{\rho^2 + a^2}(xdy - ydx), \tag{4}$$

$m$ and $a$ are arbitrary constants, and $\rho$ is given by

$$\frac{x^2 + y^2}{\rho^2 + a^2} + \frac{z^2}{\rho^2} = 1. \tag{5}$$

The surfaces of constant $\rho$ are confocal ellipsoids of revolution.

For large spatial distances, $\rho$ is given asymptotically by

$$\rho = r + 0(r^{-1}). \tag{6}$$

The metric (3), expanded in powers of $r^{-1}$ becomes

$$\begin{aligned} ds^2 = dx^2 + dy^2 + dz^2 - dt^2 + \frac{2m}{r}(dt + dr)^2 \\ + \frac{4ma}{r^3}(xdy - ydx)(dt + dr) + 0(r^{-3}). \end{aligned} \tag{7}$$

The term of order $r^{-1}$ shows, e.g., by comparing with equation (1), that $m$ is the mass of the body producing the gravitational field. The term of order $r^{-2}$ shows, by comparing with the solution of the linearized field equations, that $am$ is the angular momentum about the $z$-axis of the rotating body.

The metric of equation (3) has a true singularity on the circle $\Gamma$,

$$z = 0, \qquad R \equiv (x^2 + y^2)^{1/2} = a. \tag{8}$$

This is the analogue of the true Schwarzschild singularity at $r = 0$ in equation (1).

The analogue of the Schwarzschild sphere is now the null surface $S$ given by

$$\rho^4 + a^2 z^2 = 2m\rho^3. \tag{9}$$

In Figure 1 $S$ is plotted in the $(R, z)$-plane for the two cases, $m < a$ and $m > a$. It will be observed that $S$ has a cusp on $\Gamma$. This is not significant, since the points of $\Gamma$ are singularities. For $m > a$, $S$ splits into two disjoint parts, $S_1$ and $S_2$. As $a \to 0$, the outer surface, $S_1$, becomes the Schwarzschild sphere, while $S_2$ collapses into the origin. When $a = m$ the two components, $S_1$ and $S_2$, touch on the $z$-axis. For $a > m$ the surfaces are as in

[1] More general vacuum solutions are given by Kerr and Schild (1964).

Figure 1, *a*. When $m \to 0$ $(a \neq 0)$, $S$ shrinks to the ring $\Gamma$. In most physical situations $a \gg m$, and so neither $S_1$ nor $S_2$ has the topology of a sphere.

There is a further complication of this metric, which is not present when $a = 0$. Suppose we define $D$ as the disk bounded by $\Gamma$ given by

$$R < a, \qquad z = 0 \,. \tag{10}$$

It is represented in Figure 1 by a solid line. We shall now show that the metric in equation (3) is not even differentiable, let alone analytic, on $D$. To see this we first observe that equation (3) has two distinct real roots, $\rho_+ > 0$ and $\rho_- < 0$, for all points except $D$. In order for the metric to be continuous, we choose the root $\rho_+$ for all points. From equation (5) this gives

$$\rho_+ = \frac{a\,|\,z\,|}{\sqrt{a^2 - r^2}}, \qquad \text{near } D\,, \tag{11}$$

and so $\rho_+$ is not differentiable on $D$. Substituting equation (11) into equation (3) we can easily see that the metric itself is not differentiable on $D$.

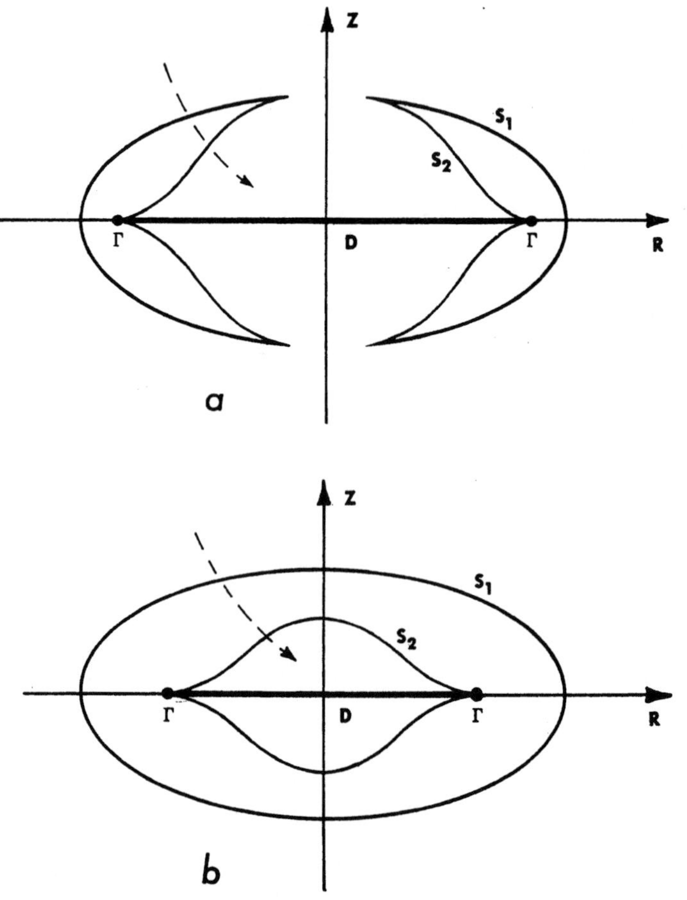


Fig. 1.—*a*, the "Schwarzschild surface" for $a > m > 0$. The solid disk, $D$, is a branch cut, bounded by the ring, $\Gamma$. *b*, the "Schwarzschild surface" for $m > a > 0$.

This behavior may be compared to that of $Z^{1/2}$ in the complex plane, with the branch point at the origin corresponding to the ring $\Gamma$. The function $\rho^2$ is real and positive at all points, except $\Gamma$, but $\rho$ is two-valued. As with $Z^{1/2}$ it is necessary to introduce two spaces, $E_1$ and $E_2$, with the topology of the points of $R^4$, less the ring $\Gamma$ and the disk $D$. The points above $D$ in $E_1$ are joined onto the points below $D$ in $E_2$, and vice versa. The points of $D$ should be thought of as a branch cut. The points immediately above the disk $D$ in $E_1$ are not close to those immediately below the disk $D$ in $E_1$. In $E_1$ we take the root $\rho_+$, whereas in $E_2$ $\rho = \rho_- \leq 0$. An observer passing through the ring $\Gamma$ will observe that $\rho$ changes sign, but the metric tensor is now analytic at all points of the space.

To interpret the space with $\rho = \rho_-$ we observe that if $\rho$ is replaced by $-\rho$ in equation (3), and the coordinate transformation $t \to -t$, $y \to -y$ is performed, then the metric of equations (3) and (4) is recovered except that the mass has the opposite sign. This means that if we take $m > 0$ in $E_1$ then an observer at a large distance from the origin in $E_2$ will consider that there is a negative mass near the origin.

It is not possible to say whether a rotating galaxy with this metric as its external field would allow much greater radiation than a non-rotating galaxy. However, all the evidence at present appears to confirm this. In particular, when $a > m$ radiation and matter will not be trapped inside $S$, as will be seen from the direction of the arrows in Figure 1, *b*. As a radiating particle collapses inside $S_1$ its radiation can be received by an observer outside the body, since it can cross $S_2$ in the indicated direction and then escape along the $z$-axis.

REFERENCES

Eddington, A. S. 1924, *Nature*, **113**, 192.
Kerr, R. P. 1963, *Phys. Rev. Letters*, **11**, 522.
Kerr, R. P., and Schild, A. 1964, Proceedings of the American Mathematical Society's Symposium on Applications of Partial Differential Equations in Mathematical Physics, New York City, April, 1964 (in press).

# Index